son
2011

CW01454598

THE CASIMIR EFFECT

Physical Manifestations of Zero-Point Energy

THE CASIMIR EFFECT
Physical Manifestations of Zero-Point Energy

K A Milton

University of Oklahoma, USA

World Scientific
New Jersey • London • Singapore • Hong Kong • Bangalore

Published by

World Scientific Publishing Co. Pte. Ltd.

5 Toh Tuck Link, Singapore 596224

USA office: 27 Warren Street, Suite 401-402, Hackensack, NJ 07601

UK office: 57 Shelton Street, Covent Garden, London WC2H 9HE

British Library Cataloguing-in-Publication Data
A catalogue record for this book is available from the British Library.

First published 2001
Reprinted 2005

ISBN 981-02-4397-9

Printed in Singapore by World Scientific Printers

To the memory of Hendrik Brugt Gerhard Casimir, who contributed so prodigiously to physics.

Preface

Preface

I first became interested in the Casimir effect [1] in 1975 when I heard the lectures of Julian Schwinger on the subject, in which he succeeded in deriving the Casimir force between parallel conducting plates without making reference to zero-point energy, a concept foreign to his non-operator version of quantum field theory, source theory. That presentation shortly appeared as a brief letter [2]. He justified this publication not merely as a rederivation of this effect in his own language, but as a resolution of the discrepancy between the finite-temperature effect first obtained by Sauer and Mehra [3, 4, 5, 6] and that obtained from the Lifshitz formula for parallel dielectrics [7, 8, 9]. Because of this discrepancy, Hargreaves [10] had called it "desirable that the whole general theory be reexamined and perhaps set up anew." Schwinger attempted just that. Unfortunately, he was unaware that the simple error in Lifshitz's paper had been subsequently corrected, so this was a non-issue. Nevertheless, this sparked an interest in the Casimir effect on Schwinger's part, which continued for the rest of his life.

Within a year or so Lester DeRaad and I, his postdocs at UCLA, joined Schwinger in reproducing the results of Lifshitz [11]; aside from a somewhat speculative treatment of surface tension, this paper contained few results that were new. Lifshitz wrote a somewhat peevish note to Schwinger complaining about the elevation of the temperature error to a significant discrepancy. But what was significant about this paper was the formulation: A general Green's dyadic procedure was developed that could be applied to a wide variety of problems. That procedure was immediately applied to a recalculation of the Casimir effect of a perfectly conducting spherical shell, which contrary to the expectation of Casimir [12] had been shown by Boyer [13] to be repulsive, not attractive. We derived a general formula, but then became stuck on its evaluation for a few months; in the meantime, the paper of Balian and Duplantier [14] appeared. DeRaad and I quickly found an even more accurate method of evaluation, and our confirmation of Boyer's result followed [15].

Thus my interest in the Casimir effect was launched. Before I left UCLA I explored the Casimir effect for a dielectric sphere, with inconclusive but seminal results [16]. My interest was rekindled a year or so later, when Ken Johnson proposed adapting the Casimir effect in a bag model of the vacuum [17]. Since he used the estimate from both interior and exterior modes, which seemed hardly applicable to the confinement situation of QCD, I proposed a better estimate based on interior contributions only [18, 19, 20], and followed it by examining the local Casimir contribution to the gluon and quark condensates [21]. I tried to improve the global estimates of these QCD effects shortly after I moved to Oklahoma, by attempting to elucidate the cutoff dependence of the interior modes [22]. I also worked out the unambiguously finite results for massless fermions interior and exterior to a perfect spherical bag [23] (in informal collaboration with Ken Johnson), and, somewhat earlier with DeRaad, computed the more difficult electromagnetic Casimir effect for a conducting cylinder [24].

In the late 1980s I was interested for a while in the Casimir effect in Kaluza-Klein spaces, particularly when the dimensionality of the compact-ified space was even [25, 26, 27], growing out of the work of Appelquist and Chodos [28] and Candelas and Weinberg [29]. A few years later I wrote two papers with Ng on the Maxwell-Chern-Simons Casimir effect [30, 31], the second of which signaled a serious problem for the Casimir effect in a two-dimensional space with a circular boundary. This was clarified shortly thereafter in a paper with Bender [32], where we computed the Casimir effect for a scalar field with Dirichlet boundary conditions on a D-dimensional sphere. Poles occur for arbitrary positive even D. I extended the work to include the TM modes, which exhibited qualitatively similar behavior [33].

It was the still inadequately understood phenomenon of sonolumines-cence [34] that sparked some of my most recent work in the field. Schwinger, in the last years of his life, suggested that the mechanism by which sound was converted into light in these repeatedly collapsing air bubbles in water had to do with the "dynamical Casimir effect" [35, 36]. After his death, I concluded he was wrong [37, 38]. But, probably the most interesting result of this work was a simple finite calculation of the regulated and renormalized van der Waals energy of a dielectric sphere. A year later Brevik, Marachevsky, and I, and others [39, 40, 41, 42], demonstrated that this coincided with the Casimir energy of a dilute dielectric ball, as formulated by me nearly two decades previously [16], suitably regulated and renormalized. At the same time I discovered that the Casimir energy of a dilute cylinder (with the speed of light the same inside and outside) vanished, as did the regulated van der Waals energy for a purely dielectric cylinder [43, 44]. The significance of these null results is still not clear.

This recounting of my personal odyssey through the Casimir world of course does no justice to the many other workers in the field, whose contributions I will attempt to more fully trace in the following. It is rather intended as a guide to the reader so my own personal biases may be discerned, biases which will be reflected in the following as well. Although I will attempt to survey the field, I will, of necessity, approach it with my own personal viewpoint. I will make some attempt to survey the literature, but I beg forgiveness from those authors whose work I slight or fail to cite. Hopefully, a document with an individual orientation will still have value in the new millennium.

Finally, I must thank the US Department of Energy for partial support of my research over the years, and my various collaborators whose contributions were invaluable. And most of all, I thank my wife, Margarita Baños-Milton, without whose support none of this would have been possible.

Kimball A. Milton
Norman, Oklahoma
April, 2001

Contents

Chapter 1

Introduction to the Casimir Effect

1.1 Van der Waals Forces

The understanding of the nature of the force between molecules has a long history. We will start our synopsis of that history with van der Waals. It was early recognized, by Herapath, Joule, Kronig, Clausius and others (for an annotated bibliography see ter Haar [45]), that the ideal gas laws of Boyle and Gay-Lussac could be explained by the kinetic theory of noninteracting point molecules. However, these laws were hardly exact. Van der Waals [46, 47] found in 1873 that significant improvements could be effected by including a finite size of the molecules and weak forces between the molecules. At the time, these forces were introduced in a completely *ad hoc* manner, by placing two parameters in the equation of state,

$$\left(p + \frac{a}{v^2}\right)(v - b) = RT. \tag{1.1}$$

Of course, it required the birth of quantum mechanics to begin to understand the origin of atomic and interatomic forces. In 1930 London [48, 49] showed that the force between molecules possessing electric dipole moments falls off with the distance r between the molecules as $1/r^6$. The simple argument goes as follows: The interaction Hamiltonian of a dipole moment $\mathbf{d}$ with an electric field $\mathbf{E}$ is $H = -\mathbf{d} \cdot \mathbf{E}$. From this, one sees that the the interaction energy between two such dipoles, labelled 1 and 2, is

$$H_{\text{int}} = \frac{(\mathbf{d}_1 \cdot \mathbf{d}_2)r^2 - 3(\mathbf{d}_1 \cdot \mathbf{r})(\mathbf{d}_2 \cdot \mathbf{r})}{r^5} \tag{1.2}$$

where $\mathbf{r}$ is the relative position vector of the two dipoles. Now in first

1

order of perturbation theory, the energy is given by $\langle H_{\text{int}}\rangle$, but this is zero because the dipoles are oriented randomly, $\langle \mathbf{d}_i\rangle = 0$. A nonzero result first emerges in second order,

$$V_{\text{eff}} = \sum_{m \neq 0} \frac{\langle 0|H_{\text{int}}|m\rangle\langle m|H_{\text{int}}|0\rangle}{E_0 - E_m} \qquad (1.3)$$

which evidently gives $V_{\text{eff}} \sim r^{-6}$. This is a short-distance electrostatic effect.

In 1947 Casimir and Polder [50] included *retardation*. They found that at large distances the interaction between the molecules goes like $1/r^7$. This result can be understood by a simple dimensional argument. For weak electric fields $\mathbf{E}$ the relation between the induced dipole moment $\mathbf{d}$ and the electric field is linear (isotropy is assumed for convenience),

$$\mathbf{d} = \alpha\mathbf{E}, \qquad (1.4)$$

where the constant of proportionality α is called the polarizability. At zero temperature, due to fluctuations in $\mathbf{d}$, the two atoms polarize each other. Because of the following dimensional properties:

$$[d] = eL, \quad [E] = eL^{-2}, \quad \Rightarrow [\alpha] = L^3, \qquad (1.5)$$

where L represents a dimension of length, we conclude that the effective potential between the two polarizable atoms has the form

$$V_{\text{eff}} \sim \frac{\alpha_1\alpha_2}{r^6}\frac{\hbar c}{r}, \qquad (1.6)$$

while at high temperatures the $1/r^6$ behavior is recovered,

$$V_{\text{eff}} \sim \frac{\alpha_1\alpha_2}{r^6}kT, \quad T \to \infty. \qquad (1.7)$$

The London result is reproduced by noting that in arguing (1.6) we implicitly assumed that $r \gg \lambda$, where λ is a characteristic wavelength associated with the polarizability, that is

$$\alpha(\omega) \approx \alpha(0) \quad \text{for} \quad \omega < \frac{c}{\lambda}. \qquad (1.8)$$

In the opposite limit,

$$V_{\text{eff}} \sim \frac{\hbar}{r^6}\int_0^\infty d\omega\, \alpha_1(\omega)\alpha_2(\omega), \quad r \ll \lambda. \qquad (1.9)$$

These results, with the precise numerical coefficients, will be derived in Chapter 3.

1.2 Casimir Effect

In 1948 Casimir [1] shifted the emphasis from action at a distance between molecules to local action of fields.* That is, the phenomenon discussed above in terms of fluctuating dipoles can equally be thought of in terms of fluctuating electric fields, in view of the linear relation between these quantities. This apparently rather trivial change of viewpoint opens up a whole new array of phenomena, which we refer to as the Casimir effect. Specifically, in 1948 Casimir considered two parallel conducting plates separated by a distance a. (See Fig. 2.1.) Although $\langle \mathbf{E} \rangle = 0$ if there is no charge on the plates, the same is not true of the square of the fields,

$$\langle \mathbf{E}^2 \rangle \neq 0, \quad \langle \mathbf{B}^2 \rangle \neq 0, \tag{1.10}$$

and so the expectation value of the energy,

$$H = \frac{1}{2} \int (d\mathbf{r}) \left(\mathbf{E}^2(\mathbf{r}) + \mathbf{B}^2(\mathbf{r}) \right), \tag{1.11}$$

is not zero. This gives rise to a measurable force on the plates. It is not possible without a detailed calculation to determine the sign of the force, however. It turns out in this circumstance to be attractive. Much of the following chapter will be devoted to a careful derivation of this force; the results, as found by Casimir, for the energy per unit area and the force per unit area are

$$\mathcal{E} = -\frac{\pi^2}{720a^3}\hbar c, \quad \mathcal{F} = -\frac{d}{da}\mathcal{E} = -\frac{\pi^2}{240a^4}\hbar c. \tag{1.12}$$

The dependence on a is, of course, completely determined by dimensional considerations. Numerically, the result is quite small,

$$\mathcal{F} = -8.11\,\mathrm{MeV\,fm}\,a^{-4} = -1.30 \times 10^{-27}\mathrm{N\,m^2}\,a^{-4}, \tag{1.13}$$

---

*This was due to a comment by Bohr in 1947, to the effect that the van der Waals force "must have something to do with zero-point energy." In April 1948 Casimir communicated a new derivation of the force between an atom and a plate, and between two atoms, based on quantum fluctuations to a meeting in Paris [51]. For further history of the development of the Casimir effect, see Refs. [52, 53].

and will be overwhelmed by electrostatic repulsion between the plates if each plate has an excess electron density n greater than $1/a^2$, from which it is clear that the experiment must be performed at the μm level. Nevertheless, there have many attempts to directly measure this effect, although somewhat inconclusively [54, 55, 56, 57, 58, 59, 60, 61, 62, 63, 64, 65]. [The cited measurements include insulators as well as conducting surfaces; the corresponding theory will be given in Chapter 3.] Until recently, the most convincing experimental evidence came from the study of thin helium films [66]; there the corresponding Lifshitz theory [7, 67, 8] was confirmed over nearly 5 orders of magnitude in the van der Waals potential (nearly two orders of magnitude in distance). However, the Casimir effect between conductors has recently been confirmed to about the 5% level by Lamoreaux [68, 69, 70], to perhaps 1% by Mohideen and Roy [71, 72, 73], and by Erdeth [74], and to about the same level very recently by Chan, Aksyuk, Kleiman, Bishop, and Capasso [75].

In general, let us define the Casimir effect as the stress on the bounding surfaces when a quantum field is confined to a finite volume of space. The boundaries may be described by real material media, with electromagnetic properties such as dielectric functions, in which case fields will exist on both sides of the material interface. The boundaries may also represent the interface between two different phases of the vacuum of a field theory such as quantum chromodynamics, in which case colored fields may only exist in the interior region. The boundaries may, on the other hand, represent the topology of space, as in higher-dimensional theories (e.g., Kaluza-Klein or strings), where the extra dimensions may be curled up into a finite geometry of a sphere, for example. In any case, the boundaries restrict the modes of the quantum fields, and give rise to measurable and important forces which may be more or less readily calculated. It is the aim of the present monograph to give a unified treatment of all these phenomena, which have implications for physics on all scales, from the substructure of quarks to the large scale structure of the universe. Although similar claims of universality of other particle physics phenomena are often made, the Casimir effect truly does have real-world applications to condensed-matter and atomic physics.

There are many ways in which the Casimir effect may be computed. Perhaps the most obvious procedure is to compute the zero-point energy in the presence of the boundaries. Although

$$\sum_{\text{modes}} \frac{1}{2}\hbar\omega \qquad (1.14)$$

is terribly divergent, it is possible to regulate the sum, subtract off the divergences (one only measures the change from the value of the sum when no boundaries are present), and compute a measurable Casimir energy. (A simple version of this procedure is given in Sec. 2.2.) However, a far superior technique is based upon the use of Green's functions. Because the Green's function represents the vacuum expectation value of the (time-ordered) product of fields, it is possible to compute the vacuum expectation value of (1.11), for example, in terms of the Green's function at coincident arguments. Once the energy U is computed as a function of the coordinates $\mathbf{X}$ of a portion of the boundary, one can compute the force on that portion of the boundary by differentiation,

$$F = -\frac{\partial}{\partial \mathbf{X}}U. \qquad (1.15)$$

Similarly, one can compute the stress-energy tensor, $T^{\mu\nu}$, from the Green's function, and thereby compute the stress on a boundary element,

$$\frac{dF}{d\sigma} = \langle T^{nn} \rangle, \qquad (1.16)$$

where n represents the normal to the surface element $d\sigma$, and where the brackets represent the vacuum expectation value.

The connection between the sum of the zero-point energies of the modes and the vacuum expectation value of the field energy may be easily given. Let us regulate the former with an oscillating exponential:

$$\frac{1}{2}\sum_a \hbar\omega_a e^{-i\omega_a\tau} = \frac{\hbar}{2}\int_{-\infty}^{\infty}\frac{d\omega}{2\pi i}e^{-i\omega\tau}\omega\sum_a\frac{2\omega}{\omega_a^2-\omega^2-i\epsilon}, \qquad (1.17)$$

where a labels the modes, and, because we assume τ goes to zero through positive values, the contour of integration in the second form may be closed in the lower half plane. For simplicity of notation let us suppose we are dealing with massless scalar modes, for which the eigenfunctions and eigenvalues satisfy

$$-\nabla^2\psi_a = \omega_a^2\psi_a. \qquad (1.18)$$

Because these are presumed normalized, we may write the second form in (1.17) as

$$\hbar \int \frac{d\omega}{2\pi i} \omega^2 e^{-i\omega\tau} \int (d\mathbf{x}) \sum_a \frac{\psi_a(\mathbf{x})\psi_a^*(\mathbf{x})}{\omega_a^2 - \omega^2 - i\epsilon}$$

$$= \frac{\hbar}{i} \int (d\mathbf{x}) \int \frac{d\omega}{2\pi} \omega^2 G(\mathbf{x}, \mathbf{x}; \omega) e^{-i\omega(t-t')} \Big|_{t \to t'}$$

$$= \int (d\mathbf{x}) \partial^0 \partial'^0 \langle \phi(x)\phi(x') \rangle \big|_{x' \to x}, \qquad (1.19)$$

where the Green's function $G(\mathbf{x}, t; \mathbf{x}', t')$ satisfies

$$\left(-\nabla^2 + \frac{\partial^2}{\partial t^2}\right) G(\mathbf{x}, t; \mathbf{x}', t') = \delta(\mathbf{x} - \mathbf{x}')\delta(t - t'), \qquad (1.20)$$

and is related to the vacuum expectation value of the time-ordered product of fields according to

$$G(\mathbf{x}, t; \mathbf{x}', t') = \frac{i}{\hbar} \langle \mathrm{T}\phi(\mathbf{x}, t)\phi(\mathbf{x}', t') \rangle. \qquad (1.21)$$

For a massless scalar field, the canonical energy-momentum tensor is

$$T^{\mu\nu} = \partial^\mu \phi \partial^\nu \phi - \frac{1}{2} g^{\mu\nu} \partial^\lambda \phi \partial_\lambda \phi. \qquad (1.22)$$

The second term involving the Lagrangian in (1.22) may be easily shown not to contribute when integrated over all space, by virtue of the equation of motion, $-\partial^2 \phi = 0$ outside the sources, so we have the result, identifying the zero-point energy with the vacuum expectation value of the field energy,

$$\frac{1}{2} \sum_a \hbar \omega_a = \int (d\mathbf{x}) \langle T^{00}(\mathbf{x}) \rangle. \qquad (1.23)$$

In the vacuum this is divergent and meaningless. What is observable is the *change* in the zero-point energy when matter is introduced. In this way we can calculate the Casimir forces.

Variational forms may also be given. For example, in Chapter 3 we will derive the following formula for the variation in the electromagnetic energy when the dielectric function is varied slightly,

$$\delta U = \frac{i\hbar}{2} \int (d\mathbf{r}) \frac{d\omega}{2\pi} \delta\epsilon(\mathbf{r}, \omega) \Gamma_{kk}(\mathbf{r}, \mathbf{r}', \omega), \qquad (1.24)$$

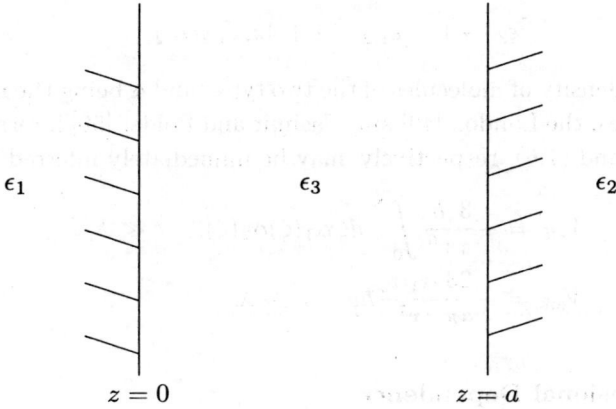

Fig. 1.1   Geometry of parallel dielectric surfaces.

where Γ is the electric Green's dyadic. From this formula one can easily compute the force between semi-infinite parallel dielectrics, as shown in Fig. 1.1. The celebrated formula of Lifshitz [7, 67, 8] is obtained in this way for the force per unit area,

$$\mathcal{F} = -\frac{1}{8\pi^2} \int_0^\infty d\zeta \int_0^\infty dk^2\, 2\kappa_3 \left[\left(\frac{\kappa_3 + \kappa_1}{\kappa_3 - \kappa_1} \frac{\kappa_3 + \kappa_2}{\kappa_3 - \kappa_2} e^{2\kappa_3 a} - 1 \right)^{-1} \right.$$
$$\left. + \left(\frac{\kappa_3' + \kappa_1'}{\kappa_3' - \kappa_1'} \frac{\kappa_3' + \kappa_2'}{\kappa_3' - \kappa_2'} e^{2\kappa_3 a} - 1 \right)^{-1} \right]. \quad (1.25)$$

Here the frequency integration has been rotated $\pi/2$ in the complex frequency plane, $\zeta = -i\omega$, ζ real, and the following abbreviations have been used,

$$k^2 = k_\perp^2, \quad \kappa^2 = k^2 + \zeta^2 \epsilon, \quad \kappa' = \frac{\kappa}{\epsilon}. \quad (1.26)$$

It is this formula which was spectacularly confirmed for a thin film of helium on a quartz substrate in a beautiful experiment by Sabisky and Anderson [66]. By taking the appropriate limit,

$$\epsilon_3 \to 1, \quad \epsilon_1 = \epsilon_2 \to \infty, \quad (1.27)$$

the Lifshitz formula (1.25) reproduces the Casimir result for parallel conductors, (1.12). Furthermore, by regarding the dielectrics to be tenuous gases,

$$\epsilon_3 \to 1, \quad \epsilon_{1,2} = 1 + 4\pi \mathcal{N}_{1,2}\alpha_{1,2}, \tag{1.28}$$

$\mathcal{N}$ being the density of molecules of the two types and α being the molecular polarizabilities, the London [49] and Casimir and Polder [50] intermolecular forces, (1.9) and (1.6), respectively, may be immediately inferred:

$$V_{\text{eff}} = -\frac{3}{\pi} \frac{\hbar}{r^6} \int_0^\infty d\zeta \, \alpha_1(\zeta)\alpha_2(\zeta), \quad r \ll \lambda, \tag{1.29}$$

$$V_{\text{eff}} = -\frac{23}{4\pi} \frac{\alpha_1\alpha_2}{r^7} \hbar c, \quad r \gg \lambda. \tag{1.30}$$

1.3 Dimensional Dependence

We have already noted above that the sign of the Casimir effect cannot be deduced until after the entire calculation is completed. This is because the starting expressions are purely formal, and require regularization and a careful subtraction of infinities before a finite force can be extracted. Nevertheless, the results quoted above demonstrate that in the one-dimensional geometries considered to this point the Casimir force is strictly *attractive*, whether one is dealing with conductors or dielectrics, and whether the helicity of the field is 0 or 1. [The same is true for spin 1/2; see Sec. 2.7.] This is certainly in accord with the interpretation of the effect as the sum of van der Waals attractions between molecules. Accordingly, Casimir suggested in 1956 [12] that the Casimir force could play the role of a Poincaré stress in stabilizing a classical model of the electron. In this way he hoped that a value for the fine structure constant could be calculated. Unfortunately, when Boyer did the calculation in 1968 [13] he found a result which was *repulsive*; Boyer's calculation was a *tour de force*, and has been independently confirmed [76, 14, 15]: The stress on a perfectly conducting spherical shell of radius a is

$$\mathcal{S}_{\text{sphere}} = \frac{\hbar c}{a^2}(0.04618\ldots). \tag{1.31}$$

The details of this calculation will be given in Chapter 4. Also there will be given the corresponding result for fermions [23]:

$$\mathcal{S}_{\text{sphere}} = \frac{\hbar c}{a^2}(0.0204\ldots). \tag{1.32}$$

These nonintuitive results immediately raise the question of what happens at intermediate dimensions. A partial answer was given in Ref. [24], where the Casimir effect was derived for a right-circular cylinder, with a small, *attractive* result for the force per unit area

$$\mathcal{F}_{\text{cylinder}} = -\frac{\hbar c}{a^4}(0.00432\ldots). \tag{1.33}$$

(See Chapter 7.)

Another context in which the dimensional dependence of the Casimir effect was studied was for Kaluza-Klein theories in $4 + N$ dimensions, where the extra dimensions were compactified into a sphere (or products of spheres). For odd N, and a single sphere S^N, the story was given by Candelas and Weinberg [29]: for example, for a scalar field, for $N = 1$ the Casimir energy is negative; then for $N = 3, 5, \ldots, 19$, the Casimir energy is positive, and for $N \geq 21$ the energy becomes increasingly negative. [See Table 10.1 and Fig. 10.2.] For even N the result is divergent; if a cutoff is introduced, the coefficient of the logarithmic divergence is negative for all N [25]. [See Table 10.2 and Fig. 10.5.] These calculations will be treated in detail in Chapter 10.

Until recently, the balance of our knowledge of the dimensional dependence of the Casimir effect referred to the force computed in parallelepiped geometries, where only interior modes are computed. Calculations of the Casimir energies *inside* rectangular cavities were first given by Lukosz [77] (see also [78, 79]) and later by Ruggiero, Zimerman, and Villani [80, 81] and by Ambjørn and Wolfram [82]. In general these results are highly questionable, because no exterior contributions can be included, because it is impossible to separate the Klein-Gordon equation, for example, in the exterior of a rectangular cavity. (Thus we will defer the discussion of these calculations until Secs. 6.1 and 7.1.2.) The exception is the case of infinite parallel plates embedded in a D-dimensional space and separated by a distance $2a$; that is, there is one longitudinal dimension and $D - 1$ transverse dimensions. The result for the force per unit area for a scalar field satisfying Dirichlet or Neumann boundary conditions as found by Ambjørn and Wolfram [82] is [see (2.35)]

$$\mathcal{F} = -a^{-D-1}2^{-2D-2}\pi^{-(D+1)/2}D\,\Gamma\left(\frac{D+1}{2}\right)\zeta(D+1), \tag{1.34}$$

which we have plotted in Fig. 1.2. Note that $\mathcal{F}$ has a simple pole (due to the

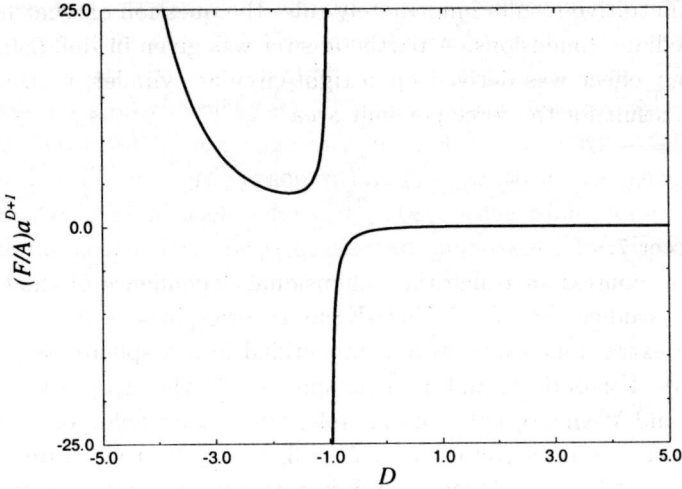

Fig. 1.2   A plot of the Casimir force per unit area $\mathcal{F}$ in (1.34) for $-5 < D < 5$ for the case of a slab geometry (two parallel plates).

gamma function) at $D = -1$. However, $\mathcal{F}$ is not infinite at the other poles of the gamma function, which are located at all the negative odd integral values of D, because the Riemann zeta function vanishes at all negative even values of its argument. One interesting and well-known special case of (1.34) is $D = 1$ [83]:

$$\mathcal{F}\big|_{D=1} = -\frac{\pi}{96a^2}, \tag{1.35}$$

where the negative sign indicates that the force is attractive. (This was originally calculated in the context of the string theory for the potential between heavy quarks.) For the case of $D = 3$ we recover precisely one-half Casimir's result (1.12) (of course, with $a \to 2a$), suggesting, perhaps misleadingly, that each electrodynamic mode contributes one half the total. But, all these cases are essentially one-dimensional.

However, more recently we have examined the Casimir force of a scalar field satisfying Dirichlet boundary conditions on a spherical shell in D dimensions [32]. The details of this calculation will be given in Chapter 9. The numerical results are shown in Fig. 9.1. We find that the

Casimir stress vanishes as $D \to \infty$ (largely because the surface area of a D-dimensional sphere tends to zero as $D \to \infty$), and also vanishes when D is a negative even integer. Remarkably, the stress has simple poles at positive even integer values of D. These results for scalar, or in waveguide terminology, TE modes, are generalized to include the TM modes as well, which behave qualitatively similarly [33]. Does this mean that we cannot make sense of the Casimir effect in two space dimensions, an arena of tremendous interest in condensed matter physics [30, 31]? Some hints of a resolution of this serious difficulty are suggested in Sec. 9.3.

1.4 Applications

It might seem to the reader that the Casimir effect is an esoteric aspect of quantum mechanics of interest only to specialists. That this is not the case should be apparent from the duality of this effect with van der Waals forces between molecules. The structure of gross matter is therefore intimately tied to the Casimir effect. But we can be more specific in citing true field-theoretic application of these phenomena.

Perhaps the first extensive reference to the Casimir effect in particle physics occurred with the development of the bag model of hadrons [84, 85, 86, 87, 88, 89]. There a hadron was modeled as a quark and an antiquark, or three quarks, confined to the interior of a (spherical) cavity. Asymptotic freedom was implemented by positing that the interior of the cavity was a chromomagnetic vacuum ($\mu = 1$), while the exterior was a perfect chromomagnetic conductor ($\mu = \infty$). Predictions could readily be made for masses and magnetic moments, in terms of a few parameters, principally, the bag constant B, the strong coupling constant α_s, the radius of the nucleon R_N, and the "zero-point energy" parameter Z, as well as quark masses. These parameters were, in fact, determined from fits to the data. In particular, the parameter Z, which occurred in the bag-model Hamiltonian as a term $-Z/a$, where a is the bag radius, was assumed to be positive. In fact, Boyer's result [13] already suggested that was in error, since the Casimir effect for a spherical shell is repulsive. However, in this situation, there are no exterior modes, so the result is less clear. Nevertheless, it has been argued that a repulsive result is expected theoretically, and that the model must be modified correspondingly [18,

19]. These issues, and other hadronic applications of the Casimir effect will
be treated in Chapter 6. See also Sec. 11.4.

We have already referred to higher-dimensional theories, such as Kaluza-
Klein models, in which extra dimensions are curled up into finite topological structures, where the scale is set by the Planck length. This gives
rise to large observable vacuum energy effects [28, 29, 25]. The hope is
that quantum fluctuations of the gravitational (and other) fields can stabilize the geometry and explain why all dimensions are not of the scale
$L_{\text{Planck}} \approx 10^{-33}$ cm. Although there is now quite an extensive literature on
this subject, progress has been slow because of the technical difficulties associated with implementing the required Vilkovisky-DeWitt formalism [90,
91]. The status of this important interface between "unified" theories and
cosmology will be related in Chapter 10.

We will come back to earth with a recounting of the Maxwell-Chern-
Simons Casimir effect in Chapter 8. In two space dimensions it is possible
to introduce a mass term for the photon without spoiling gauge invariance.
That is, in place of the Maxwell Lagrangian we can write

$$\mathcal{L} = -\frac{1}{4}F^{\mu\nu}F_{\mu\nu} + \frac{1}{4}\mu\epsilon^{\mu\alpha\beta}F_{\alpha\beta}A_\mu. \tag{1.36}$$

The mass term has the form of the Chern-Simons topological Lagrangian
which occur for the anyon fields perhaps relevant for the fractional quantum
Hall effect [92, 93, 94, 95] and for high temperature superconductivity [96,
97, 98]. Here, however, we are regarding A_μ to be a physical photon field
somehow trapped in two dimensions. We will attempt to make sense of
the Casimir effect in two dimensions by use of the procedure described in
Chapter 9, and contrast scalar and vector fields. Salient features include [30,
31]

- For parallel lines vector and scalar fields give identical attractive
  Casimir forces. This again illustrates the universal character of
  one-dimensional geometries.
- For a circular boundary vector and scalar fields give completely
  different results for the Casimir effect; in the leading approximation,
  the scalar force is repulsive, while the vector is attractive.
- The following dimensional reduction theorem holds true for the
  massless theory ($\mu = 0$): The Casimir effect for a right circular
  cylinder in (3+1) dimensions for a vector field [24], as described
  in Chapter 7, reduces, as the component of momentum along the

longitudinal axis goes to zero, $k_z \to 0$, to the sum of the Casimir effects for a scalar and for a vector field in (2+1) dimensions:

$$(3+1)_v \to (2+1)_v + (2+1)_s, \quad (k_z \to 0). \tag{1.37}$$

We will discuss the temperature dependence and possibilities of observing this phenomenon in condensed matter systems in Chapter 8.

1.5 Local Effects

To this point we have considered the Casimir effect as a global phenomenon. The observable Casimir force on a macroscopic bounding surface is a collective effect, and localization of the phenomenon would seem to be nonunique. [The nonlocalization of physical phenomena should already be familiar at the classical level in connection with radiation. For example, the question of whether a uniformly accelerated charge radiates or not is only answerable if the beginning and end of the acceleration process is specified; and even then it is only a manner of speaking to say that the radiation is associated with the beginning and end of the process. See, for example, Ref. [99].]

Nevertheless, we have already indicated that construction of the energy-momentum tensor $T^{\mu\nu}$ is an important route toward calculation of the Casimir effect. Further if one imposes conservation and tracelessness of that tensor in electromagnetism,

$$\partial_\mu T^{\mu\nu} = 0, \quad T^\mu{}_\mu = 0, \tag{1.38}$$

one can infer [100] a unique vacuum expectation value of the energy-momentum tensor for the case of parallel conducting plates located at $z = 0$ and $z = a$:

$$\langle T_{\mu\nu} \rangle = -\frac{\pi^2}{720a^4} \begin{pmatrix} 1 & 0 & 0 & 0 \\ 0 & -1 & 0 & 0 \\ 0 & 0 & -1 & 0 \\ 0 & 0 & 0 & 3 \end{pmatrix} \tag{1.39}$$

The energy density and the force per area given in (1.12) are contained in this result. Besides having great interest in their own right, these local effects could have important gravitational consequences [101]. Local effects also must be understood if one is to correctly interpret the divergences inherent in the theory. The nature of boundary divergences, a subject to which we will return later, was first studied systematically by Deutsch and Candelas [102]. These issues and consequences, particularly in hadronic physics, will be discussed in Chapter 11.

1.6 Sonoluminescence

One of the most intriguing phenomena in physics today is sonolumines-cence [34]. In the experiment, a small (radius $\sim 10^{-3}$ cm) bubble of air or other gas is injected into water, and subjected to an intense acoustic field (overpressure ~ 1 atm, frequency $\sim 2 \times 10^4$ Hz). If the parameters are carefully chosen, the repetitively collapsing bubble emits an intense flash of light at minimum radius (something like a million optical photons are emitted per flash), yet the process is sufficiently non-catastrophic that a single bubble may continue to undergo collapse and emission 20,000 times a second for many minutes, if not months. Many curious properties have been observed, such as sensitivity to small impurities, strong temperature dependence, necessity of small amounts of noble gases, possible strong iso-tope effect, etc. [34].

No convincing theoretical explanation of the light-emission process has yet been put forward. This is certainly not for want of interesting the-oretical ideas. One of the most intriguing suggestions was advocated by Schwinger [35, 36], based on a reanalysis of the Casimir effect. Specifically, he proposed that the Casimir effect be generalized to the spherical volume defined by the bubble (as we will discuss in Chapter 5), and with the static boundary conditions replaced by time-varying ones. He called this idea the dynamical Casimir effect. Unfortunately, although Schwinger began the general reformulation of the static problem in Refs. [103, 104] [most of which had been, unbeknownst to him, given earlier [16] (see Chapter 5), he did not live to complete the program. Instead, he proposed a rather naive approximation of subtracting the zero-point energy $\frac{1}{2}\sum \hbar\omega$ of the medium from that of the vacuum, leading, for a spherical bubble of radius a in a medium with index of refraction n, to a Casimir energy proportional to the volume of the bubble:

$$E_{\text{bulk}} = \frac{4\pi a^3}{3} \int \frac{(d\mathbf{k})}{(2\pi)^3} \frac{1}{2} k \left(1 - \frac{1}{n}\right). \tag{1.40}$$

Of course, this is quartically divergent. If one puts in a suitable ultraviolet cutoff, one can indeed obtain the needed 10 MeV per flash. On the other hand, one might have serious reservations about the physical meaning of such a divergent result.

In Chapter 5 we will carefully study the basis for this model for sonoluminescence. We will argue there that the leading term (1.40) is to be removed by subtracting the contribution the formalism would give if either medium filled all space. Doing so still leaves us with a cubically divergent Casimir energy; but we will argue further that this cubic divergence can plausibly be removed as a contribution to the surface energy. The remaining finite energy has been determined by a number of authors [39, 40, 41, 42] to be positive and small:

$$E_c \sim \frac{23(n-1)^2}{384\pi a}, \quad |n-1| \ll 1, \tag{1.41}$$

I wonder...

is at least ten orders of magnitude too small, and of the wrong sign, to be relevant to sonoluminescence. This result is also equivalent to the finite van der Waals self-interaction of a spherical bubble [38], as shown in Sec. 5.9.

It remains to be confirmed whether this adiabatic approximation is valid in the extreme situation present in the sonoluminescing environment. A dynamical calculation is called for, and first steps toward that theory will be sketched. That, and a discussion of the contradictory literature on this evolving subject, will be detailed in Chapter 12.

1.7 Radiative Corrections

All of the effects so far described are at the one-loop quantum level. Two-loop effects have been considered by a few authors [105, 106, 107]. Results have been given both for parallel conducting plates and a conducting spherical shell, and will be described in Chapter 13. These effects are certainly negligible in QED—the typical correction is down not merely by a factor of the fine structure constant α, but by the ratio of the (small) Compton wavelength λ_c of the electron to the geometrical size a of the macroscopic system. However, such corrections could be important in hadronic systems, where $\alpha_s \sim 1$ and $\lambda_c \sim a$; but there the relevant calculations have not been done. For the status of this important topic, see Chapter 13.

1.8   Other Topics

The above summary does not do justice to all the work carried out over many years on the theory and applications of quantum vacuum energy. Many of these topics will come up in the appropriate places in the text. For example, the Casimir effect could be relevant to the physics of cosmic strings, and a brief discussion of some of the literature on this subject will appear in Chapters 7 and 11. The Casimir energy of a closed string itself will be discussed in Appendix B.

1.9   Conclusion

It is the aim of this monograph to provide a unified, yet comprehensive, treatment of the Casimir effect in a wide variety of domains. Although from textbooks one might conclude that the Casimir effect is an esoteric subject with little practical consequence, I hope this introduction has convinced the reader of the pervasive nature of the zero-point fluctuation phenomena. These phenomena lie at the very heart of quantum mechanics, and, as noted above, what we discuss here are just the first quantum corrections to classical configurations. The subtleties and difficulties encountered in all but the simplest of the Casimir effect calculations demonstrate that we are only beginning to understand the quantum nature of the universe.

1.10   General References

Mathematical references used freely throughout this book include Whittaker and Watson [108], Gradshteyn and Rhyzik [109], Prudnikov, Brychkov, and Marichev [110], and Abramowitz and Stegun [111].

Finally, reference should be made to review articles on the Casimir effect and its applications, by Plunien, Müller, and Greiner [112] and by Mostepanenko and Trunov [113]. The latter authors have also written a book-length review of the subject [114]. Marginally related are books by Levin and Micha [115] and by Krech [116]. An excellent book is that of Milonni [117], but the orientation of that treatise is quite different.

After completion of this manuscript, a long review article has appeared by Bordag, Mohideen, and Mostepanenko [118], which is quite complementary to the present volume, being in the end primarily concerned with the experimental situation.

Chapter 2
Casimir Force Between Parallel Plates

2.1 Introduction

It is often stated that zero-point energy in quantum field theory is not observable, and for this reason the theory should be defined by normal ordering. That such a conclusion is incorrect was recognized by Casimir in 1948 when he showed that zero-point fluctuations in electromagnetic fields gave rise to an attractive force between parallel, perfectly conducting plates [1]. His result, at zero temperature, for the force per unit area between such plates separated by a distance a is

$$\mathcal{F}_{\text{em}} = -\frac{\pi^2 \hbar c}{240 a^4}. \tag{2.1}$$

Experiments have confirmed this *Casimir Effect*. Our aim in this chapter is to rederive Casimir's result using careful Green's function techniques which should lay to rest any uneasiness concerning control of infinities in the problem. The formalism developed here will be applied in subsequent chapters to derive Casimir forces in more complicated geometries and topologies, and make application to fundamental physics issues from hadrons to cosmology.

In this chapter, we will first, in Sec. 2.2, provide a simple, unphysical, derivation of the Casimir effect between two idealized plates using dimensional regularization. The Green's function approach in the case of a scalar field satisfying Dirichlet boundary conditions will then be given in Sec. 2.3. Here we calculate the force on one of the plates by looking both at the normal-normal component of the stress tensor, and by computing the Casimir energy. In Sec. 2.4 we consider a massive scalar field. The nonzero temperature Casimir effect is examined in Sec. 2.5, with specific attention

19

to the high- and low-temperature limits. The full electromagnetic case is treated at last in Sec. 2.6, where we introduce a Green's dyadic formulation. The work of the chapter is concluded with Sec. 2.7, where the Casimir force on parallel surfaces due to fluctuations in a massless fermionic field satisfying bag-model boundary conditions is treated.

The Casimir effect evolved out of the earlier work, published the same year, 1948, by Casimir and Polder, who considered the retarded dispersive forces between polarizable atoms, the constituents of dielectric media [50]. (As mentioned, this result was verified by a cavity calculation involving zero-point energy [51].) In particular, Casimir and later, and more explicitly, Lifshitz recognized that the "Casimir" forces between bodies having different dielectric constants can be interpreted, in the limit of tenuous media, to arise from the retarded $(1/r^7)$ and the short-range $(1/r^6)$ van der Waals potentials between the molecules which make up the bodies, and that these van der Waals forces are a result of quantum fluctuations. We will discuss these questions and their experimental consequences in Chapter 3.

2.2   Dimensional Regularization

We begin by presenting a simple, "modern," derivation of the Casimir effect in its original context, the electromagnetic force between parallel, uncharged, perfectly conducting plates. No attempt at rigor will be given, for the same formulæ will be derived by a consistent Green's function technique in the following section. Nevertheless, the procedure illustrated here correctly produces the finite, observable force starting from a divergent formal expression, without any explicit subtractions, and is therefore of great utility in practice.

For simplicity we consider a massless scalar field ϕ confined between two parallel plates separated by a distance a. (See Fig. 2.1.) Assume that the field satisfies Dirichlet boundary conditions on the plates, that is

$$\phi(z = 0) = \phi(z = a) = 0. \tag{2.2}$$

The Casimir force between the plates results from the zero-point energy per unit transverse area

$$\mathcal{E} = \frac{1}{2} \sum \hbar\omega = \frac{1}{2} \sum_{n=1}^{\infty} \int \frac{d^2k}{(2\pi)^2} \sqrt{k^2 + \frac{n^2\pi^2}{a^2}}, \tag{2.3}$$

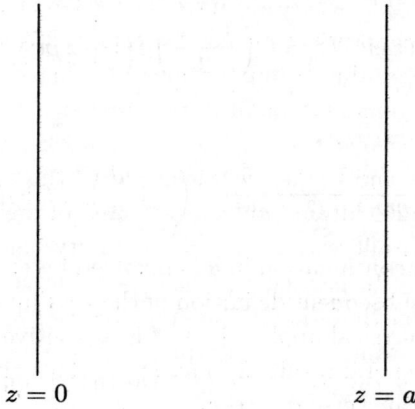

$$z = 0 \qquad\qquad z = a$$

Fig. 2.1 Geometry of parallel, infinitesimal plates.

where we have set $\hbar = c = 1$, and introduced normal modes labeled by the positive integer n and the transverse momentum k.

To evaluate (2.3) we employ dimensional regularization. That is, we let the transverse dimension be d, which we will subsequently treat as a continuous, complex variable. It is also convenient to employ the Schwinger proper-time representation for the square root:

$$\mathcal{E} = \frac{1}{2} \sum_n \int \frac{d^d k}{(2\pi)^d} \int_0^\infty \frac{dt}{t} t^{-1/2} e^{-t(k^2 + n^2\pi^2/a^2)} \frac{1}{\Gamma(-\frac{1}{2})}, \qquad (2.4)$$

where we have used the Euler representation for the gamma function. We next carry out the Gaussian integration over k:

$$\mathcal{E} = -\frac{1}{4\sqrt{\pi}} \frac{1}{(4\pi)^{d/2}} \sum_n \int_0^\infty \frac{dt}{t} t^{-1/2-d/2} e^{-tn^2\pi^2/a^2}. \qquad (2.5)$$

Finally, we again use the Euler representation, and carry out the sum over n by use of the definition of the Riemann zeta function:

$$\mathcal{E} = -\frac{1}{4\sqrt{\pi}} \frac{1}{(4\pi)^{d/2}} \left(\frac{\pi}{a}\right)^{d+1} \Gamma\left(-\frac{d+1}{2}\right) \zeta(-d-1). \qquad (2.6)$$

When d is an odd integer, this expression is indeterminate, but we can use the reflection property

$$\Gamma\left(\frac{z}{2}\right)\zeta(z)\pi^{-z/2} = \Gamma\left(\frac{1-z}{2}\right)\zeta(1-z)\pi^{(z-1)/2} \tag{2.7}$$

to rewrite (2.6) as

$$\mathcal{E} = -\frac{1}{2^{d+2}\pi^{d/2+1}}\frac{1}{a^{d+1}}\Gamma\left(1+\frac{d}{2}\right)\zeta(2+d). \tag{2.8}$$

We note that analytic continuation in d is involved here: (2.5) is only valid if $\operatorname{Re} d < -1$ and the subsequent definition of the zeta function is only valid if $\operatorname{Re} d < -2$. In the physical applications, d is a positive integer.

We evaluate this general result (2.8) at $d = 2$. This gives for the energy per unit area in the transverse direction

$$\mathcal{E} = -\frac{\pi^2}{1440}\frac{1}{a^3}, \tag{2.9}$$

where we have recalled that $\zeta(4) = \pi^4/90$. The force per unit area between the plates is obtained by taking the negative derivative of u with respect to a:

$$\mathcal{F}_s = -\frac{\partial}{\partial a}\mathcal{E} = -\frac{\pi^2}{480}\frac{1}{a^4}. \tag{2.10}$$

The above result (2.10) represents the Casimir force due to a scalar field. It is tempting (and, in this case, is correct) to suppose that to obtain the force due to electromagnetic field fluctuations between parallel conducting plates, we simply multiply by a factor of 2 to account for the two polarization states of the photon. Doing so reproduces the classic result of Casimir (2.1):

$$\mathcal{F}_{\text{em}} = -\frac{\pi^2}{240}\frac{1}{a^4}. \tag{2.11}$$

A correct derivation of this result will be given in Sec. 2.6.

2.3　Scalar Green's Function

We now rederive the result of Sec. 2.2 by a physical and rigorous Green's function approach. The equation of motion of a massless scalar field ϕ

produced by a source K is

$$-\partial^2 \phi = K, \tag{2.12}$$

from which we deduce the equation satisfied by the corresponding Green's function

$$-\partial^2 G(x, x') = \delta(x - x'). \tag{2.13}$$

For the geometry shown in Fig. 2.1, we introduce a reduced Green's function $g(z, z')$ according to the Fourier transformation

$$G(x, x') = \int \frac{d^d k}{(2\pi)^d} e^{i\mathbf{k}\cdot(\mathbf{x}-\mathbf{x}')} \int \frac{d\omega}{2\pi} e^{-i\omega(t-t')} g(z, z'), \tag{2.14}$$

where we have suppressed the dependence of g on $\mathbf{k}$ and ω, and have allowed z on the right hand side to represent the coordinate perpendicular to the plates. The reduced Green's function satisfies

$$\left(-\frac{\partial^2}{\partial z^2} - \lambda^2\right) g(z, z') = \delta(z - z'), \tag{2.15}$$

where $\lambda^2 = \omega^2 - k^2$. Equation (2.15) is to be solved subject to the boundary conditions (2.2), or

$$g(0, z') = g(a, z') = 0. \tag{2.16}$$

We solve (2.15) by the standard discontinuity method. The form of the solution is

$$g(z, z') = \begin{cases} A \sin \lambda z, & 0 < z < z' < a, \\ B \sin \lambda(z - a), & a > z > z' > 0, \end{cases} \tag{2.17}$$

which makes use of the boundary condition on the plates (2.16). According to (2.15), g is continuous at $z = z'$, but its derivative has a discontinuity:

$$A \sin \lambda z' - B \sin \lambda(z' - a) = 0, \tag{2.18a}$$
$$A\lambda \cos \lambda z' - B\lambda \cos \lambda(z' - a) = 1. \tag{2.18b}$$

The solution to this system of equations is

$$A = -\frac{1}{\lambda} \frac{\sin \lambda(z' - a)}{\sin \lambda a}, \tag{2.19a}$$
$$B = -\frac{1}{\lambda} \frac{\sin \lambda z'}{\sin \lambda a}, \tag{2.19b}$$

which implies that the reduced Green's function is

$$g(z, z') = -\frac{1}{\lambda \sin \lambda a} \sin \lambda z_< \sin \lambda(z_> - a), \qquad (2.20)$$

where $z_>$ ($z_<$) is the greater (lesser) of z and z'.

From knowledge of the Green's function we can calculate the force on the bounding surfaces from the energy-momentum or stress tensor. For a scalar field, the stress tensor* is

$$T_{\mu\nu} = \partial_\mu \phi \partial_\nu \phi + g_{\mu\nu} \mathcal{L}, \qquad (2.21)$$

where the Lagrange density is

$$\mathcal{L} = -\frac{1}{2} \partial_\lambda \phi \partial^\lambda \phi. \qquad (2.22)$$

What we require is the vacuum expectation value of $T_{\mu\nu}$ which can be obtained from the Green's function according to

$$\langle \phi(x)\phi(x') \rangle = \frac{1}{i} G(x, x'), \qquad (2.23)$$

a time-ordered product being understood in the vacuum expectation value. By virtue of the boundary condition (2.2) we compute the normal-normal component of the stress tensor for a given ω and k (denoted by a lowercase letter) on the boundaries to be

$$\langle t_{zz} \rangle = \frac{1}{2i} \partial_z \partial_{z'} g(z, z')|_{z \to z' = 0, a} = \frac{i}{2} \lambda \cot \lambda a. \qquad (2.24)$$

We now must integrate on the transverse momentum and the frequency to get the force per unit area. The latter integral is best done by performing a complex frequency rotation,

$$\omega \to i\zeta, \quad \lambda \to i\sqrt{k^2 + \zeta^2} \equiv i\kappa. \qquad (2.25)$$

Thus, the force per unit area is given by

$$\mathcal{F} = -\frac{1}{2} \int \frac{d^d k}{(2\pi)^d} \int \frac{d\zeta}{2\pi} \kappa \coth \kappa a. \qquad (2.26)$$

This integral does not exist.

---

*The ambiguity in defining the stress tensor has no effect. We can add to $T_{\mu\nu}$ an arbitrary multiple of $(\partial_\mu \partial_\nu - g_{\mu\nu} \partial^2)\phi^2$ [119, 120]. But the zz component of this tensor on the surface vanishes by virtue of (2.2). Locally, however, there is an effect. See Chapter 11.

What we do now is regard the right boundary at $z = a$, for example, to be a perfect conductor of infinitesimal thickness, and consider the flux of momentum to the right of that surface. To do this, we find the Green's function which vanishes at $z = a$, and has outgoing boundary conditions as $z \to \infty$, $\sim e^{ikz}$. A calculation just like that which led to (2.20) yields for $z, z' > a$,

$$g(z, z') = \frac{1}{\lambda} \sin \lambda (z_< - a) e^{i\lambda(z_> - a)}. \qquad (2.27)$$

The corresponding normal-normal component of the stress tensor at $z = a$ is

$$\langle t_{zz} \rangle|_{z=z'=a} = \frac{1}{2i} \partial_z \partial_{z'} g(z, z')|_{z=z'=a} = \frac{\lambda}{2}. \qquad (2.28)$$

So, from the discontinuity in t_{zz}, that is, the difference between (2.24) and (2.28), we find the force per unit area on the conducting surface:

$$\mathcal{F} = -\frac{1}{2} \int \frac{d^d k}{(2\pi)^d} \int \frac{d\zeta}{2\pi} \kappa (\coth \kappa a - 1). \qquad (2.29)$$

We evaluate this integral using polar coordinates:

$$\mathcal{F} = -\frac{A_{d+1}}{(2\pi)^{d+1}} \int_0^\infty \kappa^d \, d\kappa \frac{\kappa}{e^{2\kappa a} - 1}. \qquad (2.30)$$

Here A_n is the surface area of a unit sphere in n dimensions, which is most easily found by integrating the multiple Gaussian integral

$$\int_{-\infty}^\infty d^n x \, e^{-x^2} = \pi^{n/2} \qquad (2.31)$$

in polar coordinates. The result is

$$A_n = \frac{2\pi^{n/2}}{\Gamma(n/2)}. \qquad (2.32)$$

When we substitute this into (2.30) and use the identity

$$\Gamma(2z) = (2\pi)^{-1/2} 2^{2z - 1/2} \Gamma(z) \Gamma\left(z + \frac{1}{2}\right) \qquad (2.33)$$

as well as one of the defining equations for the Riemann zeta function,

$$\int_0^\infty dy \frac{y^{s-1}}{e^y - 1} = \Gamma(s)\zeta(s), \qquad (2.34)$$

we find for the force per unit transverse area

$$\mathcal{F} = -(d+1)2^{-d-2}\pi^{-d/2-1}\frac{\Gamma\left(1+\frac{d}{2}\right)\zeta(d+2)}{a^{d+2}}. \tag{2.35}$$

Evidently, (2.35) is the negative derivative of the Casimir energy (2.8) with respect to the separation between the plates:

$$\mathcal{F} = -\frac{\partial \mathcal{E}}{\partial a}; \tag{2.36}$$

this result has now been obtained by a completely well-defined approach. The force per unit area, (2.35), is plotted in Fig. 1.2, where $a \to 2a$ and $d = D - 1$.

We can also derive the same result by computing the energy from the energy-momentum tensor[†]. The relevant component is[‡]

$$T_{00} = \frac{1}{2}(\partial_0\phi\partial_0\phi + \partial_1\phi\partial_1\phi + \partial_2\phi\partial_2\phi + \partial_3\phi\partial_3\phi), \tag{2.37}$$

so when the vacuum expectation value is taken, we find from (2.20)

$$\begin{aligned}
\langle t_{00} \rangle &= -\frac{1}{2i\lambda}\frac{1}{\sin\lambda a}[(\omega^2 + k^2)\sin\lambda z \sin\lambda(z-a) \\
&\quad + \lambda^2\cos\lambda z\cos\lambda(z-a)] \\
&= -\frac{1}{2i\lambda\sin\lambda a}[\omega^2\cos\lambda a - k^2\cos\lambda(2z-a)]. \tag{2.38}
\end{aligned}$$

We now must integrate this over z to find the energy per area between the plates. Integration of the second term in (2.38) gives a constant, independent of a, which will not contribute to the force. The first term gives

$$\int_0^a dz \,\langle t_{00} \rangle = -\frac{\omega^2 a}{2i\lambda}\cot\lambda a. \tag{2.39}$$

As above, we now integrate over ω and k, after we perform the complex frequency rotation. We obtain

$$\mathcal{E} = -\frac{a}{2}\int\frac{d^d k}{(2\pi)^d}\int\frac{d\zeta}{2\pi}\frac{\zeta^2}{\kappa}\coth\kappa a. \tag{2.40}$$

[†] Again, the ambiguity in the stress tensor is without effect, because the extra term here is $\nabla^2\phi^2$, which upon integration over space becomes a vanishing surface integral.

[‡] As noted after (1.22), we would get the same integrated energy if we dropped the second, Lagrangian, term in T_{00} there, that is, used $T_{00} = \partial_0\phi\partial_0\phi$.

If we introduce polar coordinates so that $\zeta = \kappa \cos \theta$, we see that this differs from (2.26) by the factor of $a \langle \cos^2 \theta \rangle$. Here

$$\langle \cos^2 \theta \rangle = \frac{\int_0^\pi \cos^2 \theta \sin^{d-1} \theta \, d\theta}{\int_0^\pi \sin^{d-1} \theta \, d\theta} = \frac{1}{d+1}, \tag{2.41}$$

which uses the integral

$$\int_0^\pi \sin^{d-1} \theta \, d\theta = \int_{-1}^1 (1-x^2)^{(d-2)/2} dx = 2^{d-1} \frac{\Gamma\left(\frac{d}{2}\right)^2}{\Gamma(d)}. \tag{2.42}$$

Thus, we again recover (2.8).

For the sake of completeness, we note that it is also possible to use the eigenfunction expansion for the reduced Green's function. That expansion is

$$g(z, z') = \frac{2}{a} \sum_{n=1}^\infty \frac{\sin(n\pi z/a) \sin(n\pi z'/a)}{n^2 \pi^2 / a^2 - \lambda^2}. \tag{2.43}$$

When we insert this into the stress tensor we encounter

$$\partial_z \partial_{z'} g(z, z')|_{z=z'=0,a} = \frac{2}{a} \sum_{n=1}^\infty \frac{n^2 \pi^2 / a^2}{n^2 \pi^2 / a^2 - \lambda^2}. \tag{2.44}$$

We subtract and add λ^2 to the numerator of this divergent sum, and omit the divergent part, which is simply a constant in λ. As we will discuss more fully later, such terms correspond to δ functions in space and time (contact terms), and should be omitted, since we are considering the *limit* as the space-time points coincide. We evaluate the resulting finite sum by use of the following expression for the cotangent:

$$\cot \pi x = \frac{1}{\pi x} + \frac{2x}{\pi} \sum_{k=1}^\infty \frac{1}{x^2 - k^2}. \tag{2.45}$$

So in place of (2.24) we obtain

$$\langle t_{zz} \rangle = \frac{i}{2} \lambda \left(\cot \lambda a - \frac{1}{\lambda a} \right), \tag{2.46}$$

which agrees with (2.24) apart from an additional contact term.

In passing, we note that in the case of periodic boundary conditions, (2.43) is replaced by

$$g(z, z') = \frac{1}{a} \sum_{n=-\infty}^{\infty} \frac{e^{in2\pi(z-z')/a}}{(2n\pi/a)^2 - \lambda^2}, \tag{2.47}$$

so we see immediately that

$$\mathcal{F}_P(a) = \mathcal{F}_D(a/2), \tag{2.48}$$

for the forces referring to periodic and Dirichlet boundary conditions. The corresponding energies are related by

$$\mathcal{E}_P(a) = 2\mathcal{E}_D(a/2). \tag{2.49}$$

See Ref. [82].

Yet another method was proposed by Schwinger in Ref. [103], based on the proper-time representation of the effective action,

$$W = -\frac{i}{2} \int_{s_0 \to 0}^{\infty} \frac{ds}{s} e^{-isH}, \tag{2.50}$$

see Ref. [121]. He used it there in attempting to construct the Casimir energy of a dielectric sphere, which we shall discuss in Chapter 5, but it may be easily applied to the calculation of the Casimir effect between parallel plates—see Refs. [122, 123, 124, 125, 126].

Bordag, Hennig, and Robaschik [127] consider the Casimir effect between plates described by δ-function potentials. If g is the strength of the potential, as $ag \to \infty$ we recover the Casimir energy for Dirichlet plates, (2.9).

2.4 Massive Scalar

It is easy to modify the discussion of Sec. 2.3 to include a mass μ for the scalar field. The reduced Green's function now satisfies the equation

$$\left(-\frac{\partial^2}{\partial z^2} - \lambda^2\right) g(z, z') = \delta(z - z'), \tag{2.51}$$

where

$$\lambda^2 = \omega^2 - k^2 - \mu^2, \tag{2.52}$$

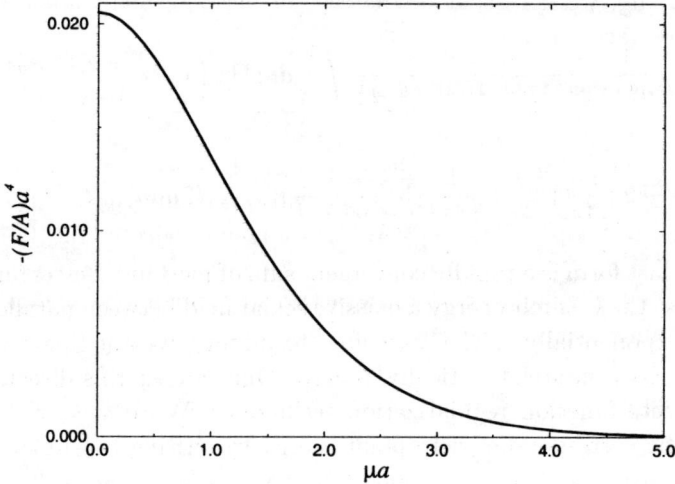

Fig. 2.2 Scalar Casimir force per unit area, $\mathcal{F}$, between parallel plates as a function of mass for $d = 2$.

instead of (2.15), so the reduced Green's function between the plates has just the form (2.20). The calculation proceeds just as in Sec. 2.3, and we find, in place of (2.30)

$$\mathcal{F} = -\frac{A_{d+1}}{(2\pi)^{d+1}} \int_0^\infty \kappa^d \, d\kappa \frac{\sqrt{\kappa^2 + \mu^2}}{e^{2a\sqrt{\kappa^2+\mu^2}} - 1}. \qquad (2.53)$$

When we substitute the value of A_{d+1} given by (2.32), and introduce a dimensionless integration variable, we find for the force per unit area

$$\mathcal{F} = -\frac{1}{2^{2(d+1)}\pi^{(d+1)/2}\Gamma\left(\frac{d+1}{2}\right)a^{d+2}} \int_{2\mu a}^\infty dx \, x^2 \frac{(x^2 - 4\mu^2 a^2)^{(d-1)/2}}{e^x - 1}. \qquad (2.54)$$

For $d = 2$ this function is plotted in Fig. 2.2. Ambjørn and Wolfram [82] considered the case of massive fields in D dimensions, which had been first treated by Hays in the two-dimensional case [128]. They expressed the result in terms of the corresponding energy, which can be presented in alternative forms (there is a misprint in (2.18) of Ref. [82] for the first form here)

$$\mathcal{E} = \frac{1}{a^{d+1}} \frac{1}{2^{d+1}\pi^{(d+1)/2}\Gamma\left(\frac{d+1}{2}\right)} \int_0^\infty dt \, t^d \ln\left(1 - e^{-2\sqrt{t^2+\mu^2 a^2}}\right)$$

$$(2.55\mathrm{a})$$

$$= -2\left(\frac{\mu a}{4\pi}\right)^{d/2+1} \frac{1}{a^{d+1}} \sum_{n=1}^\infty \frac{1}{n^{d/2+1}} K_{d/2+1}(2\mu a n),\qquad (2.55\mathrm{b})$$

where the last form is a rapidly convergent sum of modified Bessel functions.

Because the Casimir energy a massive scalar field between parallel plates vanishes exponentially as the mass goes to infinity, we anticipate that the Casimir energy, nonrelativistically, is zero. One can see this directly using a simple zeta-function regularization technique. Write $\omega = p^2/2\mu$, and evaluate the corresponding zero-point energy by writing $(s \to 1)$

$$\frac{1}{2}\sum\omega = \frac{1}{4\mu}\sum_{n=1}^\infty \int \frac{d^d k}{(2\pi)^d}\left(k^2 + \frac{n^2\pi^2}{a^2}\right)^s$$

$$= \frac{1}{2^{d+2}\pi^{(1+d)/2}\mu a^{d+2s}} \frac{\Gamma(1/2+s+d/2)\zeta(1+2s+d)}{\Gamma(-s)};\quad (2.56)$$

[see (2.4)]. Notice that this reduces to $1/2\mu$ times (2.8) for $s = 1/2$. Evidently, the $s \to 1$ limit vanishes for all $d > -2$.

2.5   Finite Temperature

We next turn to a consideration of the Casimir effect at nonzero temperature. In this case, fluctuations arise not only from quantum mechanics but from thermal effects. In fact, as we will shortly see, the high-temperature limit is a purely classical phenomenon. Finite temperature effects were first discussed by Lifshitz [7], but then considered more fully by Fierz, Sauer, and Mehra [129, 3, 4]. (Fierz's early calculation referred only to the energy, and not the free energy or force.) Hargreaves [10] analyzed the discrepancy between the results of Lifshitz [7] and Sauer [3], which turned out to be the result of transcription errors in the former paper [130]. An excellent treatment of the parallel plate problem using the stress-tensor approach and the method of images was given by Brown and Maclay [100], for both zero and finite temperatures. A multiple-scattering formulation was presented by Balian and Duplantier [131, 14].

Formally, we can easily obtain the expression for the Casimir force between parallel plates at nonzero temperature. In (2.29) we replace the imaginary frequency by

$$\zeta \to \zeta_n = \frac{2\pi n}{\beta}, \tag{2.57}$$

where $\beta = 1/kT$, T being the absolute temperature. Correspondingly, the frequency integral is replaced by a sum on the integer n:

$$\int_{-\infty}^{\infty} \frac{d\zeta}{2\pi} \to \frac{1}{\beta} \sum_{n=-\infty}^{\infty} . \tag{2.58}$$

Thus, (2.29) is replaced by

$$\mathcal{F}^T = -\frac{1}{2\beta} \int \frac{d^d k}{(2\pi)^d} \sum_{n=-\infty}^{\infty} \frac{2\kappa_n}{e^{2\kappa_n a} - 1}, \tag{2.59}$$

where $\kappa_n = \sqrt{k^2 + (2\pi n/\beta)^2}$.

We first consider the high-temperature limit. When $T \to \infty$ ($\beta \to 0$), apart from exponentially small corrections (considered in Sec. 3.2), the contribution comes from the $n = 0$ term in the sum in (2.59). That integral is easily worked out in polar coordinates using (2.32) and (2.34). The result is

$$\mathcal{F}^{T\to\infty} \sim -kT \frac{d}{(2\sqrt{\pi}a)^{d+1}} \Gamma\left(\frac{d+1}{2}\right) \zeta(d+1). \tag{2.60}$$

In particular, for two and three dimensions, $d = 1$ and $d = 2$, respectively,

$$d = 1: \quad \mathcal{F}^{T\to\infty} \sim -kT \frac{\pi}{24a^2}, \tag{2.61a}$$

$$d = 2: \quad \mathcal{F}^{T\to\infty} \sim -kT \frac{\zeta(3)}{8\pi a^3}. \tag{2.61b}$$

Note that if we apply the same procedure to the energy expression (2.40) we find that the energy vanishes in the high temperature limit, because $\zeta_0 = 0$. Accordingly, the entropy approaches a constant. This is as expected from thermodynamics, as discussed by Ref. [132].

This high-temperature limit should be classical. Indeed, we can derive this same result from the classical limit of statistical mechanics. The

Helmholtz free energy for massless bosons is

$$F = -kT \ln Z, \quad \ln Z = -\sum_i \ln(1 - e^{-\beta p_i}), \qquad (2.62)$$

from which the pressure on the plates can be obtained by differentiation:

$$p = -\frac{\partial F}{\partial V}. \qquad (2.63)$$

We make the momentum-space sum explicit for our $d+1$ spatial geometry:

$$F = kTV \int \frac{d^d k}{(2\pi)^{d+1}} \frac{\pi}{a} \sum_{n=-\infty}^{\infty} \ln(1 - \exp(-\beta\sqrt{k^2 + n^2\pi^2/a^2})). \qquad (2.64)$$

Now, for high temperature, $\beta \to 0$, we expand the exponential, and keep the first order term in β. We can write the result as

$$F = VkT \frac{1}{2a} \frac{d}{ds} \int \frac{d^d k}{(2\pi)^d} \sum_{n=-\infty}^{\infty} \frac{1}{2}\beta^{2s} \left(\frac{n^2\pi^2}{a^2} + k^2\right)^s \Bigg|_{s=0}, \qquad (2.65)$$

where we have used the identity

$$\ln \xi = \frac{d}{ds}\xi^s \Bigg|_{s=0}. \qquad (2.66)$$

This trick allows us to proceed as in Sec. 2.2. After the k integration is done, the s derivative acts only on $1/\Gamma(-s)$:

$$\frac{d}{ds}\frac{1}{\Gamma(-s)}\Bigg|_{s=0} = -1, \qquad (2.67)$$

so we easily find the result from (2.7)

$$F = -V\frac{kT}{(2a\sqrt{\pi})^{d+1}}\zeta(d+1)\Gamma\left(\frac{1+d}{2}\right). \qquad (2.68)$$

The pressure, the force per unit area on the plates, is obtained by applying the following differential operator to the free energy:

$$-\frac{\partial}{\partial V} = -\frac{1}{A}\frac{\partial}{\partial a}, \qquad (2.69)$$

where $V = Aa$; A being the d-dimensional area of the plates. The result of this operation coincides with (2.60).

The low-temperature limit ($T \to 0$ or $\beta \to \infty$) is more complicated because $\mathcal{F}^T$ is not analytic at $T = 0$. The most convenient way to proceed is to resum the series in (2.59) by means of the Poisson sum formula, which says that if $c(\alpha)$ is the Fourier transform of $b(x)$,

$$c(\alpha) = \frac{1}{2\pi} \int_{-\infty}^{\infty} b(x) e^{-i\alpha x}\, dx, \qquad (2.70)$$

then the following identity holds:

$$\sum_{n=-\infty}^{\infty} b(n) = 2\pi \sum_{n=-\infty}^{\infty} c(2\pi n). \qquad (2.71)$$

Here, we take

$$b(n) = \int \frac{d^d k}{(2\pi)^d} \frac{2\kappa_n}{e^{2\kappa_n a} - 1}. \qquad (2.72)$$

Introducing polar coordinates for $\mathbf{k}$, changing from k to the dimensionless integration variable $z = 2a\kappa_x$, and interchanging the order of x and z integration, we find for the Fourier transform[§]

$$c(\alpha) = \frac{A_d}{2^{2d}\pi^{d+1}a^{d+1}} \int_0^{\infty} \frac{dz\, z^2}{e^z - 1} \int_0^{\beta z/4\pi a} dx\, \cos \alpha x \left(z^2 - \left(\frac{4\pi a x}{\beta} \right)^2 \right)^{(d-2)/2}. \qquad (2.73)$$

The x integral in (2.73) is easily expressed in terms of a Bessel function: It is

$$\frac{\beta z^{d-1}}{4\pi a} \int_0^1 du\, \cos \left(\frac{\alpha \beta z}{4\pi a} u \right) (1 - u^2)^{(d-2)/2}$$

$$= \alpha^{(1-d)/2} z^{(d-1)/2} \left(\frac{8\pi a}{\beta} \right)^{(d-3)/2} \sqrt{\pi}\, \Gamma \left(\frac{d}{2} \right) J_{(d-1)/2} \left(\frac{\alpha \beta z}{4\pi a} \right). \qquad (2.74)$$

We thus encounter the z integral

$$I(s) = \int_0^{\infty} \frac{dz\, z^{(d+3)/2}}{e^z - 1} J_{(d-1)/2}(sz), \qquad (2.75)$$

where $s = \alpha\beta/4\pi a$.

The zero-temperature limit comes entirely from $\alpha = 0$:

$$\mathcal{F}^{T=0} = -\frac{\pi}{\beta} c(0). \qquad (2.76)$$

[§]It is obvious that $d = 2$ is an especially simple case. The calculation then is described in Sec. 3.2.

So we require the small-x behavior of the Bessel function,

$$J_{(d-1)/2}(x) \sim \left(\frac{x}{2}\right)^{(d-1)/2} \frac{1}{\Gamma\left(\frac{d+1}{2}\right)}, \quad x \to 0, \tag{2.77}$$

whence

$$I(s) \sim \left(\frac{s}{2}\right)^{(d-1)/2} \frac{2^{d+1}}{\sqrt{\pi}} \frac{d+1}{2} \Gamma\left(\frac{d}{2}+1\right) \zeta(d+2), \quad s \to 0. \tag{2.78}$$

Inserting this into (2.73) we immediately recover the zero-temperature result (2.35).

We now seek the leading correction to this. We rewrite $I(s)$ as

$$\begin{aligned}
I(s) &= \frac{1}{s^{(d+5)/2}} \int_0^\infty \frac{dy\, y^{(d+3)/2}}{1 - e^{-y/s}} e^{-y/s} J_{(d-1)/2}(y) \\
&= \frac{1}{s^{(d+5)/2}} \int_0^\infty dy\, y^{(d+3)/2} e^{-y/s} \sum_{n=0}^\infty e^{-ny/s} J_{(d-1)/2}(y), \quad (2.79)
\end{aligned}$$

where we have employed the geometric series. The Bessel-function integral has an elementary form:

$$\int_0^\infty dy\, J_q(y) e^{-by} = \frac{(\sqrt{b^2+1}-b)^q}{\sqrt{b^2+1}} \tag{2.80}$$

is the fundamental integral, and the form we want can be written as

$$\int_0^\infty dy\, J_q(y) y^p e^{-by} = (-1)^p \left(\frac{d}{db}\right)^p \frac{(\sqrt{b^2+1}-b)^q}{\sqrt{b^2+1}}, \tag{2.81}$$

provided p is a nonnegative integer. (For the application here, this means d is odd, but we will be able to analytically continue the final result to arbitrary d.) Then we can write $I(s)$ in terms of the series

$$I(s) = \frac{(-1)^{(d+3)/2}}{s^{(d+5)/2}} \sum_{l=1}^\infty f(l), \quad l = n+1, \tag{2.82}$$

where

$$f(l) = \left(\frac{d}{dl/s}\right)^{(d+3)/2} \frac{\left[\sqrt{\left(\frac{l}{s}\right)^2+1} - \frac{l}{s}\right]^{(d-1)/2}}{\sqrt{\left(\frac{l}{s}\right)^2+1}}. \tag{2.83}$$

We evaluate the sum in (2.82) by means of the Euler-Maclaurin summation formula, which has the following formal expression:

$$\sum_{l=1}^{\infty} f(l) = \int_1^{\infty} dl\, f(l) + \frac{1}{2}[f(\infty) + f(1)]$$

$$+ \sum_{k=1}^{\infty} \frac{1}{(2k)!} B_{2k}[f^{(2k-1)}(\infty) - f^{(2k-1)}(1)]. \qquad (2.84)$$

Here, B_n represents the nth Bernoulli number. Since we are considering $\alpha \neq 0$ (the $\alpha = 0$ term was dealt with in the previous paragraph), the low-temperature limit corresponds to the limit $s \to \infty$. It is easy then to see that $f(\infty)$ and all its derivatives there vanish. The function f at 1 has the general form

$$f(1) = \left(\frac{d}{d\epsilon}\right)^{(d+3)/2} \frac{\left[\sqrt{\epsilon^2 + 1} - \epsilon\right]^{(d-1)/2}}{\sqrt{\epsilon^2 + 1}}, \qquad \epsilon = 1/s \to 0. \qquad (2.85)$$

By examining the various possibilities for odd d, $d = 1$, $d = 3$, $d = 5$, and so on, we find the result

$$f(1) = -(-1)^{(d-1)/2} d!! = -(-1)^{(d-1)/2} 2^{(d+1)/2} \pi^{-1/2} \Gamma\left(\frac{d}{2} + 1\right). \qquad (2.86)$$

Because it is easily seen that before the limit $\epsilon \to 0$ is taken, $f(1)$ is an even function of ϵ, it follows that the odd derivatives of f evaluated at 1 that appear in (2.84) vanish. Finally, the integral in (2.84) is that of a total derivative:

$$\int_1^{\infty} dl\, f(l) = -\frac{1}{\epsilon} \left(\frac{d}{d\epsilon}\right)^{(d+1)/2} \frac{\left[\sqrt{\epsilon^2 + 1} - \epsilon\right]^{(d-1)/2}}{\sqrt{\epsilon^2 + 1}} = -f(1). \qquad (2.87)$$

Thus, the final expression for $I(s)$ is

$$I(s) \sim \frac{1}{s^{(d+5)/2}} 2^{(d-1)/2} \pi^{-1/2} \Gamma\left(\frac{d}{2} + 1\right). \qquad (2.88)$$

Note that a choice of analytic continuation has been made so as to avoid oscillatory behavior in d.

We return to (2.73). It may be written as

$$c(\alpha) = \frac{2\pi^{d/2}}{\Gamma(d/2)} \frac{1}{2^{2d}\pi^{d+1}a^{d+1}} \alpha^{(1-d)/2} \left(\frac{8\pi a}{\beta}\right)^{(d-3)/2} \sqrt{\pi}\, \Gamma(d/2) I\left(\frac{\alpha\beta}{4\pi a}\right)$$

$$= 2^{-d-1}\pi^{-d/2-1}\left(\frac{4\pi a}{\beta}\right)^{d+1}\frac{1}{a^{d+1}}\Gamma\left(\frac{d}{2}+1\right)\frac{1}{\alpha^{d+2}}. \tag{2.89}$$

The correction to the zero-temperature result (2.35) is obtained from

$$\mathcal{F}_{\text{corr}}^{T\to0} = -\frac{2\pi}{\beta}\sum_{n=1}^{\infty}c(2\pi n)$$

$$= -\pi^{-d/2-1}\Gamma(d/2+1)\zeta(d+2)\beta^{-d-2}. \tag{2.90}$$

Thus, the force per unit area in the low-temperature limit has the form

$$\mathcal{F}^{T\to0} \approx -\frac{(d+1)2^{-d-2}\pi^{-d/2-1}}{a^{d+2}}\Gamma(d/2+1)\zeta(d+2)\left[1+\frac{1}{d+1}\left(\frac{2a}{\beta}\right)^{d+2}\right], \tag{2.91}$$

of which the $d=1$ and $d=2$ cases are familiar:

$$d=1: \quad \mathcal{F}^{T\to0}\approx -\frac{1}{8\pi a^3}\zeta(3)\left[1+4\frac{a^3}{\beta^3}\right], \tag{2.92a}$$

$$d=2: \quad \mathcal{F}^{T\to0}\approx -\frac{\pi^2}{480a^4}\left[1+\frac{16}{3}\frac{a^4}{\beta^4}\right]. \tag{2.92b}$$

These equations are incomplete in that they omit exponentially small terms; for example, in the last square bracket, we should add the term

$$-\frac{240}{\pi}\frac{a}{\beta}e^{-\pi\beta/a}. \tag{2.93}$$

We will discuss such corrections in Sec. 3.2.

Mitter and Robaschik [133] considered the Casimir effect between two plates where the temperature between the plates T is different from the external temperature T'. If $T' < T$ the difference in thermal pressure can balance the Casimir attraction.

2.6 Electromagnetic Casimir Force

We now turn to the situation originally treated by Casimir: the force between parallel conducting plates due to quantum fluctuations in the electromagnetic field. An elegant procedure,[¶] which can be applied to much more

---

[¶] One advantage of this scheme is its explicit gauge invariance, as contrasted with methods making use of the Green's functions for the potentials.

complicated geometries, involves the introduction of the Green's dyadic, defined as the response function between the (classical) electromagnetic field and the polarization source [11]:

$$\mathbf{E}(x) = \int d^4x' \, \mathbf{\Gamma}(x, x') \cdot \mathbf{P}(x').$$ (2.94)

In the following we will use the Fourier transform of $\mathbf{\Gamma}$ in frequency:

$$\mathbf{\Gamma}(x, x') = \int \frac{d\omega}{2\pi} e^{-i\omega(t-t')} \mathbf{\Gamma}(\mathbf{r}, \mathbf{r}'; \omega),$$ (2.95)

which satisfies the Maxwell equations

$$\mathbf{\nabla} \times \mathbf{\Gamma} = i\omega\mathbf{\Phi},$$ (2.96a)

$$-\mathbf{\nabla} \times \mathbf{\Phi} - i\omega\mathbf{\Gamma} = i\omega\mathbf{1}\delta(\mathbf{r} - \mathbf{r}').$$ (2.96b)

The second Green's dyadic appearing here is solenoidal,

$$\mathbf{\nabla} \cdot \mathbf{\Phi} = 0,$$ (2.96c)

as is $\mathbf{\Gamma}$ if a multiple of a δ function is subtracted:

$$\mathbf{\nabla} \cdot \mathbf{\Gamma}' = 0, \quad \mathbf{\Gamma}' = \mathbf{\Gamma} + \mathbf{1}\delta(\mathbf{r} - \mathbf{r}').$$ (2.96d)

The system of first-order equations (2.96a), (2.96b) can be easily converted to second-order form:

$$(\nabla^2 + \omega^2)\mathbf{\Gamma}' = -\mathbf{\nabla} \times (\mathbf{\nabla} \times \mathbf{1})\delta(\mathbf{r} - \mathbf{r}'),$$ (2.97a)

$$(\nabla^2 + \omega^2)\mathbf{\Phi} = i\omega\mathbf{\nabla} \times \mathbf{1}\delta(\mathbf{r} - \mathbf{r}').$$ (2.97b)

The system of equations (2.97a), (2.97b) is quite general. We specialize to the case of parallel plates by introducing the transverse Fourier transform:

$$\mathbf{\Gamma}'(\mathbf{r}, \mathbf{r}'; \omega) = \int \frac{d^2k}{(2\pi)^2} e^{i\mathbf{k}\cdot(\mathbf{r}_\perp - \mathbf{r}'_\perp)} \mathbf{g}(z, z'; \mathbf{k}, \omega).$$ (2.98)

The equations satisfied by the various Cartesian components of $\mathbf{\Gamma}$ may be easily worked out once it is recognized that

$$[\mathbf{\nabla} \times (\mathbf{\nabla} \times \mathbf{1})]_{ij} = \partial_i\partial_j - \delta_{ij}\partial^2.$$ (2.99)

In terms of the Fourier transforms, these equations are

$$\left(\frac{\partial^2}{\partial z^2} - k^2 + \omega^2\right) g_{zz} = -k^2 \delta(z - z'), \tag{2.100a}$$

$$\left(\frac{\partial^2}{\partial z^2} - k^2 + \omega^2\right) g_{zx} = -ik_x \frac{\partial}{\partial z} \delta(z - z'), \tag{2.100b}$$

$$\left(\frac{\partial^2}{\partial z^2} - k^2 + \omega^2\right) g_{zy} = -ik_y \frac{\partial}{\partial z} \delta(z - z'), \tag{2.100c}$$

$$\left(\frac{\partial^2}{\partial z^2} - k^2 + \omega^2\right) g_{xx} = \left(-k_y^2 + \frac{\partial^2}{\partial z^2}\right) \delta(z - z'), \tag{2.100d}$$

$$\left(\frac{\partial^2}{\partial z^2} - k^2 + \omega^2\right) g_{yy} = \left(-k_x^2 + \frac{\partial^2}{\partial z^2}\right) \delta(z - z'), \tag{2.100e}$$

$$\left(\frac{\partial^2}{\partial z^2} - k^2 + \omega^2\right) g_{xy} = k_x k_y \delta(z - z'). \tag{2.100f}$$

We solve these equations subject to the boundary condition that the transverse components of the electric field vanish on the conducting surfaces, that is,

$$\mathbf{n} \times \mathbf{\Gamma'}|_{z=0,a} = 0, \tag{2.101}$$

where $\mathbf{n}$ is the normal to the surface. That means any x or y components vanish at $z = 0$ or at $z = a$. Therefore, g_{xy} is particularly simple. By the standard discontinuity method, we immediately find [cf. (2.20)]

$$g_{xy} = g_{yx} = \frac{k_x k_y}{\lambda \sin \lambda a}(ss), \tag{2.102}$$

where

$$(ss) = \sin \lambda z_< \sin \lambda(z_> - a). \tag{2.103}$$

To find g_{xx} we simply subtract a δ function:

$$g'_{xx} = g_{xx} - \delta(z - z'). \tag{2.104}$$

Then, we again find at once

$$g'_{xx} = \frac{k_x^2 - \omega^2}{\lambda \sin \lambda a}(ss), \tag{2.105a}$$

and similarly

$$g'_{yy} = \frac{k_y^2 - \omega^2}{\lambda \sin \lambda a}(ss).$$ (2.105b)

To determine the boundary condition on g_{zz}, we recall the solenoidal condition on $\mathbf{\Gamma}'$, (2.96d), which implies that

$$\left.\frac{\partial}{\partial z}g_{zz}\right|_{z=0,a} = 0.$$ (2.106)

This then leads straightforwardly to the conclusion

$$g_{zz} = -\frac{k^2}{\lambda \sin \lambda a}(cc),$$ (2.107)

where

$$(cc) = \cos \lambda z_< \cos \lambda(z_> - a).$$ (2.108)

The remaining components have the property that the functions are discontinuous, while, apart from a δ function, their derivatives are continuous:

$$g_{zx} = -\frac{ik_x}{\sin \lambda a}(cs),$$ (2.109a)

$$g_{zy} = -\frac{ik_y}{\sin \lambda a}(cs),$$ (2.109b)

where

$$(cs) = \begin{cases} \cos \lambda z \sin \lambda(z' - a), & z < z', \\ \sin \lambda z' \cos \lambda(z - a), & z > z'. \end{cases}$$ (2.110)

Similarly,

$$g_{xz} = \frac{ik_x}{\sin \lambda a}(sc),$$ (2.111a)

$$g_{yz} = \frac{ik_y}{\sin \lambda a}(sc),$$ (2.111b)

where

$$(sc) = \begin{cases} \sin \lambda z \cos \lambda(z' - a), & z < z', \\ \cos \lambda z' \sin \lambda(z - a), & z > z', \end{cases}$$ (2.112)

which just reflects the symmetry

$$\mathbf{\Gamma}'(\mathbf{r}, \mathbf{r}') = \mathbf{\Gamma}'(\mathbf{r}', \mathbf{r}). \tag{2.113}$$

The normal-normal component of the electromagnetic stress tensor is

$$T_{zz} = \frac{1}{2}(E^2 + H^2) - (E_z^2 + H_z^2). \tag{2.114}$$

The vacuum expectation value is obtained by the replacements

$$\langle \mathbf{E}(x)\mathbf{E}(x')\rangle = \frac{1}{i}\mathbf{\Gamma}(x, x'), \tag{2.115a}$$

$$\langle \mathbf{H}(x)\mathbf{H}(x')\rangle = -\frac{1}{i}\frac{1}{\omega^2}\mathbf{\nabla} \times \mathbf{\Gamma}(x, x') \times \overleftarrow{\mathbf{\nabla}}'. \tag{2.115b}$$

In terms of the Fourier transforms, we have

$$\begin{aligned}
\langle t_{zz}\rangle = \frac{1}{2i\omega^2}\Big[&-(\omega^2 - k^2)g_{zz} + (\omega^2 - k_y^2)g_{xx} + (\omega^2 - k_x^2)g_{yy} \\
&+ ik_y\left(\frac{\partial}{\partial z}g_{yz} - \frac{\partial}{\partial z'}g_{zy}\right) + ik_x\left(\frac{\partial}{\partial z}g_{xz} - \frac{\partial}{\partial z'}g_{zx}\right) \\
&+ k_xk_y(g_{xy} + g_{yx}) + \frac{\partial}{\partial z}\frac{\partial}{\partial z'}(g_{xx} + g_{yy})\Big].
\end{aligned} \tag{2.116}$$

When the appropriate Green's functions are inserted into the above, enormous simplification occurs on the surface, and we are left with

$$\langle t_{zz}\rangle|_{z=0,a} = i\lambda \cot \lambda a, \tag{2.117}$$

which indeed is twice the scalar result (2.24), as claimed at the end of Sec. 2.2.

2.6.1 *Variations*

The force between a perfectly conducting plate and a perfectly permeable one was worked out by Boyer [134] and studied more recently in Refs. [135, 136]. A repulsive result, $-\frac{7}{8}$ times that for a scalar field with Dirichlet boundary conditions, (2.9), is found at zero temperature. It is extremely interesting that this answer differs only by a sign from the fermionic Casimir force we will derive in the next section. Kenneth and Nussinov [137] derive the Casimir effect between plates which conduct in single, different, directions. As expected, when the conductivities are parallel, the energy is 1/2 that for ordinary conductors.

Fluctuations in the Casimir stress have been considered by Barton [138]. If τ is the observation time, then for $\tau \gg a$,

$$\frac{\Delta \mathcal{F}}{\mathcal{F}} = \mathcal{O}\left(\frac{a}{\tau}\right)^3, \tag{2.118}$$

far beyond experimental reach. The stress correlation function was analyzed by Barton in Ref. [139]. See also Refs. [140, 141, 142].

How real is Casimir energy? Just as real as any other form, as demonstrated by Jaekel and Reynaud [143], who consider the mechanical and inertial properties of Casimir energy and conclude that "vacuum fluctuations result in mechanical effects which conform with general principles of mechanics."

Mention should also be made of the *Scharnhorst* effect, in which light speeds greater than the vacuum speed of light are possible in a parallel plate capacitor, as an induced consequence of the Casimir effect [144, 145, 146]. It is interesting that Schwinger in 1990 wrote a manuscript, which may never have been submitted to a journal, that claimed, in contradiction with the above referenced results, that the effect was nonuniform, dispersive, and persisted if only a single plate was present [147].

2.7   Fermionic Casimir Force

We conclude this Chapter with a discussion of the force on parallel surfaces due to fluctuations in a massless Dirac fermionic field. For this simple geometry, the primary distinction between this case and what has gone before lies in the boundary conditions. The boundary conditions appropriate to the Dirac equation are the so-called bag-model boundary conditions. That is, if n^μ represents an outward normal at a boundary surface, the condition on the Dirac field ψ there is

$$(1 + in \cdot \gamma)\psi = 0. \tag{2.119}$$

For the situation of parallel plates at $z = 0$ and $z = a$, this means

$$(1 \mp i\gamma^3)\psi = 0 \tag{2.120}$$

at $z = 0$ and $z = a$, respectively. In the following, we will choose a representation of the Dirac matrices in which $i\gamma_5$ is diagonal, in 2×2 block form,

$$i\gamma_5 = \begin{pmatrix} 1 & 0 \\ 0 & -1 \end{pmatrix} \tag{2.121a}$$

while

$$\gamma^0 = \begin{pmatrix} 0 & -i \\ i & 0 \end{pmatrix}, \tag{2.121b}$$

from which the explicit form of all the other Dirac matrices follow from $\gamma = i\gamma^0\gamma_5\boldsymbol{\sigma}$.

The effect of fermionic fluctuations was first investigated by Johnson [87], and quoted in Ref. [86]. (The bag model and its boundary conditions were introduced in [84].)

2.7.1 *Summing Modes*

It is easiest, but not rigorous, to sum modes as in Sec. 2.2. We introduce a Fourier transform in time and the transverse spatial directions,

$$\psi(x) = \int \frac{d\omega}{2\pi} e^{-i\omega t} \int \frac{d^2k}{(2\pi)^2} e^{i\mathbf{k}\cdot\mathbf{x}} \psi(z; \mathbf{k}, \omega), \tag{2.122}$$

so that the Dirac equation for a massless fermion $-i\gamma\partial\psi = 0$ becomes, in the coordinate system in which $\mathbf{k}$ lies along the x axis,

$$\left(-\omega \mp i\frac{\partial}{\partial z}\right) u_\pm \pm k v_\pm = 0, \tag{2.123a}$$

$$\pm k u_\pm + \left(-\omega \pm i\frac{\partial}{\partial z}\right) v_\pm = 0, \tag{2.123b}$$

where the subscripts indicate the eigenvalues of $i\gamma_5$ and and u and v are eigenvectors of σ^3 with eigenvalue $+1$ or -1, respectively. This system of equations is to be solved to the boundary conditions (2.120), or

$$u_+ + u_-|_{z=0} = 0, \tag{2.124a}$$

$$v_+ - v_-|_{z=0} = 0, \tag{2.124b}$$

$$u_+ - u_-|_{z=a} = 0, \tag{2.124c}$$

$$v_+ + v_-|_{z=a} = 0. \tag{2.124d}$$

The solution is straightforward. Each component satisfies

$$\left(\frac{\partial^2}{\partial z^2} + \lambda^2\right)\psi = 0, \tag{2.125}$$

where $\lambda^2 = \omega^2 - k^2$, so each component is expressed as follows:

$$u_+ + u_- = A \sin \lambda z, \tag{2.126a}$$
$$v_+ - v_- = B \sin \lambda z, \tag{2.126b}$$
$$u_+ - u_- = C \sin \lambda(z - a), \tag{2.126c}$$
$$v_+ + v_- = D \sin \lambda(z - a). \tag{2.126d}$$

Inserting these into the Dirac equation (2.123a) and (2.123b), we find, first, a condition on λ:

$$\cos \lambda a = 0, \tag{2.127}$$

or

$$\lambda a = \left(n + \frac{1}{2}\right)\pi, \tag{2.128}$$

where n is an integer. We then find two independent solutions for the coefficients:

$$A \neq 0, \tag{2.129a}$$
$$B = 0, \tag{2.129b}$$
$$C = \frac{i\omega}{\lambda}(-1)^n A, \tag{2.129c}$$
$$D = \frac{ik}{\lambda}(-1)^n A, \tag{2.129d}$$

and

$$A = 0, \tag{2.130a}$$
$$B \neq 0, \tag{2.130b}$$
$$C = \frac{k}{i\lambda}(-1)^n B, \tag{2.130c}$$
$$D = \frac{\omega}{i\lambda}(-1)^n B. \tag{2.130d}$$

Thus, when we compute the zero-point energy, we must sum over odd integers, noting that there are two modes, and remembering the characteristic minus sign for fermions: Instead of (2.3), the Casimir energy is

$$
\begin{aligned}
u &= -2\frac{1}{2}\sum_{n=0}^{\infty}\frac{d^2k}{(2\pi)^2}\sqrt{k^2 + \frac{(n+1/2)^2\pi^2}{a^2}}\\
&= \frac{1}{2\sqrt{\pi}}\frac{1}{4\pi}\sum_{n=0}^{\infty}\int_0^{\infty}\frac{dt}{t}t^{-3/2}e^{-t(n+1/2)^2\pi^2/a^2}\\
&= \frac{1}{8\pi^{3/2}}\Gamma\left(-\frac{3}{2}\right)\sum_{n=0}^{\infty}\frac{(n+1/2)^3\pi^3}{a^3}\\
&= -\frac{\pi^2}{6a^3}\frac{7}{8}\zeta(-3),
\end{aligned}
\tag{2.131}
$$

which is $\frac{7}{8}\times 2$ times the scalar result (2.9) because $\zeta(-3) = -B_4/4 = 1/120$. (The factor of 2 refers to the two spin modes of the fermion.)

2.7.2　Green's Function Method

Again, a more controlled calculation starts from the equation satisfied by the Dirac Green's function,

$$
\gamma\frac{1}{i}\partial G(x,x') = \delta(x-x'),
\tag{2.132}
$$

subject to the boundary condition

$$
(1 + i\mathbf{n}\cdot\boldsymbol{\gamma})G|_{z=0,a} = 0.
\tag{2.133}
$$

We introduce a reduced, Fourier-transformed, Green's function,

$$
G(x,x') = \int\frac{d\omega}{2\pi}e^{-i\omega(t-t')}\int\frac{d^2k}{(2\pi)^2}e^{i\mathbf{k}\cdot(\mathbf{x}-\mathbf{x}')}g(z,z';\mathbf{k},\omega),
\tag{2.134}
$$

which satisfies

$$
\left(-\gamma^0\omega + \boldsymbol{\gamma}\cdot\mathbf{k} + \gamma^3\frac{1}{i}\frac{\partial}{\partial z}\right)g(z,z') = \delta(z-z').
\tag{2.135}
$$

Introducing the representation for the gamma matrices given above, we find that the components of g corresponding to the $+1$ or -1 eigenvalues of $i\gamma_5$,

$$
g = \begin{pmatrix} g_{++} & g_{+-} \\ g_{-+} & g_{--} \end{pmatrix},
\tag{2.136}
$$

satisfy the coupled set of equations

$$\left(-\omega \pm \boldsymbol{\sigma} \cdot \mathbf{k} \mp i\sigma^3 \frac{\partial}{\partial z}\right) g_{\pm\pm} = 0, \tag{2.137a}$$

$$\left(-\omega \pm \boldsymbol{\sigma} \cdot \mathbf{k} \mp i\sigma^3 \frac{\partial}{\partial z}\right) g_{\pm\mp} = \mp i\delta(z - z'). \tag{2.137b}$$

We then resolve each of these components into eigenvectors of σ^3:

$$g_{\pm\pm} = \begin{pmatrix} u_{\pm\pm}^{(+)} & u_{\pm\pm}^{(-)} \\ v_{\pm\pm}^{(+)} & v_{\pm\pm}^{(-)} \end{pmatrix} \tag{2.138}$$

and similarly for $g_{\pm\mp}$. These components satisfy the coupled equations

$$\left(-\omega \mp i\frac{\partial}{\partial z}\right) u_{\pm\pm}^{(\pm)} \pm k v_{\pm\pm}^{(\pm)} = 0, \tag{2.139a}$$

$$\pm k u_{\pm\pm}^{(\pm)} + \left(-\omega \pm i\frac{\partial}{\partial z}\right) v_{\pm\pm}^{(\pm)} = 0, \tag{2.139b}$$

$$\left(-\omega \mp i\frac{\partial}{\partial z}\right) u_{\pm\mp}^{(+)} \pm k v_{\pm\mp}^{(+)} = \mp i\delta(z - z'), \tag{2.139c}$$

$$\left(-\omega \mp i\frac{\partial}{\partial z}\right) u_{\pm\mp}^{(-)} \pm k v_{\pm\mp}^{(-)} = 0, \tag{2.139d}$$

$$\pm k u_{\pm\mp}^{(+)} + \left(-\omega \pm i\frac{\partial}{\partial z}\right) v_{\pm\mp}^{(+)} = 0, \tag{2.139e}$$

$$\pm k u_{\pm\mp}^{(-)} + \left(-\omega \pm i\frac{\partial}{\partial z}\right) v_{\pm\mp}^{(-)} = \mp i\delta(z - z'), \tag{2.139f}$$

which aside from the inhomogeneous terms are replicas of (2.123a) and (2.123b). These equations are to be solved subject to the boundary conditions

$$u_{++}^{(\pm)} - u_{-+}^{(\pm)}|_{z=a} = 0, \tag{2.140a}$$

$$u_{++}^{(\pm)} + u_{-+}^{(\pm)}|_{z=0} = 0, \tag{2.140b}$$

$$u_{+-}^{(\pm)} - u_{--}^{(\pm)}|_{z=a} = 0, \tag{2.140c}$$

$$u_{+-}^{(\pm)} + u_{--}^{(\pm)}|_{z=0} = 0, \tag{2.140d}$$

$$v_{++}^{(\pm)} + v_{-+}^{(\pm)}|_{z=a} = 0, \tag{2.140e}$$

$$v_{++}^{(\pm)} - v_{-+}^{(\pm)}|_{z=0} = 0, \tag{2.140f}$$

$$v^{(\pm)}_{+-} + v^{(\pm)}_{--}|_{z=a} = 0, \tag{2.140g}$$

$$v^{(\pm)}_{+-} - v^{(\pm)}_{--}|_{z=0} = 0. \tag{2.140h}$$

The solution is straightforward. We find

$$u^{(+)}_{++} = u^{(+)*}_{--} = v^{(-)*}_{++} = v^{(-)}_{--}$$

$$= \frac{1}{2\cos\lambda a} \left[\cos\lambda(z+z'-a) + \frac{i\omega}{\lambda}\sin\lambda(z+z'-a) \right], \tag{2.141a}$$

$$v^{(+)}_{++} = v^{(+)}_{--} = u^{(-)*}_{++} = u^{(-)*}_{--}$$

$$= \frac{ik}{2\lambda}\frac{1}{\cos\lambda a}\sin\lambda(z+z'-a), \tag{2.141b}$$

$$u^{(+)}_{-+} = u^{(+)*}_{+-} = -v^{(-)*}_{-+} = -v^{(-)}_{+-}$$

$$= \frac{1}{2\cos\lambda a} \left[\epsilon(z-z')\cos\lambda(z_> - z_< - a) \right.$$

$$\left. - \frac{i\omega}{\lambda}\sin\lambda(z_> - z_< - a) \right], \tag{2.141c}$$

$$v^{(+)}_{-+} = v^{(+)}_{+-} = u^{(-)}_{-+} = u^{(-)}_{+-}$$

$$= \frac{ik}{2\lambda}\frac{1}{\cos\lambda a}\sin\lambda(z_> - z_< - a), \tag{2.141d}$$

where

$$\epsilon(z-z') = \begin{cases} 1 & \text{if } z > z', \\ -1 & \text{if } z < z'. \end{cases} \tag{2.142}$$

We now insert these Green's functions into the vacuum expectation value of the energy-momentum tensor. The latter is

$$T^{\mu\nu} = \frac{1}{2}\psi\gamma^0\frac{1}{2}\left(\gamma^\mu\frac{1}{i}\partial^\nu + \gamma^\nu\frac{1}{i}\partial^\mu\right)\psi + g^{\mu\nu}\mathcal{L}, \tag{2.143}$$

$$\mathcal{L} = -\frac{1}{2}\psi\gamma\frac{1}{i}\partial\psi. \tag{2.144}$$

We take the vacuum expectation value by the replacement

$$\psi\psi\gamma^0 \to \frac{1}{i}G, \tag{2.145}$$

where G is the fermionic Green's function computed above. Because we are interested in the *limit* as $x' \to x$ we can ignore the Lagrangian term in the energy-momentum tensor, leaving us with

$$\langle T^{33} \rangle = \frac{1}{2} \mathrm{Tr}\, \gamma^3 \partial_3 G(x, x')|_{x' \to x}, \tag{2.146}$$

so for a given frequency and transverse momentum,

$$
\begin{aligned}
\langle t^{33} \rangle &= \frac{i}{2} \frac{\partial}{\partial z} \mathrm{tr}\, \sigma^3 (g_{-+} + g_{+-})|_{z' \to z} \\
&= \frac{i}{2} \frac{\partial}{\partial z} [u_{-+}^{(+)} + u_{+-}^{(+)} - (v_{-+}^{(-)} + v_{+-}^{(-)})]|_{z' \to z}.
\end{aligned} \tag{2.147}
$$

When we insert the solution found above (2.141c), we obtain

$$\langle t^{33} \rangle = 2i \frac{\partial}{\partial z} \mathrm{Re}\, u_{-+}^{(+)}. \tag{2.148}$$

Carrying out the differentiation and setting $z = z'$ we find instead of (2.24),

$$\langle t^{33} \rangle = i\lambda \tan \lambda a, \tag{2.149}$$

where again we ignore the δ-function term.

We now follow the same procedure given in Sec. 2.3: The force per unit area is

$$
\begin{aligned}
\mathcal{F} &= \int \frac{d^2 k}{(2\pi)^2} \int \frac{d\omega}{2\pi} i\lambda \tan \lambda a \\
&= \int \frac{d^2 k}{(2\pi)^2} \int \frac{d\zeta}{2\pi} \kappa \tanh \kappa a \\
&= \frac{1}{2\pi^2} \int_0^\infty d\kappa\, \kappa^3 \left[1 - \frac{2}{e^{2\kappa a} + 1} \right]
\end{aligned} \tag{2.150}
$$

As in (2.26) we omit the 1 in the last square bracket: The same term is present in the vacuum stress outside the plates, so cancels out when we compute the discontinuity across the plates. We are left with, then,

$$\mathcal{F} = -\frac{1}{16\pi^2 a^4} \int_0^\infty \frac{dx\, x^3}{e^x + 1}. \tag{2.151}$$

But

$$\int_0^\infty \frac{x^{s-1}\, dx}{e^x + 1} = (1 - 2^{1-s})\zeta(s)\Gamma(s), \tag{2.152}$$

so here we find

$$\mathcal{F}_f = -\frac{7\pi^2}{1920a^4},$$ (2.153)

which is, indeed, $\frac{7}{4}$ times the scalar force given in (2.10).

Chapter 3

Casimir Force Between Parallel Dielectrics

3.1 The Lifshitz Theory

The formalism given in Sec. 2.6 can be readily extended to dielectric bodies [11]. The starting point is the effective action in the presence of an external polarization source $\mathbf{P}$:

$$W = \int (dx)[\mathbf{P} \cdot (-\dot{\mathbf{A}} - \boldsymbol{\nabla}\phi) + \epsilon \mathbf{E} \cdot (-\dot{\mathbf{A}} - \boldsymbol{\nabla}\phi)$$

$$-\mathbf{H} \cdot (\boldsymbol{\nabla} \times \mathbf{A}) + \frac{1}{2}(H^2 - \epsilon E^2)], \tag{3.1}$$

which, upon variation with respect to $\mathbf{H}$, $\mathbf{E}$, $\mathbf{A}$, and ϕ, yields the appropriate Maxwell's equations. Thus, because W is stationary with respect to these field variations, its response to a change in dielectric constant is explicit:

$$\delta_\epsilon W = \int (dx)\delta\epsilon \frac{1}{2} E^2. \tag{3.2}$$

Comparison of $i\delta_\epsilon W$ with the second iteration of the source term in the vacuum persistence amplitude,

$$e^{iW} = \cdots + \frac{1}{2}\left[i \int (dx)\mathbf{E} \cdot \mathbf{P}\right]^2 + \cdots, \tag{3.3}$$

allows us to identify the effective product of polarization sources,

$$i\mathbf{P}(x)\mathbf{P}(x')\big|_{\text{eff}} = \mathbf{1}\delta\epsilon \, \delta(x - x'). \tag{3.4}$$

Thus, the numerical value of the action according to (2.94),

$$W = \frac{1}{2} \int (dx) \mathbf{P}(x) \cdot \mathbf{E}(x) = \frac{1}{2} \int (dx)(dx') \mathbf{P}(x) \cdot \mathbf{\Gamma}(x, x') \cdot \mathbf{P}(x'), \quad (3.5)$$

implies the following change in the action when the dielectric constant is varied slightly,

$$\delta W = -\frac{i}{2} \int (dx) \delta\epsilon(x) \Gamma_{kk}(x, x), \qquad (3.6)$$

where the repeated indices on the dyadic indicate a trace. In view of (3.2), this is equivalent to the vacuum-expectation-value replacement (2.115a).

For the geometry of parallel dielectric slabs, shown in Fig. 1.1, where the dielectric constants in the three regions are

$$\epsilon(z) = \begin{cases} \epsilon_1, & z < 0, \\ \epsilon_3, & 0 < z < a, \\ \epsilon_2, & a < z, \end{cases} \qquad (3.7)$$

the components of the Green's dyadics may be expressed in terms of the TE (transverse electric or H) modes and the TM (transverse magnetic or E) modes,[*] given by the reduced scalar Green's functions satisfying

$$\left(-\frac{\partial^2}{\partial z^2} + k^2 - \omega^2 \epsilon \right) g^H(z, z') = \delta(z - z'), \qquad (3.8a)$$

$$\left(-\frac{\partial}{\partial z} \frac{1}{\epsilon} \frac{\partial}{\partial z'} + \frac{k^2}{\epsilon} - \omega^2 \right) g^E(z, z') = \delta(z - z'), \qquad (3.8b)$$

where, quite generally, $\epsilon = \epsilon(z)$, $\epsilon' = \epsilon(z')$. The nonzero components of the Fourier transform $\mathbf{g}$ given by (2.98) are easily found to be (in the coordinate system where $\mathbf{k}$ lies along the $+x$ axis)

$$g_{xx} = -\frac{1}{\epsilon} \delta(z - z') + \frac{1}{\epsilon} \frac{\partial}{\partial z} \frac{1}{\epsilon'} \frac{\partial}{\partial z'} g^E, \qquad (3.9a)$$

$$g_{yy} = \omega^2 g^H, \qquad (3.9b)$$

$$g_{zz} = -\frac{1}{\epsilon} \delta(z - z') + \frac{k^2}{\epsilon\epsilon'} g^E, \qquad (3.9c)$$

$$g_{xz} = i \frac{k}{\epsilon\epsilon'} \frac{\partial}{\partial z} g^E, \qquad (3.9d)$$

---

[*]We have changed the notation from that of the original reference [11]. The TE modes are denoted by H, the TM modes are denoted by E, to be consistent with the notation used later in the book.

$$g_{zx} = -i \frac{k}{\epsilon \epsilon'} \frac{\partial}{\partial z'} g^E. \tag{3.9e}$$

The trace required in the change of the action (3.6) is obtained by taking the limit $z' \to z$, and consequently omitting delta functions:

$$g_{kk} = \left(\omega^2 g^H + \frac{k^2}{\epsilon \epsilon'} g^E + \frac{1}{\epsilon} \frac{\partial}{\partial z} \frac{1}{\epsilon'} \frac{\partial}{\partial z'} g^E \right) \bigg|_{z=z'}. \tag{3.10}$$

This appears in the change of the energy when the second interface is displaced by an amount δa,

$$\delta \epsilon(z) = -\delta a (\epsilon_2 - \epsilon_3) \delta(z - a), \tag{3.11}$$

namely (A is the transverse area)

$$\frac{\delta E}{A} = \frac{i}{2} \int \frac{d\omega}{2\pi} \frac{(d\mathbf{k})}{(2\pi)^2} dz \, \delta \epsilon(z) g_{kk}(z, z; \mathbf{k}, \omega) = -\delta a \, \mathcal{F}, \tag{3.12}$$

where the force per unit area is

$$\mathcal{F} = \frac{i}{2} \int \frac{d\omega}{2\pi} \frac{(d\mathbf{k})}{(2\pi)^2} (\epsilon_2 - \epsilon_3) g_{kk}(a, a; \mathbf{k}, \omega). \tag{3.13}$$

Because g^H, g^E and $\frac{1}{\epsilon} \frac{\partial}{\partial z} \frac{1}{\epsilon'} \frac{\partial}{\partial z'} g^E$ are all continuous, while $\epsilon \epsilon'$ is not, we interpret the trace of $\mathbf{g}$ in (3.12) symmetrically; we let z and z' approach the interface from opposite sides, so the term $\frac{k^2}{\epsilon \epsilon'} g^E \to \frac{k^2}{\epsilon_1 \epsilon_2} g^E$. Subsequently, we may evaluate the Green's function on a single side of the interface. In terms of the notation (for $\epsilon = 1$, $\kappa = -i\lambda$ in the notation used in the previous chapter)

$$\kappa^2 = k^2 - \omega^2 \epsilon, \tag{3.14}$$

which is positive after a complex frequency rotation is performed (it is automatically positive for finite temperature), the magnetic (TE) Green's function is in the region $z, z' > a$,

$$g^H(z, z') = \frac{1}{2\kappa_2} \left(e^{-\kappa_2 |z - z'|} + r e^{-\kappa_2 (z + z' - 2a)} \right), \tag{3.15}$$

where the reflection coefficient is

$$r = \frac{\kappa_2 - \kappa_3}{\kappa_2 + \kappa_3} + \frac{4\kappa_2 \kappa_3}{\kappa_3^2 - \kappa_2^2} d^{-1}, \tag{3.16}$$

with the denominator here being

$$d = \frac{\kappa_3 + \kappa_1}{\kappa_3 - \kappa_1} \frac{\kappa_3 + \kappa_2}{\kappa_3 - \kappa_2} e^{2\kappa_3 a} - 1. \tag{3.17}$$

The electric (TM) Green's function g^E has the same form but with the replacement

$$\kappa \to \kappa/\epsilon \equiv \kappa', \tag{3.18}$$

except in the exponentials; the corresponding denominator is denoted by d'. [It is easy to see that g^H reduces to (2.27) when $r = -1$; the results in Sec. 2.6 follow from (3.9a)–(3.9e) in the coordinate system adopted here.]

Evaluating these Green's functions just outside the interface, we find for the force on the surface per unit area

$$\mathcal{F} = \frac{i}{2} \int \frac{d\omega}{2\pi} \frac{(d\mathbf{k})}{(2\pi)^2} \left\{ \left[\kappa_3 - \kappa_2 + 2\kappa_3 d^{-1} \right] + \left[\kappa_3 - \kappa_2 + 2\kappa_3 d'^{-1} \right] \right\}, \tag{3.19}$$

where the first bracket comes from the TE modes, and the second from the TM modes. The first term in each bracket, which does not make reference to the separation a between the surfaces, is seen to be a change in the volume energy of the system. These terms correspond to the electromagnetic energy required to replace medium 2 by medium 3 in the displacement volume. They constitute the so-called bulk energy contribution. (It will be discussed further in Chapter 12.) The remaining terms are the Casimir force. We rewrite the latter by making a complex rotation in the frequency,

$$\omega \to i\zeta, \quad \zeta \text{ real}, \quad \text{so} \quad \kappa^2 = k^2 + \epsilon \zeta^2. \tag{3.20}$$

This gives for the force per unit area at zero temperature

$$\mathcal{F}_{\text{Casimir}}^{T=0} = -\frac{1}{8\pi^2} \int_0^\infty d\zeta \int_0^\infty dk^2 \, 2\kappa_3 \left(d^{-1} + d'^{-1} \right). \tag{3.21}$$

From this, we can obtain the finite temperature expression immediately by the substitution

$$\zeta^2 \to \zeta_n^2 = 4\pi^2 n^2/\beta^2, \tag{3.22}$$

$$\int_0^\infty \frac{d\zeta}{2\pi} \to \frac{1}{\beta} \sum_{n=0}^\infty{}', \tag{3.23}$$

the prime being a reminder to count the $n = 0$ term with half weight. These results agree with those of Lifshitz et al. [7, 67, 8]. The connection between Casimir's ideas of zero-point energy and Lifshitz' theory of retarded dispersion forces appears in Boyer's paper [148].[†]

Note that the same result (3.21) may be easily rederived by computing the normal-normal component of the stress tensor on the surface, T_{zz}, provided two constant stresses are removed, terms which would be present if either medium filled all space. The difference between these two constant stresses,

$$T_{zz}^{\text{vol}} = -i \int \frac{(d\mathbf{k})}{(2\pi)^2} \frac{d\omega}{2\pi} (\kappa_2 - \kappa_3), \tag{3.24}$$

precisely corresponds to the deleted volume energy in the previous calculation.

3.2 Applications

Various applications can be made from this general formula (3.21). In particular, if we set the intermediate material to be vacuum, $\epsilon_3 = 1$, and set $\epsilon_1 = \epsilon_2 = \infty$, so that $\kappa_1 = \kappa_2 = \infty$, $\kappa_1' = \kappa_2' = 0$, we recover the Casimir force (2.11) between parallel, perfectly conducting plates. More generally, we can let the intermediate material have a dispersive permittivity, so that we obtain

$$\begin{aligned} \mathcal{F} &= -\frac{1}{8\pi^2} \int_0^\infty d\zeta \int_{\zeta^2\epsilon}^\infty d\kappa^2 \frac{4\kappa}{e^{2\kappa a} - 1} \\ &= -\frac{1}{\pi^2} \int_0^\infty d\zeta \frac{(\sqrt{\epsilon}\zeta)^3}{e^{2\sqrt{\epsilon}\zeta a} - 1} = -\frac{\pi^2}{240\sqrt{\epsilon}a^4}, \end{aligned} \tag{3.25}$$

which, until the last step, still admits of dispersion. This last expression is an obvious generalization of Casimir's result to a dielectric-filled capacitor. Note that the corresponding energy per unit area is

$$\mathcal{E} = \frac{1}{2\pi^2} \int_0^\infty d\zeta \, \epsilon\zeta^2 \ln\left(1 - e^{-2a\sqrt{\epsilon}\zeta}\right), \tag{3.26}$$

[†]The nonretarded part of the Lifshitz formula, for $d \ll \lambda$, the "principal absorption wavelength of the material," was rederived in 1968 by van Kampen, Nijboer, and Schram [149], using the "argument principle" described in Appendix A to evaluate the zero-point energy.

which is a considerably simpler expression than, but equivalent to, that given in Ref. [150]. (See their note added in proof.) [Cf. (2.55a).]

3.2.1 *Temperature Dependence for Conducting Plates*

As for the temperature dependence, we note that we must take this limit with special care for the static situation $\omega = 0$. In order to enforce correctly the electrostatic boundary conditions, we adopt the prescription that we take the limit $\epsilon \to \infty$ before we set $\omega = 0$. Doing so for the temperature-dependent version of (3.21) gives

$$\mathcal{F}^T = -\frac{1}{\pi\beta} \sum_{n=0}^{\infty} {}' \int_0^{\infty} dk^2 \frac{\kappa_n}{e^{2\kappa_n a} - 1}, \tag{3.27}$$

where $\kappa_n^2 = k^2 + (2\pi n/\beta)^2$, This is exactly twice the scalar result given in (2.59) for $d = 2$. Notice that if we had simply let $\epsilon_{1,2} \to \infty$, the first denominator structure in (3.21) for $n = 0$ would not contribute, which, among other consequences, would imply an incorrect $T \to 0$ limit.[‡] Defining

$$y = 2\kappa_n a, \quad t = 4\pi a/\beta, \tag{3.28}$$

we find the Casimir force for arbitrary temperature to be

$$\mathcal{F}^T = -\frac{1}{4\pi\beta a^3} \sum_{n=0}^{\infty} {}' \int_{nt}^{\infty} y^2 \, dy \frac{1}{e^y - 1}. \tag{3.29}$$

As noted in the previous chapter, the high-temperature, $t \gg 1$, limit is particularly easy to obtain, for then the $n = 0$ term is the only one which is not exponentially small. Including the first of these exponentially small corrections (from $n = 1$) we find for large T

$$\mathcal{F}^T \sim -\frac{1}{4\pi\beta a^3}\zeta(3) - \frac{1}{2\pi\beta a^3}\left(1 + t + \frac{t^2}{2}\right)e^{-t}, \quad \beta \ll 4\pi a. \tag{3.30}$$

[‡]Remarkably, this seemingly obvious prescription has become controversial. Boström and Sernelius [151, 152] claim that the $n = 0$ mode of the TE mode should be omitted. This would give rise to a significant temperature correction in the experimentally accessible region, while none is seen—see below. That this claim is incorrect has been demonstrated by Lamoreaux [153]. See also Svetovoy and Lokhanin [154, 155], who obtain a large linear temperature correction, again in contradiction with experiment. A sensible explanation of the ambiguities which led to these erroneous results appears in Bordag et al. [156].

[The term which is not exponentially small is twice that given in (2.61b).] This coincides with the result first found by Sauer [3] and Mehra [5, 4, 6]. See also Levin and Rytov [157]. Our form (3.21) is especially suited to obtain the high temperature limit, in contradistinction to the forms obtained by Sauer and Mehra and by Brown and Maclay [100].

The low-temperature limit was also worked out in the previous chapter, but for a general value of d, the transverse dimension. There are significant simplifications when $d = 2$. We use the Poisson sum formula (2.71) for functions related by a Fourier transformation (2.70). Here we take

$$b(n) = \int_{|n|t}^{\infty} y^2 \, dy \frac{1}{e^y - 1}, \tag{3.31}$$

which has the Fourier transform for $\alpha \neq 0$

$$\begin{aligned}
c(\alpha) &= \frac{1}{\pi} \int_0^{\infty} dx \cos \alpha x \int_{xt}^{\infty} y^2 \, dy \frac{1}{e^y - 1} \\
&= -\frac{1}{\pi \alpha} \frac{d^2}{dz^2} \left[\frac{\pi}{2} \coth \pi z - \frac{1}{2z} \right] \bigg|_{z = \alpha/t} \\
&= -\frac{1}{\pi \alpha} \left[4\pi^3 \frac{e^{-2\pi\alpha/t}(1 + e^{-2\pi\alpha/t})}{(1 - e^{-2\pi\alpha/t})^3} - \frac{t^3}{\alpha^3} \right].
\end{aligned} \tag{3.32}$$

Here we have interchanged the order of integration and used the fact that

$$\int_0^{\infty} \frac{dy}{e^y - 1} \sin zy = \frac{\pi}{2} \coth \pi z - \frac{1}{2z}, \tag{3.33}$$

which may be easily derived from (2.45). The evaluation of $c(0)$ is easily accomplished directly, or by expanding $\coth \pi z$ in the above, yielding

$$c(0) = \frac{\pi^3}{15t}. \tag{3.34}$$

We therefore find an alternative form for the sum in (3.29)

$$2 \sum_{n=0}^{\infty} {}'b(n) = \frac{2\pi^4}{15t} + \frac{t^3}{360} - 8\pi^2 \sum_{n=1}^{\infty} \frac{1}{n} \frac{e^{-4\pi^2 n/t}(1 + e^{-4\pi^2 n/t})}{(1 - e^{-4\pi^2 n/t})^3}, \tag{3.35}$$

which, apart from a factor, expresses the general temperature dependence of the Casimir force. From this form, it is very easy to obtain the low-temperature limit

$$\mathcal{F}^T \sim -\frac{\pi^2}{240a^4}\left[1 + \frac{16}{3}\frac{a^4}{\beta^4} - \frac{240}{\pi}\frac{a}{\beta}e^{-\pi\beta/a}\right], \quad \beta \gg 4\pi a. \qquad (3.36)$$

again in agreement with Sauer and Mehra [3, 5, 4, 6], and with that found by Lifshitz [7, 8] when the transcription error there is corrected. The result is, of course, twice that given in (2.92b), including the exponential correction (2.93). We recognize the second term here as the blackbody radiation pressure arising from thermal fluctuations above the plate, $z > a$, (external fluctuations),[§] so we write

$$\mathcal{F}^T = -\frac{\pi^2}{45}(kT)^4 - \frac{\partial}{\partial a}\left(\frac{F}{A}\right), \qquad (3.37)$$

where the corresponding (internal) free energy per unit area is ($\xi = akT$)

$$\frac{F}{A} = -\frac{1}{a^3}f(\xi), \quad f(\xi) = \frac{\pi^2}{720} + X\xi^3 + \xi^2\left(1 + \frac{\xi}{\pi}\right)e^{-\pi/\xi}, \quad \xi \ll 1, \quad (3.38)$$

where the constant X is undetermined by (3.36). Under the inversion symmetry discovered by Ravndal and Tollefsen [158], a generalization of that found by Brown and Maclay [100], this low temperature result can be extended to the high-temperature limit by the inversion formula

$$f(\xi) = (2\xi)^4 f(1/4\xi), \qquad (3.39)$$

so here

$$f(\xi) = \frac{\pi^2}{45}\xi^4 + \frac{1}{4}X\xi + \xi\left(\frac{1}{4\pi} + \xi\right)e^{4\pi\xi}, \quad \xi \gg 1. \qquad (3.40)$$

The corresponding force is, from (3.37),

$$\mathcal{F}^T \sim -\frac{X}{2a^3}kT - \frac{kT}{4\pi a^3}(t^2 + 2t + 2)e^{-t}, \quad t \gg 1, \qquad (3.41)$$

where $t = 4\pi akT$. We see that the Stefan's law contribution cancelled between the interior and exterior modes, and that the first term expresses the correct linear behavior shown in (3.30) with $X = \zeta(3)/2\pi$. The exponentially small term in (3.30) is reproduced, which shows the (limited) efficacy of this inversion symmetry.

[§]See footnote 7 of Ref. [100].

Recently, these results have been rederived by semiclassical orbit theory by Schaden and Spruch [159].

3.2.2 *Finite Conductivity*

Another interesting result, important for the recent experiments [71, 72], is the correction for an imperfect conductor, where for frequencies above the infrared, an adequate representation for the dielectric constant is [99]

$$\epsilon(\omega) = 1 - \frac{\omega_p^2}{\omega^2}, \tag{3.42}$$

where the plasma frequency is[¶]

$$\omega_p^2 = \frac{4\pi e^2 N}{m}, \tag{3.43}$$

where e and m are the charge and mass of the electron, and N is the number density of free electrons in the conductor. A simple calculation shows, at zero temperature [10, 11],

$$\mathcal{F} \approx -\frac{\pi^2}{240a^4}\left[1 - \frac{8}{3\sqrt{\pi}}\frac{1}{ea}\left(\frac{\mu}{N}\right)^{1/2}\right]. \tag{3.44}$$

If we define a penetration parameter, or skin depth, by $\delta = 1/\omega_p$, we can write the result out to second order as [160, 114]

$$\mathcal{F} \approx -\frac{\pi^2}{240a^4}\left(1 - \frac{16}{3}\frac{\delta}{a} + 24\frac{\delta^2}{a^2}\right). \tag{3.45}$$

Recently, Lambrecht, Jaekel, and Reynaud [161] analyzed the Casimir force between mirrors with arbitrary frequency-dependent reflectivity, and found that it is always smaller than that between perfect reflectors.

3.2.3 *van der Waals Forces*

Now suppose the central slab consists of a tenuous medium and the surrounding medium is vacuum, so that the dielectric constant in the slab

---

[¶]Although we have used rationalized Heaviside-Lorentz units in our electromagnetic action formalism, that is without effect, in that the one-loop Casimir effect is independent of electromagnetic units. For considerations where the electric charge and polarizability appear, it seems more convenient to use unrationalized Gaussian units.

differs only slightly from unity,

$$\epsilon - 1 \ll 1. \qquad (3.46)$$

Then, with a simple change of variable,

$$\kappa = \zeta p, \qquad (3.47)$$

we can recast the Lifshitz formula (3.21) into the form

$$\mathcal{F} \approx -\frac{1}{32\pi^2} \int_0^\infty d\zeta\, \zeta^3 \int_1^\infty \frac{dp}{p^2} [\epsilon(\zeta) - 1]^2 [(2p^2 - 1)^2 + 1] e^{-2\zeta pa}. \qquad (3.48)$$

If the separation of the surfaces is large compared to the characteristic wavelength characterizing ϵ, $a\zeta_c \gg 1$, we can disregard the frequency dependence of the dielectric constant, and we find

$$\mathcal{F} \approx -\frac{23(\epsilon - 1)^2}{640\pi^2 a^4}. \qquad (3.49)$$

For short distances, $a\zeta_c \ll 1$, the approximation is

$$\mathcal{F} \approx -\frac{1}{32\pi^2} \frac{1}{a^3} \int_0^\infty d\zeta (\epsilon(\zeta) - 1)^2. \qquad (3.50)$$

These formulas are identical with the well-known forces found for the complementary geometry in Ref. [11].

Now we wish to obtain these results from the sum of van der Waals forces, derivable from a potential of the form

$$V = -\frac{B}{r^\gamma}. \qquad (3.51)$$

We do this by computing the energy ($\mathcal{N}$ = density of molecules)

$$E = -\frac{1}{2} B\mathcal{N}^2 \int_0^a dz \int_0^a dz' \int (d\mathbf{r}_\perp)(d\mathbf{r}_\perp') \frac{1}{[(\mathbf{r}_\perp - \mathbf{r}_\perp')^2 + (z - z')^2]^{\gamma/2}}. \qquad (3.52)$$

If we disregard the infinite self-interaction terms (analogous to dropping the volume energy terms in the Casimir calculation), we get [11, 38]

$$\mathcal{F} = -\frac{\partial}{\partial a} \frac{E}{A} = -\frac{2\pi B\mathcal{N}^2}{(2 - \gamma)(3 - \gamma)} \frac{1}{a^{\gamma - 3}}. \qquad (3.53)$$

So then, upon comparison with (3.49), we set $\gamma = 7$ and in terms of the polarizability,

$$\alpha = \frac{\epsilon - 1}{4\pi \mathcal{N}}, \tag{3.54}$$

we find

$$B = \frac{23}{4\pi}\alpha^2, \tag{3.55}$$

or, equivalently, we recover the retarded dispersion potential of Casimir and Polder [50],

$$V = -\frac{23}{4\pi}\frac{\alpha^2}{r^7}, \tag{3.56}$$

whereas for short distances we recover from (3.50) the London potential [49],

$$V = -\frac{3}{\pi}\frac{1}{r^6}\int_0^\infty d\zeta\, \alpha(\zeta)^2, \tag{3.57}$$

which are the quantitative forms of (1.6) and (1.9), given in (1.30) and (1.29), respectively.

3.2.4 *Force between Polarizable Molecule and a Dielectric Plate*

As a final application of these ideas, we will calculate the energy of interaction between a molecule of polarizability $\alpha(\omega)$ and a dielectric slab. This energy is given by (3.6) with

$$\delta\epsilon(\mathbf{r}, \omega) = 4\pi\alpha(\omega)\delta(\mathbf{r} - \mathbf{R}), \tag{3.58}$$

which expresses the change in the dielectric constant when a molecule is inserted in the vacuum at $\mathbf{R}$. We will suppose that the dielectric slab occupies the region of space $z < 0$ with vacuum above it. The appropriate Green's functions here, referring to a single interface, are trivially obtained from those discussed in Sec. 3.1. In region 2, g^H has the form of (3.15) with the reflection coefficient r given by

$$1 + r = \frac{2\kappa}{\kappa + \kappa_1}, \tag{3.59}$$

where

$$\kappa^2 = k^2 - \omega^2, \quad \kappa_1^2 = k^2 - \epsilon_1 \omega^2, \tag{3.60}$$

which is obtained from (3.16) by taking the limits $a \to \infty$, $\epsilon_2 = 1$, $\epsilon_3 = \epsilon_1$. The energy is then ($R_3 = z$)

$$E = \frac{i}{2} \int \frac{d\omega}{2\pi} \frac{(d\mathbf{k})}{(2\pi)^2} 4\pi\alpha(\omega) g_{kk}(z, z; \mathbf{k}, \omega), \tag{3.61}$$

where, from (3.10),

$$
\begin{aligned}
g_{kk} &= \omega^2 g^H + \left(k^2 + \frac{\partial}{\partial z} \frac{\partial}{\partial z'} \right) g^E \bigg|_{z' \to z} \\
&= \frac{\omega^2}{\kappa} + \frac{1}{2\kappa} \left[\omega^2 \frac{\kappa - \kappa_1}{\kappa + \kappa_1} + (2k^2 - \omega^2) \frac{\epsilon_1 \kappa - \kappa_1}{\epsilon_1 \kappa + \kappa_1} \right] e^{-2\kappa z}. \tag{3.62}
\end{aligned}
$$

The necessary contact term here is easily deduced from the physical requirement that the energy of interaction go to zero as the separation gets large, $z \to \infty$, which effectively removes the ω^2/κ term in g_{kk}. Therefore, the interaction energy between the molecule and the dielectric slab separated by a distance z is

$$
\begin{aligned}
E = -\frac{1}{16\pi^2} \int_0^\infty d\zeta \, 4\pi\alpha(\zeta) \int_0^\infty dk^2 \frac{1}{\kappa} \\
\times \left[-\zeta^2 \frac{\kappa - \kappa_1}{\kappa + \kappa_1} + (2k^2 + \zeta^2) \frac{\epsilon_1 \kappa - \kappa_1}{\epsilon_1 \kappa + \kappa_1} \right] e^{-2\kappa z}. \tag{3.63}
\end{aligned}
$$

One application of this result refers to the attraction of a molecule by a perfectly conducting plate. We merely take $\epsilon_1 \to \infty$ and then easily find

$$E = -\frac{3\alpha}{8\pi z^4}, \tag{3.64}$$

a result first calculated by Casimir and Polder [50]. This result was experimentally verified by Sukenik, Boshier, Cho, Sandoghar, and Hinds [162]. (Actually, they measured the force on an atom between two plates, the general theory of which was given by Barton [163].) A second, particularly interesting possibility occurs when the molecule is of the same type as those composing the dielectric slab. When the common dielectric constant is close to unity, the energy of interaction, to lowest order in $\epsilon - 1$, is

$$E \approx -\frac{1}{16\pi^2} \frac{1}{\mathcal{N}} \int_0^\infty d\zeta \, (\epsilon - 1)^2 \int_0^\infty \frac{dk^2}{\kappa^2} \left(k^4 + k^2 \zeta^2 + \frac{1}{2} \zeta^4 \right) e^{-2\kappa z}, \quad (3.65)$$

where we have expressed the polarizability of the molecule in terms of the dielectric constant according to (3.54). Lifshitz et al. [7, 8, 9] have considered the limiting behavior of large separations (small ζ) where ϵ can be considered to be constant.

A recent proposal by Ford and Svaiter [164] suggests focusing the vacuum modes of a quantized field by a parabolic mirror, thereby enhancing the Casimir-Polder force on an atom, which would be drawn into the focus of the mirror. The approach used in that paper is a semiclassical approximation, based on geometrical optics. It is related to the calculations of Schaden and Spruch [165, 166] who used a semiclassical approximation and geometrical optics to calculate Casimir energies between pairs of conductors, plates, a plate and a sphere, spheres, and concentric spheres, in the approximation that the separations of the objects are small compared to all radii of the objects. They also provide a rigorous derivation of the proximity theorem result of Derjaguin [167, 168, 54, 55, 56], which is discussed in the next section.

3.3 Experimental Verification of the Casimir Effect

Attempts to measure the Casimir effect between solid bodies date back to the middle 1950s. The early measurements were, not surprisingly, somewhat inconclusive [55, 57, 58, 59, 60, 61, 62, 63, 64, 65]. The Lifshitz theory (3.21), for zero temperature, was, however, confirmed accurately in the experiment of Sabisky and Anderson in 1973 [66]. So there could be no serious doubt of the reality of zero-point fluctuation forces. For a review of the earlier experiments, see Ref. [169].

New technological developments allowed for dramatic improvements in experimental techniques in recent years, and thereby permitted nearly direct confirmation of the Casimir force between parallel conductors. First, in 1996 Lamoreaux used a electromechanical system based on a torsion pendulum to measure the force between a conducting plate and a sphere [68, 69]. The force per unit area is, of course, no longer given by (1.12) or (1.13), but may be obtained from that result by the proximity force theorem [170] which here says that the attractive force F between a sphere of radius R and a flat surface is simply the circumference of the sphere times the energy per unit area for parallel plates, or, from (1.12),

$$F = 2\pi R \mathcal{E}(d) = -\frac{\pi^3}{360} \frac{R}{d} \frac{\hbar c}{d^2}, \quad R \gg d, \tag{3.66}$$

where d is the distance between the plate and the sphere at the point of closest approach, and R is the radius of curvature of the sphere at that point. The proof of (3.66) is quite simple. If $R \gg d$, each element of the sphere may be regarded as parallel to the plane, so the potential energy of the sphere is

$$V(d) = \int_0^\pi 2\pi R \sin\theta R \, d\theta \, \mathcal{E}(d + R(1 - \cos\theta)) = 2\pi R \int_{-R}^R dx \, \mathcal{E}(d + R - x). \tag{3.67}$$

To obtain the force between the sphere and the plate, we differentiate with respect to d:

$$\begin{aligned} F = -\frac{\partial V}{\partial d} &= 2\pi R \int_{-R}^R dx \, \frac{\partial}{\partial x} \mathcal{E}(d + R - x) \\ &= 2\pi R[\mathcal{E}(d) - \mathcal{E}(d + 2R)] \approx 2\pi R \mathcal{E}(d), \quad d \ll R, \end{aligned} \tag{3.68}$$

provided that $\mathcal{E}(a)$ falls off with a. This result was already given in Refs. [54, 55, 57]. The proximity theorem itself dates back to a paper by Derjaguin in 1934 [167, 168].

Lamoreaux in 1997 [68, 69] claimed an agreement with this theoretical value at the 5% level, although it seems that finite conductivity was not included correctly, nor were roughness corrections incorporated [171]. Further, Lambrecht and Reynaud [172] analyzed the effect of conductivity and found discrepancies with Lamoreaux [70], and therefore stated that it is too early to claim agreement between theory and experiment. See also Refs. [173, 174].

An improved experimental measurement was reported in 1998 by Mohideen and Roy [71], based on the use of an atomic force microscope. They included finite conductivity, roughness, and temperature corrections, although the latter remains beyond experimental reach.[||] Spectacular agreement with theory at the 1% level was attained. Improvements were subsequently reported [72, 73]. (The nontrivial effects of corrugations in the sur-

---

[||] The low temperature correction for the force between a perfectly conducting sphere and

face were examined in Ref. [175].) Erdeth [74] used template-stripped surfaces, and measured the Casimir forces with similar devices at separations of 20–100 nm. Rather complete analyses of the roughness, conductivity, and temperature correction to this experment have now been published [176, 156, 177].

Very recently, a new measurement of the Casimir force (3.66) has been announced by a group at Bell Labs [75], using a micromachined torsional device, by which they measure the attraction between a polysilicon plate and a spherical metallic surface. Both surfaces are plated with a 200 nm film of gold. The authors include finite conductivity [172, 178] and surface roughness corrections [179, 180], and obtain agreement with theory at better than 0.5% at the smallest separations of about 75 nm. However, potential corrections of greater than 1% exist, so that limits the level of verification of the theory. Their experiment suggests novel nanoelectromechanical applications.

The recent intense experimental activity is very encouraging to the development of the field. Coming years, therefore, promise ever increasing experimental input into a field that has been dominated by theory for five decades.

a perfectly conducting plate is [156, 68, 69]

$$F^T = -\frac{\pi^3}{360}\frac{R}{d^3}\left[1 + \frac{360\zeta(3)}{\pi^3}(Td)^3 - 16(Td)^4\right]. \tag{3.69}$$

For the closest separations yet measured, $d \sim 100$ nm, this correction is only $\sim 10^{-5}$ at room temperature.

Chapter 4

Casimir Effect with Perfect Spherical Boundaries

4.1 Electromagnetic Casimir Self-Stress on a Spherical Shell

The zero-point fluctuations due to parallel plates, either conducting or insulating, give rise to an attractive force, which seems intuitively understandable in view of the close connection with the attractive van der Waals interactions. However, one's intuition fails when more complicated geometries are considered.

In 1956 Casimir proposed that the zero-point force could be the Poincaré stress stabilizing a semiclassical model of an electron [12]. For definiteness, take a naive model of an electron as a perfectly conducting shell of radius a carrying a total charge e. The Coulomb repulsion must be balanced by some attractive force; Casimir proposed that that could be provided by the vacuum fluctuation energy, so that the effective energy of the configuration would be

$$E = \frac{e^2}{2a} - \frac{Z}{a}\hbar c, \tag{4.1}$$

where the Casimir energy is characterized by a pure number Z. The would open the way for a semiclassical calculation of the fine-structure constant, for stability results if $E = 0$ or

$$\alpha = \frac{e^2}{\hbar c} = 2Z. \tag{4.2}$$

Unfortunately as Tim Boyer was to discover a decade later after a heroic calculation [13], the Casimir force in this case is *repulsive*, $Z = -0.04618$. The sign results from delicate cancellations between interior and exterior

65

modes, and between TE and TM modes, so it appears impossible to predict the sign *a priori.*

Boyer's calculation was rather complicated, involving finding the zeroes of Bessel functions. Boyer's expression was subsequently evaluated with greater precision by Davies [76]. In the late 1970s two independent calculations appeared confirming this surprising result. Balian and Duplantier [14] used a multiple scattering formalism to obtain a quite tractable form for the Casimir energy for both zero and finite temperature, while Milton, DeRaad, and Schwinger [15] exploited the Green's function technique earlier developed for the parallel plate geometry. We will describe the latter approach here. In particular the Green's dyadic formalism of Sec. 2.6 may be used, except now the modes must be described by vector spherical harmonics, defined by [99, 181, 182, 183, 184]

$$\mathbf{X}_{lm} = [l(l+1)]^{-1/2}\mathbf{L}Y_{lm}(\theta, \phi), \tag{4.3}$$

where $\mathbf{L}$ is the orbital angular momentum operator,

$$\mathbf{L} = \frac{1}{i}\mathbf{r} \times \boldsymbol{\nabla}. \tag{4.4}$$

Notice that we may take $l \geq 1$, because spherically symmetric solutions to Maxwell's equations do not exist for $\omega \neq 0$. The vector spherical harmonics satisfy the orthonormality condition

$$\int d\Omega \, \mathbf{X}_{l'm'}^* \cdot \mathbf{X}_{lm} = \delta_{ll'}\delta_{mm'}, \tag{4.5}$$

as well as the sum rule

$$\sum_{m=-l}^{l} |\mathbf{X}_{lm}(\theta, \phi)|^2 = \frac{2l+1}{4\pi}. \tag{4.6}$$

The divergenceless dyadics $\boldsymbol{\Gamma}'$ and $\boldsymbol{\Phi}$ may be expanded in terms of vector spherical harmonics as

$$\boldsymbol{\Gamma}' = \sum_{lm}\left(f_l\mathbf{X}_{lm} + \frac{i}{\omega}\boldsymbol{\nabla} \times g_l\mathbf{X}_{lm}\right), \tag{4.7a}$$

$$\boldsymbol{\Phi} = \sum_{lm}\left(\tilde{g}_l\mathbf{X}_{lm} - \frac{i}{\omega}\boldsymbol{\nabla} \times \tilde{f}_l\mathbf{X}_{lm}\right), \tag{4.7b}$$

where the second suppressed tensor index is carried by the coefficient functions $f_l, g_l, \tilde{f}_l, \tilde{g}_l$.

Inserting this expansion into the first-order equations (2.96a), (2.96b), and using the properties of the vector spherical harmonics, we straightforwardly find [15] that the Green's dyadic may be expressed in terms of two scalar Green's functions, the electric and the magnetic:

$$\mathbf{\Gamma}(\mathbf{r}, \mathbf{r}'; \omega) = \sum_{lm} \{\omega^2 F_l(r, r') \mathbf{X}_{lm}(\Omega) \mathbf{X}_{lm}^*(\Omega')$$

$$- \mathbf{\nabla} \times [G_l(r, r') \mathbf{X}_{lm}(\Omega) \mathbf{X}_{lm}^*(\Omega')] \times \overleftarrow{\mathbf{\nabla}}'\}$$

$$+ \delta\text{-function terms}, \tag{4.8}$$

where the expression "δ-function terms" refers to terms proportional to spatial delta functions. These terms may be omitted, as we are interested in the *limit* in which the two spatial points approach coincidence. These scalar Green's functions satisfy the differential equation

$$\left(\frac{1}{r}\frac{d^2}{dr^2}r - \frac{l(l+1)}{r^2} + \omega^2\right) \left\{ \begin{array}{c} F_l(r, r') \\ G_l(r, r') \end{array} \right\} = -\frac{1}{r^2}\delta(r - r'), \tag{4.9}$$

subject to the boundary conditions that they be finite at the origin (the center of the sphere), which picks out the spherical Bessel function of the first kind, j_l, there, and correspond to outgoing spherical waves at infinity,* which selects out the spherical Hankel function of the first kind, $h_l^{(1)}$. On the surface of the sphere, we must have

$$F_l(a, r') = 0, \quad \frac{\partial}{\partial r}rG_l(r, r')\bigg|_{r=a} = 0, \tag{4.10}$$

so that F is the TE (H), and G is the TM (E), Green's function. The result is that

$$\left\{ \begin{array}{c} G_l \\ F_l \end{array} \right\} = G_l^0 + \left\{ \begin{array}{c} \tilde{G}_l \\ \tilde{F}_l \end{array} \right\}, \tag{4.11}$$

where G_l^0 is the vacuum Green's function ($k = |\omega|$),

$$G_l^0(r, r') = ikj_l(kr_<)h_l^{(1)}(kr_>), \tag{4.12}$$

---

*The terminology refers to the associated Helmholtz equation, so the behavior at spatial infinity is e^{ikr}/r. The time dependence is $e^{-i\omega t}$, where $k = |\omega|$. Thus, in field-theoretic terms, we are using the usual causal or Feynman Green's function.

and in the interior and the exterior of the sphere respectively,

$$r, r' < a : \quad \left\{ \begin{matrix} \tilde{G}_l \\ \tilde{F}_l \end{matrix} \right\} = -A_{G,F} i k j_l(kr) j_l(kr'), \tag{4.13a}$$

$$r, r' > a : \quad \left\{ \begin{matrix} \tilde{G}_l \\ \tilde{F}_l \end{matrix} \right\} = -B_{G,F} i k h_l^{(1)}(kr) h_l^{(1)}(kr'), \tag{4.13b}$$

where the coefficients are

$$A_F = B_F^{-1} = \frac{h_l^{(1)}(ka)}{j_l(ka)}, \tag{4.14a}$$

$$A_G = B_G^{-1} = \frac{[kah_l^{(1)}(ka)]'}{[kaj_l(ka)]'}, \tag{4.14b}$$

the prime signifying differentiation with respect to the argument ka. From the electromagnetic energy density we may derive the following formula for the energy of the system

$$E = \int (d\mathbf{r}) \frac{1}{2i} \int_{-\infty}^{\infty} \frac{d\omega}{2\pi} e^{-i\omega(t-t')}$$

$$\times \sum_{l=1}^{\infty} \sum_{m=-l}^{l} \{ k^2 [\tilde{F}_l(r, r') + \tilde{G}_l(r, r')] \mathbf{X}_{lm}(\Omega) \cdot \mathbf{X}_{lm}^*(\Omega')$$

$$- \nabla \times \mathbf{X}_{lm}(\Omega) \cdot [\tilde{F}_l(r, r') + \tilde{G}_l(r, r')] \cdot \mathbf{X}_{lm}^*(\Omega') \times \overleftarrow{\nabla}' \} \Big|_{\mathbf{r}=\mathbf{r}'} . \tag{4.15}$$

Note here that the vacuum term in the Green's functions has been removed, since that corresponds to the zero-point energy that would be present in this formalism if no bounding surface were present. Here we are putting the two spatial points coincident, while we leave a temporal separation, $\tau = t - t'$, which is only to be set equal to zero at the end of the calculation, and therefore serves as a kind of regulator. The integration over the solid angle and the sum on m may be easily carried out, with the result

$$E = \frac{1}{2i} \sum_{l=1}^{\infty} (2l+1) \int_{-\infty}^{\infty} \frac{d\omega}{2\pi} e^{-i\omega\tau} \int_0^{\infty} r^2 dr \left(2k^2 [\tilde{F}_l + \tilde{G}_l](r, r) \right.$$

$$\left. + \frac{1}{r^2} \frac{d}{dr} r \left\{ \frac{d}{dr'} r' \left[\tilde{F}_l(r, r') + \tilde{G}_l(r, r') \right] \right\}_{r'=r} \right) . \tag{4.16}$$

The integral of the derivative term here is equal to zero, which can be seen by explicit calculation. The radial integral over Bessel functions is simply done using recurrence relations. (For further details, see Chapter 5.) The result is, in Minkowski spacetime, with $z = ka$, and $h_l(z) = h_l^{(1)}(z)$,

$$E = \frac{i}{2a} \sum_{l=1}^{\infty} (2l + 1) \int \frac{d(\omega a)}{2\pi} e^{-i\omega\tau} z \left\{ \frac{(zj_l)'}{zj_l} + \frac{(zj_l)''}{(zj_l)'} + \frac{(zh_l)'}{zh_l} + \frac{(zh_l)''}{(zh_l)'} \right\}.$$

(4.17)

Now it may be verified that the integrand in (4.17) has the following analytic properties in the complex variable $\zeta = k$:

- The singularities lie in the lower half plane or on the real axis.[†] Consequently, the integration contour C in ω lies just above the real axis for $\omega > 0$, and just below the real axis for $\omega < 0$.
- For Im $\zeta > 0$, the integrand goes to zero as $1/|\zeta|^2$. (This is a weaker condition than specified in Ref. [15].) This convergent behavior is the result of including both interior and exterior contributions.

Then we may write the energy of the sphere as

$$E = \int_C \frac{d\omega}{2\pi} e^{-i\omega\tau} g(|\omega|),$$

(4.18)

where the integrand satisfies the dispersion relation

$$g(|\omega|) = \frac{1}{\pi i} \int_{-\infty}^{\infty} d\zeta \frac{\zeta}{\zeta^2 - \omega^2 - i\epsilon} g(\zeta),$$

(4.19)

because the singularities of $g(\zeta)$ occur only for Im $\zeta \leq 0$. Now we can carry out the ω integral in (4.18) to obtain

$$E = \frac{1}{2\pi} \int_{-\infty}^{\infty} d\zeta \frac{\zeta}{|\zeta|} e^{-i|\zeta||\tau|} g(\zeta).$$

(4.20)

Finally, we rewrite the result in Euclidean space by making the Euclidean transformation $i|\tau| \to |\tau_4| > 0$, so that we have the representation

$$\frac{1}{2|\zeta|} e^{-|\zeta||\tau_4|} = \int_{-\infty}^{\infty} \frac{dk_4}{2\pi} \frac{e^{ik_4\tau_4}}{k_4^2 + \zeta^2}.$$

(4.21)

[†]If a large external sphere is added, as Hagen [185] advocates, the singularities arising from modes in the annulus become real. This is because there is then no energy radiated to infinity. However, this has no effect on the stress on the inner sphere [186].

Thus the Euclidean transform of the energy is

$$E \to E_E = i \int_{-\infty}^{\infty} \frac{dk_4}{2\pi} e^{ik_4 \tau_4} g(i|k_4|). \tag{4.22}$$

In effect, then, the Euclidean transformation is given by the recipe $\omega \to ik_4$, $|\omega| \to i|k_4|$, $\tau \to i\tau_4$. In particular, the energy (4.17) is transformed into the expression

$$\begin{aligned}
E_E &= -\frac{1}{2\pi a} \sum_{l=1}^{\infty} (2l+1) \frac{1}{2} \int_{-\infty}^{\infty} dy\, e^{i\delta y} x \left(\frac{s_l'}{s_l} + \frac{s_l''}{s_l'} + \frac{e_l'}{e_l} + \frac{e_l''}{e_l'} \right) \\
&= -\frac{1}{2\pi a} \sum_{l=1}^{\infty} (2l+1) \frac{1}{2} \int_{-\infty}^{\infty} dy\, e^{i\delta y} x \frac{d}{dx} \ln(1-\lambda_l^2), \tag{4.23}
\end{aligned}$$

where

$$\lambda_l = [s_l(x)e_l(x)]' \tag{4.24}$$

is written in terms of Ricatti-Bessel functions of imaginary argument,

$$s_l(x) = \sqrt{\frac{\pi x}{2}} I_{l+1/2}(x),$$

$$e_l(x) = \sqrt{\frac{2x}{\pi}} K_{l+1/2}(x). \tag{4.25}$$

In the above we have used the value of the Wronskian,

$$W(e_l, s_l) \equiv e_l(x)s_l'(x) - s_l(x)e_l'(x) = 1. \tag{4.26}$$

Here, as a result of the Euclidean rotation,

$$x = |y|, \quad y = \frac{1}{i}ka \text{ is real, as is } \delta = \frac{1}{i}\frac{\tau}{a} \to 0. \tag{4.27}$$

The same formula may be derived by computing the stress on the surface through use of the stress tensor [15], the force per unit area being given by the discontinuity

$$\mathcal{F} = T_{rr}\Big|_{r=a+}^{r=a-} = -\frac{1}{4\pi a^2} \frac{\partial}{\partial a} E, \tag{4.28}$$

E being given by (4.17), or by (4.23) after the Euclidean rotation.

A very rapidly convergent evaluation of this formula can be achieved by using the uniform asymptotic expansions for the Bessel functions:

$$s_l(x) \sim \frac{1}{2} \frac{z^{1/2}}{(1+z^2)^{1/4}} e^{\nu\eta} \left[1 + \sum_{k=1}^{\infty} \frac{u_k(t)}{\nu^k} \right], \qquad (4.29a)$$

$$e_l(x) \sim \frac{z^{1/2}}{(1+z^2)^{1/4}} e^{-\nu\eta} \left[1 + \sum_{k=1}^{\infty} \frac{(-1)^k u_k(t)}{\nu^k} \right], \quad l \to \infty, \quad (4.29b)$$

where

$$x = \nu z, \quad \nu = l + 1/2, \quad t = (1+z^2)^{-1/2}, \quad \eta = t^{-1} + \ln \frac{z}{1+t^{-1}}, \quad (4.30)$$

and the $u_k(t)$ are polynomials in t of definite parity and of order $3k$ [111], the first few of which are

$$u_1(t) = \frac{1}{24}(3t - 5t^3), \qquad (4.31a)$$

$$u_2(t) = \frac{1}{1152}(81t^2 - 462t^4 + 285t^6), \qquad (4.31b)$$

$$u_3(t) = \frac{1}{414720}(30375t^3 - 369603t^5 + 765765t^7 - 425425t^9), \quad (4.31c)$$

$$u_4(t) = \frac{1}{39813120}(4465125t^4 - 94121676t^6 + 349922430t^8$$
$$- 446185740t^{10} + 185910725t^{12}). \qquad (4.31d)$$

If we now write

$$E = -\frac{1}{2a} \sum_{l=1}^{\infty} J(l, \delta), \qquad (4.32)$$

we easily find from the leading approximation,

$$(2\nu)^2 \ln(1 - \lambda_l^2) \sim -\frac{1}{(1+z^2)^3}, \qquad (4.33)$$

so that

$$J(l, 0) \sim \frac{3}{32}, \quad l \to \infty. \qquad (4.34)$$

In order to obtain a finite sum, therefore, we must keep $\delta \neq 0$ until the end of the calculation. By adding and subtracting the leading approximation to the logarithm, we can write

$$J(l, \delta) = R_l + S_l(\delta),$$ (4.35)

where

$$R_l = -\frac{1}{2\pi} \int_0^\infty dz \left[(2l+1)^2 \ln(1 - \lambda_l^2) + \frac{1}{(1+z^2)^3} \right] = J(l, 0) - \frac{3}{32},$$ (4.36)

and

$$S_l(\delta) = -\frac{1}{4\pi} \int_{-\infty}^\infty z \, dz \, e^{i\delta\nu z} \frac{d}{dz} \frac{1}{(1+z^2)^3}.$$ (4.37)

By use of the Euler-Maclaurin sum formula (2.84), we can work out the sum

$$\sum_{l=1}^\infty S_l(\delta) = -\frac{3}{32},$$ (4.38)

precisely the negative of the value of a single term at $\epsilon = 0$![‡] The sum of the remainder, $\sum_l R_l$, is easily evaluated numerically, and changes this result by less than 2%. Thus the result for the Casimir energy for a spherical conducting shell is found to be

$$E = \frac{0.092353}{2a}.$$ (4.40)

This agrees with the result found in 1968 by Boyer [13], evaluated more precisely by Davies [76], and confirmed by a completely different method by Balian and Duplantier in 1978 [14]. Recently, this result has been re-confirmed, using a zeta function method, by Leseduarte and Romeo [187, 188]. Reconsiderations using direct mode summation have also appeared [189, 190, 191].

It is, of course, possible to derive the result using potentials and ghost fields. Unlike in our manifestly gauge-invariant approach, gauge invariance must then be verified. See Ref. [192, 193].

‡This result may be formally obtained by zeta-function regularization:

$$\sum_{l=1}^\infty \nu^s = (2^{-s} - 1)\zeta(-s) - 2^{-s}, \quad s < -1,$$ (4.39)

so if we formally extrapolate to $s = 0$, the angular momentum sum of unity becomes -1.

Eberlein [194] considered the fluctuations of the force on a sphere. If the observation time τ is large compared to the radius of the sphere, $\tau \gg a$,

$$\frac{\Delta \mathcal{F}}{\mathcal{F}} \sim \left(\frac{a}{\tau}\right)^5, \tag{4.41}$$

which is two orders of magnitude smaller than for parallel plates, as seen in (2.118).

Bordag, Elizalde, Kirsten, and Leseduarte [195] examined a scalar field with mass μ subject to a spherical boundary. Divergences were encountered, which were removed by renormalizing constants in a classical Hamiltonian,

$$H_{\text{classical}} = pV + \sigma S + Fa + k + \frac{h}{a}, \tag{4.42}$$

where $V = 4\pi a^3/3$, $S = 4\pi a^2$. Although this would seem to make it impossible to determine the Casimir energy, which is of the form of h/a, a renormalization prescription was imposed that only the contributions corresponding to $\mu \to \infty$ were to removed. Doing so left mass corrections which did not decrease exponentially, as they did for parallel plates, as discussed in Sec. 2.4. Clearly there are issues here yet to be resolved. The completely finite result for a massless scalar will be derived in Chapter 9.

4.1.1   *Temperature Dependence*

Balian and Duplantier [14] also considered the temperature dependence of the electromagnetic Casimir effect for a sphere. They computed the free energy in both the low and high temperature limits, with the results

$$F \sim \frac{0.04618}{a} - (\pi a)^3 \frac{(kT)^4}{15}, \quad kT \ll 1/a, \tag{4.43a}$$

$$F \sim -\frac{kT}{4}\left(\ln kTa + 0.769\right) - \frac{1}{3840kTa^2}, \quad kT \gg 1/a. \tag{4.43b}$$

In view of the relation between the force and the energy [which follows from applying the substitution (2.58) on the zero-temperature expression (4.28) for the force] the corresponding expressions for the energy are

$$E^T = -a\frac{\partial}{\partial a}F = \begin{cases} 0.04618/a + \pi^3 a^3 (kT)^4/5, & kTa \ll 1, \\ kT/4 - 1/(1920kTa^2), & kTa \gg 1. \end{cases} \tag{4.44}$$

Note that, unlike the situation for parallel plates, discussed after (2.61b), E^T does not vanish in the $T \to \infty$ limit. See Ref. [132, 196, 197].

We sketch the derivation of these results in the leading uniform asymptotic approximation, where the approximation (4.33) holds. Then we may write the approximation for the energy at finite temperature as a double sum:[§]

$$E \sim -\frac{3\beta^5}{(2\pi a)^6} \sum_{l=1}^{\infty} \sum_{n=1}^{\infty} \frac{n^2 \nu^5}{[n^2 + (\beta\nu/2\pi a)^2]^4}. \qquad (4.45)$$

As $\beta \to 0$ we can approximate the sum over l by an integral,

$$E \sim -\frac{3}{\beta} \sum_{n=1}^{\infty} n^2 \int_0^{\infty} dx \frac{x^5}{(n^2 + x^2)^4}$$

$$= -\frac{1}{2\beta} \sum_{n=1}^{\infty} n^0 = -\frac{1}{2\beta}\zeta(0) = \frac{kT}{4}, \qquad (4.46)$$

where in the last step we adopted a zeta-function evaluation. Alternatively, we could keep the $e^{i2\pi na\delta/\beta}$ point-splitting factor in the n sum, which then evaluates as

$$\sum_{l=1}^{\infty} e^{2\pi i n a\delta/\beta} = -\frac{1}{2} - \frac{1}{2i}\cot\frac{\pi a\delta}{\beta}, \qquad (4.47)$$

the real part of which is correctly $-1/2$.

For low temperature, $\beta \gg a$, we instead replace the sum on n in (4.45) by an integral,

$$E \sim -\frac{3}{2\pi a} \sum_{l=1}^{\infty} \nu^5 \int_0^{\infty} dx \frac{x^2}{(x^2 + \nu^2)^4} e^{i\delta x}$$

$$= -\frac{3}{2\pi a} \sum_{l=1}^{\infty} \frac{\pi}{192}[6 + 6\delta\nu - 2(\delta\nu)^3]e^{-\nu\delta}$$

$$= \frac{3}{64a}, \qquad (4.48)$$

where the l sum may be carried out directly, or as in Ref. [15]. There are no power of T corrections in this approximation.

[§]This resembles the double-sum representation found by Brown and Maclay for parallel plates [100].

To obtain the latter, it is necessary to use the exact expression (4.23), which for finite T becomes

$$E^T = -\frac{1}{2\pi a} \sum_{l=1}^{\infty} (2l+1) \frac{2\pi a}{\beta} \sum_{n=0}^{\infty} f_{n,l}, \tag{4.49}$$

where

$$f_{n,l} = x_n \frac{d}{dx_n} \ln\left(1 - \lambda_l^2(x_n)\right), \quad x_n = \frac{2\pi a n}{\beta}, \tag{4.50}$$

where we note that $f_{0,l} = 0$. We may evaluate this by use of the Euler-Maclaurin sum formula (2.84). Now the correction to the zero temperature result (4.40) comes from the neighborhood of $n = 0$, where

$$\frac{d}{dn} f_{n,l}\bigg|_{n=0} = 0, \quad \frac{d^3}{dn^3} f_{n,l}\bigg|_{n=0} = -6\left(\frac{2\pi a}{\beta}\right)^3 \delta_{l1}, \tag{4.51}$$

so that

$$E^T_{\text{corr}} = -\frac{6}{4!} B_4 \frac{3}{2\pi a} \left(\frac{2\pi a}{\beta}\right)^4 = \frac{(\pi a)^3 (kT)^4}{5}, \quad kT \ll 1/a. \tag{4.52}$$

4.2 Fermion Fluctuations

The corresponding calculation for a massless spin-1/2 particle subject to bag model boundary conditions (2.119) on a spherical surface,

$$(1 + i\mathbf{n} \cdot \gamma)G\bigg|_S = 0, \tag{4.53}$$

was carried out by Ken Johnson [198] and by Milton [23]. The result is also a repulsive stress, of less than one-half the magnitude of the electromagnetic result. (Recall that for parallel plates, the reduction factor was 7/8.)

In this case we wish to solve the Green's function equation

$$\left(\gamma \frac{1}{i} \partial\right) G(x, x') = \delta(x - x') \tag{4.54}$$

subject to the boundary condition (4.53). In the same representation for the gamma matrices used before in Sec. 2.7, this may be easily achieved in terms of the total angular momentum eigenstates ($\mathbf{J} = \mathbf{L} + (1/2)\boldsymbol{\sigma}$):

$$Z_{JM}^{l=J\pm1/2}(\Omega) = \left(\frac{l+1/2\mp M}{2l+1}\right)^{1/2} Y_{lM-1/2}(\Omega)|+\rangle$$
$$\mp \left(\frac{l+1/2\pm M}{2l+1}\right)^{1/2} Y_{lM+1/2}(\Omega)|-\rangle. \quad (4.55)$$

These may be interchanged by the radial spin operator

$$\boldsymbol{\sigma}\cdot\hat{\mathbf{r}}Z_{JM}^{l=J\pm1/2} = Z_{JM}^{l=J\mp1/2}. \quad (4.56)$$

These harmonics satisfy the addition theorem, the analog of (4.6)

$$\sum_{M=-J}^{J} \operatorname{tr} Z_{JM}^{J\pm1/2}(\Omega)Z_{JM}^{J\pm1/2}(\Omega)^* = \frac{2J+1}{4\pi}. \quad (4.57)$$

From this point, it is straightforward to derive the fermionic Green's function (the details are given in Ref. [19]). It differs from the free Dirac Green's function $G^{(0)}$ by

$$G = G^{(0)} + \tilde{G}, \quad (4.58)$$

where, using a matrix notation for the two-dimensional spin space spanned by $Z_{JM}^{J\pm1/2}$,

$$\tilde{G}_{\pm\mp} = -ik\sum_J \frac{h_{J+1/2}(ka)j_{J+1/2}(ka) - h_{J-1/2}(ka)j_{J-1/2}(ka)}{[j_{J+1/2}(ka)]^2 - [j_{J-1/2}(ka)]^2}$$
$$\times \begin{pmatrix} \mp i\omega j_{J+1/2}(kr)j_{J+1/2}(kr') & kj_{J+1/2}(kr)j_{J-1/2}(kr') \\ -kj_{J-1/2}(kr)j_{J+1/2}(kr') & \mp i\omega j_{J-1/2}(kr)j_{J-1/2}(kr') \end{pmatrix}, \quad (4.59a)$$

$$\tilde{G}_{\pm\pm} = -ik\sum_J \frac{1/k^2a^2}{[j_{J+1/2}(ka)]^2 - [j_{J-1/2}(ka)]^2}$$
$$\times \begin{pmatrix} -ik\omega j_{J+1/2}(kr)j_{J+1/2}(kr') & \mp\omega j_{J+1/2}(kr)j_{J-1/2}(kr') \\ \mp\omega j_{J-1/2}(kr)j_{J+1/2}(kr') & ikj_{J-1/2}(kr)j_{J-1/2}(kr') \end{pmatrix}. \quad (4.59b)$$

Here, as in Sec. 2.7.2, the subscripts denote the eigenvalues of $i\gamma_5$.

Once the Green's function is found, it can be used in the usual way to compute the vacuum expectation value of the stress tensor, which in the

Dirac case is given by (2.143), which leads directly to (unlike in Sec. 2.7, a factor of 2 is included for the charge trace)

$$\langle T_{rr}\rangle = \frac{\partial}{\partial r}\mathrm{tr}\,\boldsymbol{\gamma}\cdot\hat{\mathbf{r}}G(x,x')\Big|_{x'\to x}. \qquad (4.60)$$

The discontinuity of the stress tensor across the surface of the sphere gives the energy according to

$$4\pi a^2[\langle T_{rr}\rangle(a-) - \langle T_{rr}\rangle(a+)] = -\frac{\partial}{\partial a}E(a). \qquad (4.61)$$

A quite straightforward calculation (the details are given in Refs. [19, 23]) gives the result for the sum of exterior and exterior modes, again, in terms of modified spherical Bessel functions:

$$E = \frac{2}{\pi a}\sum_{l=0}^{\infty}(l+1)\int_0^{\infty} dx\, x\cos x\delta\,\frac{d}{dx}\ln[(e_l^2 + e_{l+1}^2)(s_l^2 + s_{l+1}^2)]. \qquad (4.62)$$

The argument of the logarithm may also be written in an alternative form

$$(e_l^2 + e_{l+1}^2)(s_l^2 + s_{l+1}^2) = 1 + \lambda_l^2, \qquad (4.63)$$

where

$$\lambda_l = \frac{d}{dx}\left[\left(l + \frac{1}{2}\right) - \frac{x}{2}\frac{d}{dx}\right]\frac{e_l s_l}{x}. \qquad (4.64)$$

This expression may again be numerically evaluated through use of the uniform asymptotic approximants, with the result

$$E = \frac{0.0204}{a}. \qquad (4.65)$$

Somewhat less precision was obtained because, in this case, the leading uniform asymptotic approximation vanished. This result has been verified, to perhaps one more significant figure,

$$E = \frac{0.02037}{a}, \qquad (4.66)$$

by Elizalde, Bordag, and Kirsten [199]. See also Ref. [200].

Chapter 5

The Casimir Effect of a Dielectric Ball: The Equivalence of the Casimir Effect and van der Waals Forces

A natural generalization of the considerations of the previous Chapter is to allow the spherical shell to be replaced by a dielectric ball, with permittivity ϵ. The Casimir energy, or self-stress, for such a situation was first considered by me in 1980 [16]. This is a rather more subtle situation than the situation considered above, because when the speed of light is different on the two sides of the boundary, the zero-point energy is not finite. However, as we shall see, it is possible to extract an unambiguous finite part, at least in the dilute approximation, by regulating the divergences, and renormalizing physical parameters. In this Chapter we will consider the most general situation, in which a ball of radius a, composed of a material having permittivity ϵ' and permeability μ', is embedded in a uniform medium having permittivity ϵ and permeability μ. Dispersion is included by allowing these electromagnetic parameters to depend on the frequency ω. This configuration allows us to apply the results to the situation of sonoluminescence, for example, where a bubble of air ($\epsilon' \approx 1$, $\mu' = 1$) is inserted into a standing acoustic wave in water ($\epsilon > 1$, $\mu = 1$). This application will be discussed in Chapter 12.

note the applicat

5.1 Green's Dyadic Formulation

We use the Green's dyadic formulation of Chapter 4, as modified for dielectric materials. In terms of Green's dyadics, Maxwell's equations become in a region where ϵ and μ are constant and there are no free charges or currents [cf. (2.96a)–(2.96d)]

$$\boldsymbol{\nabla} \times \boldsymbol{\Gamma} = i\omega\boldsymbol{\Phi} \ , \qquad \boldsymbol{\nabla} \cdot \boldsymbol{\Phi} = 0,$$

$$\frac{1}{\mu}\boldsymbol{\nabla} \times \boldsymbol{\Phi} = -i\omega\epsilon\boldsymbol{\Gamma}' \ , \qquad \boldsymbol{\nabla} \cdot \boldsymbol{\Gamma}' = 0, \tag{5.1}$$

in which $\boldsymbol{\Gamma}' = \boldsymbol{\Gamma} + \mathbf{1}/\epsilon$, where $\mathbf{1}$ includes a spatial delta function. The two solenoidal Green's dyadics given here satisfy the following second-order equations:

$$(\nabla^2 + \omega^2\epsilon\mu)\boldsymbol{\Gamma}' = -\frac{1}{\epsilon}\boldsymbol{\nabla} \times (\boldsymbol{\nabla} \times \mathbf{1}), \tag{5.2a}$$

$$(\nabla^2 + \omega^2\epsilon\mu)\boldsymbol{\Phi} = i\omega\mu\boldsymbol{\nabla} \times \mathbf{1}. \tag{5.2b}$$

These can be expanded in terms of vector spherical harmonics (4.3) as follows

$$\boldsymbol{\Gamma}'(\mathbf{r}, \mathbf{r}') = \sum_{lm}\left(f_l(r, \mathbf{r}')\mathbf{X}_{lm}(\Omega) + \frac{i}{\omega\epsilon\mu}\boldsymbol{\nabla} \times g_l(r, \mathbf{r}')\mathbf{X}_{lm}(\Omega)\right), \tag{5.3a}$$

$$\boldsymbol{\Phi}(\mathbf{r}, \mathbf{r}') = \sum_{lm}\left(\tilde{g}_l(r, \mathbf{r}')\mathbf{X}_{lm}(\Omega) - \frac{i}{\omega}\boldsymbol{\nabla} \times \tilde{f}_l(r, \mathbf{r}')\mathbf{X}_{lm}(\Omega)\right). \tag{5.3b}$$

When these are substituted in Maxwell's equations (5.1) we obtain, first,

$$g_l = \tilde{g}_l, \qquad f_l = \tilde{f}_l + \frac{1}{\epsilon}\frac{1}{r^2}\delta(r - r')\mathbf{X}_{lm}^*(\Omega'), \tag{5.4}$$

and then the second-order equations

$$(D_l + \omega^2\mu\epsilon)g_l(r, \mathbf{r}') = i\omega\mu\int d\Omega'' \, \mathbf{X}_{lm}^*(\Omega'') \cdot \boldsymbol{\nabla}'' \times \mathbf{1}, \tag{5.5a}$$

$$(D_l + \omega^2\mu\epsilon)f_l(r, \mathbf{r}') = -\frac{1}{\epsilon}\int d\Omega'' \, \mathbf{X}_{lm}^*(\Omega'') \cdot \boldsymbol{\nabla}'' \times (\boldsymbol{\nabla}'' \times \mathbf{1})$$

$$= \frac{1}{\epsilon}D_l\frac{1}{r^2}\delta(r - r')\mathbf{X}_{lm}^*(\Omega'), \tag{5.5b}$$

where the spherical Bessel operator is

$$D_l = \frac{\partial^2}{\partial r^2} + \frac{2}{r}\frac{\partial}{\partial r} - \frac{l(l+1)}{r^2}. \tag{5.6}$$

These equations can be solved in terms of Green's functions satisfying

$$(D_l + \omega^2\epsilon\mu)F_l(r, r') = -\frac{1}{r^2}\delta(r - r'). \tag{5.7}$$

Let us specialize to the case of a sphere of radius a centered on the origin, with properties ϵ', μ' in the interior and ϵ, μ outside. Then the solutions to (5.7) have the form

$$F_l(r, r') = \begin{cases} ik'j_l(k'r_<)[h_l(k'r_>) - Aj_l(k'r_>)], & r, r' < a, \\ ikh_l(kr_>)[j_l(kr_<) - Bh_l(kr_<)], & r, r' > a, \end{cases} \tag{5.8}$$

where

$$k = |\omega|\sqrt{\mu\epsilon}, \qquad k' = |\omega|\sqrt{\mu'\epsilon'}, \tag{5.9}$$

and $h_l = h_l^{(1)}$ is the spherical Hankel function of the first kind. Specifically, we have

$$\tilde{f}_l(r, \mathbf{r}') = \omega^2 \mu F_l(r, r') \mathbf{X}_{lm}^*(\Omega'), \tag{5.10a}$$

$$g_l(r, \mathbf{r}') = -i\omega\mu \mathbf{\nabla}' \times G_l(r, r') \mathbf{X}_{lm}^*(\Omega'), \tag{5.10b}$$

where F_l and G_l are Green's functions of the form (5.8) with the constants A and B determined by the boundary conditions given below. Given F_l, G_l, the fundamental Green's dyadic is given by the generalization of (4.8),

$$\mathbf{\Gamma}'(\mathbf{r}, \mathbf{r}') = \sum_{lm} \Big\{ \omega^2 \mu F_l(r, r') \mathbf{X}_{lm}(\Omega) \mathbf{X}_{lm}^*(\Omega')$$

$$- \frac{1}{\epsilon} \mathbf{\nabla} \times G_l(r, r') \mathbf{X}_{lm}(\Omega) \mathbf{X}_{lm}^*(\Omega') \times \overleftarrow{\mathbf{\nabla}}'$$

$$+ \frac{1}{\epsilon} \frac{1}{r^2} \delta(r - r') \mathbf{X}_{lm}(\Omega) \mathbf{X}_{lm}^*(\Omega') \Big\}. \tag{5.11}$$

Because of the boundary conditions that

$$\mathbf{E}_\perp, \quad \epsilon E_r, \quad B_r, \quad \frac{1}{\mu}\mathbf{B}_\perp \tag{5.12}$$

be continuous at $r = a$, we find for the constants A and B in the two Green's functions in (5.11)

$$A_F = \frac{\sqrt{\epsilon\mu'}\,\tilde{e}_l(x')\tilde{e}_l'(x) - \sqrt{\epsilon'\mu}\,\tilde{e}_l(x)\tilde{e}_l'(x')}{\Delta_l}, \tag{5.13a}$$

$$B_F = \frac{\sqrt{\epsilon\mu'}\,\tilde{s}_l(x')\tilde{s}_l'(x) - \sqrt{\epsilon'\mu}\,\tilde{s}_l(x)\tilde{s}_l'(x')}{\Delta_l}, \tag{5.13b}$$

$$A_G = \frac{\sqrt{\epsilon'\mu}\,\tilde{e}_l(x')\tilde{e}_l'(x) - \sqrt{\epsilon\mu'}\,\tilde{e}_l(x)\tilde{e}_l'(x')}{\tilde{\Delta}_l}, \tag{5.13c}$$

$$B_G = \frac{\sqrt{\epsilon'\mu}\tilde{s}_l(x')\tilde{s}_l'(x) - \sqrt{\epsilon\mu'}\tilde{s}_l(x)\tilde{s}_l'(x')}{\tilde{\Delta}_l}. \tag{5.13d}$$

Here we have introduced $x = ka$, $x' = k'a$, the Riccati-Bessel functions

$$\tilde{e}_l(x) = x h_l(x), \qquad \tilde{s}_l(x) = x j_l(x), \tag{5.14}$$

and the denominators

$$\Delta_l = \sqrt{\epsilon\mu'}\tilde{s}_l(x')\tilde{e}_l'(x) - \sqrt{\epsilon'\mu}\tilde{s}_l'(x')\tilde{e}_l(x),$$
$$\tilde{\Delta}_l = \sqrt{\epsilon'\mu}\tilde{s}_l(x')\tilde{e}_l'(x) - \sqrt{\epsilon\mu'}\tilde{s}_l'(x')\tilde{e}_l(x), \tag{5.15}$$

and have denoted differentiation with respect to the argument by a prime.

5.2   Stress on the Sphere

We can calculate the stress (force per unit area) on the sphere by computing the discontinuity of the radial-radial component of the stress tensor:

$$\mathcal{F} = \langle T_{rr}\rangle(a-) - \langle T_{rr}\rangle(a+), \tag{5.16}$$

where

$$T_{rr} = \frac{1}{2}[\epsilon(E_\perp^2 - E_r^2) + \mu(H_\perp^2 - H_r^2)]. \tag{5.17}$$

The vacuum expectation values of the product of field strengths are given directly by the Green's dyadics computed in Section 5.1; according to (2.115a) and (2.115b),

$$i\langle \mathbf{E}(\mathbf{r})\mathbf{E}(\mathbf{r}')\rangle = \mathbf{\Gamma}(\mathbf{r},\mathbf{r}'), \tag{5.18a}$$

$$i\langle \mathbf{B}(\mathbf{r})\mathbf{B}(\mathbf{r}')\rangle = -\frac{1}{\omega^2}\boldsymbol{\nabla} \times \mathbf{\Gamma}(\mathbf{r},\mathbf{r}') \times \overleftarrow{\boldsymbol{\nabla}}', \tag{5.18b}$$

where here and in the following we ignore δ functions because we are interested in the *limit* as $\mathbf{r}' \to \mathbf{r}$. It is then rather immediate to find for the stress on the sphere (the *limit* $t' \to t$ is assumed)

$$\mathcal{F} = \frac{1}{2ia^2}\int_{-\infty}^{\infty}\frac{d\omega}{2\pi}e^{-i\omega(t-t')}\sum_{l=1}^{\infty}\frac{2l+1}{4\pi}$$

$$\times \left\{(\epsilon'-\epsilon)\left[\frac{k^2}{\epsilon}a^2 F_l + \left(\frac{l(l+1)}{\epsilon'} + \frac{1}{\epsilon}\frac{\partial}{\partial r}r\frac{\partial}{\partial r'}r'\right)G_l\right]\right\}\bigg|_{r=r'=a+}$$

$$+ (\mu' - \mu) \left[\frac{k^2}{\mu} a^2 G_l + \left(\frac{l(l+1)}{\mu'} + \frac{1}{\mu} \frac{\partial}{\partial r} r \frac{\partial}{\partial r'} r' \right) F_l \right] \Bigg|_{r=r'=a+} \Bigg\}$$

$$= \frac{i}{2a^4} \int_{-\infty}^{\infty} \frac{dy}{2\pi} e^{-iy\delta} \sum_{l=1}^{\infty} \frac{2l+1}{4\pi} x \frac{d}{dx} \ln D_l, \qquad (5.19)$$

where $y = \omega a$, $\delta = (t - t')/a$, and, up to a multiplicative constant,

$$D_l = (\tilde{s}_l(x')\tilde{e}'_l(x) - \tilde{s}'_l(x')\tilde{e}_l(x))^2 - \xi^2 (\tilde{s}_l(x')\tilde{e}'_l(x) + \tilde{s}'_l(x')\tilde{e}_l(x))^2. \quad (5.20)$$

Here the parameter ξ is

$$\xi = \frac{\sqrt{\frac{\epsilon'}{\epsilon} \frac{\mu}{\mu'}} - 1}{\sqrt{\frac{\epsilon'}{\epsilon} \frac{\mu}{\mu'}} + 1}. \qquad (5.21)$$

This is not yet the answer. We must remove the term which would be present if either medium filled all space (the same was done in the case of parallel dielectrics: see Chapter 3). [This issue will be discussed more fully in Chapter 12.] The corresponding Green's function is given by the first term in (5.8)

$$F_l^{(0)} = \begin{cases} ik' j_l(k' r_<) h_l(k' r_>), & r, r' < a \\ ik j_l(k r_<) h_l(k r_>), & r, r' > a \end{cases} \qquad (5.22)$$

The resulting stress is

$$\mathcal{F}^{(0)} = \frac{1}{a^3} \int_{-\infty}^{\infty} \frac{d\omega}{2\pi} e^{-i\omega\tau} \sum_{l=1}^{\infty} \frac{2l+1}{4\pi} \Big\{ x'[\tilde{s}'_l(x')\tilde{e}'_l(x') - \tilde{e}_l(x')\tilde{s}''_l(x')]$$

$$- x[\tilde{s}'_l(x)\tilde{e}'_l(x) - \tilde{e}_l(x)\tilde{s}''_l(x)] \Big\}. \qquad (5.23)$$

The final formula for the stress is obtained by subtracting (5.23) from (5.19):

$$\mathcal{F} = -\frac{1}{2a^4} \int_{-\infty}^{\infty} \frac{dy}{2\pi} e^{iy\delta} \sum_{l=1}^{\infty} \frac{2l+1}{4\pi} \Big\{ x \frac{d}{dx} \ln D_l$$

$$+ 2x'[s'_l(x')e'_l(x') - e_l(x')s''_l(x')] - 2x[s'_l(x)e'_l(x) - e_l(x)s''_l(x)] \Big\},$$

$$(5.24)$$

where we have now performed a Euclidean rotation, as discussed more fully in the previous chapter,

$$y \to iy, \quad x \to ix, \quad \tau = t - t' \to i(x_4 - x_4') \quad [\delta = (x_4 - x_4')/a],$$
$$\tilde{s}_l(x) \to s_l(x), \quad \tilde{e}_l(x) \to e_l(x), \tag{5.25}$$

where the Ricatti-Bessel functions of imaginary argument are given in (4.25).

5.3   Total Energy

In a similar way we can directly calculate the Casimir energy of the configuration, starting from the energy density

$$u = \frac{1}{2}\langle \epsilon E^2 + \mu H^2 \rangle. \tag{5.26}$$

In terms of the Green's dyadic, the total energy is*

$$E = \int (d\mathbf{r})\, u$$
$$= \frac{1}{2i} \int r^2 dr\, d\Omega \left[\epsilon \mathrm{Tr}\, \mathbf{\Gamma}(\mathbf{r},\mathbf{r}) - \frac{1}{\omega^2 \mu} \mathrm{Tr}\, \boldsymbol{\nabla} \times \mathbf{\Gamma}(\mathbf{r},\mathbf{r}') \times \overleftarrow{\boldsymbol{\nabla}}' \Big|_{\mathbf{r}=\mathbf{r}'} \right] \tag{5.27a}$$
$$= \frac{1}{2i} \int_{-\infty}^{\infty} \frac{d\omega}{2\pi} e^{-i\omega(t-t')} \sum_{l=1}^{\infty} (2l+1) \int_0^{\infty} r^2\, dr$$
$$\times \left\{ 2k^2 [F_l(r,r) + G_l(r,r)] + \frac{1}{r^2}\frac{d}{dr} r \left(\frac{\partial}{\partial r'} r'[F_l + G_l](r,r') \right)_{r'=r} \right\}, \tag{5.27b}$$

where there is no explicit appearance of ϵ or μ. (The last expression looks just like (4.16) for a conducting shell in vacuum. Here, however, the value of k depends on which medium we are in.) As in Ref. [15] and Chapter 4 we can easily show that the total derivative term integrates to zero. We are left with

$$E = \frac{1}{2i} \int_{-\infty}^{\infty} \frac{d\omega}{2\pi} e^{-i\omega\tau} \sum_{l=1}^{\infty} (2l+1) \int_0^{\infty} r^2\, dr\, 2k^2 [F_l(r,r) + G_l(r,r)]. \tag{5.28}$$

---

*Here we ignore dispersion. For the stress, it is sufficient to insert $\epsilon(\omega)$, for example, while in the energy one should write $\frac{d}{d\omega}\omega\epsilon(\omega)$—see Ref. [99], p. 76. This answers to a certain extent the objection of Candelas [201].

However, again we should subtract off that contribution which the formalism would give if either medium filled all space. That means we should replace F_l and G_l by

$$\tilde{F}_l, \tilde{G}_l = \begin{cases} -ik'A_{F,G}j_l(k'r)j_l(k'r'), & r, r' < a \\ -ikB_{F,G}h_l(kr)h_l(kr'), & r, r' > a \end{cases} \tag{5.29}$$

so then (5.28) says

$$E = -\sum_{l=1}^{\infty}(2l+1)\int_{-\infty}^{\infty}\frac{d\omega}{2\pi}e^{-i\omega\tau}\left\{\int_0^a r^2 dr\, k'^3(A_F + A_G)j_l^2(k'r)\right.$$
$$\left. + \int_a^{\infty} r^2 dr\, k^3(B_F + B_G)h_l^2(kr)\right\}. \tag{5.30}$$

The radial integrals may be done by using the following indefinite integral for any spherical Bessel function j_l:

$$\int dx\, x^2 j_l^2(x) = \frac{x}{2}[((xj_l)')^2 - j_l(xj_l)' - xj_l(xj_l)'']. \tag{5.31}$$

But we must remember to add the contribution of the total derivative term in (5.27b) which no longer vanishes when the replacement (5.29) is made. The result is precisely that expected from the stress (5.24),

$$E = 4\pi a^3 \mathcal{F}, \qquad \mathcal{F} = \frac{1}{4\pi a^2}\left(-\frac{\partial}{\partial a}\right)E, \tag{5.32}$$

where the derivative is the naive one, that is, the cutoff δ has no effect on the derivative.

It is useful here to make contact with the formalism introduced by Schwinger [103, 104]. In terms of an imaginary frequency ζ and a parameter w, he derived the following simple formula for the energy from the proper-time formalism

$$E = -\frac{1}{2\pi}\int_0^{\infty}d\zeta\int_0^{\infty}dw\,\mathrm{Tr}_s G, \tag{5.33}$$

where the trace refers to space, the Green's function is

$$G = \frac{1}{w + H}, \tag{5.34}$$

and the Hamiltonian appropriate to the two modes is (for a nonmagnetic material)

$$H = \begin{cases} \text{TE}: & \partial_0 \epsilon \partial_0 - \nabla^2, \\ \text{TM}: & \partial_0^2 - \nabla \cdot (1/\epsilon)\nabla. \end{cases} \tag{5.35}$$

Consider the TE part (the TM part is similar, but not explicitly considered by Schwinger). In terms of Green's function satisfying (5.7), we have

$$E = \frac{1}{2\pi} \int_0^\infty d\zeta \int_0^\infty dw \sum_{l=1}^\infty (2l+1) \int_0^\infty dr\, r^2 F_l(r,r;\zeta^2\epsilon + w), \tag{5.36}$$

where the third argument of the Green's function reflects the substitution in (5.7) of $\omega^2\epsilon \to -\zeta^2\epsilon - w$. We now introduce polar coordinates by writing

$$\zeta^2\epsilon + w = \rho^2, \quad d\zeta\, dw = \frac{1}{\sqrt{\epsilon}} 2\rho^2 \cos\theta\, d\rho\, d\theta, \tag{5.37}$$

and integrate over θ from 0 to $\pi/2$. The result coincides with the first term in (5.28).

5.4 Fresnel Drag

As may easily inferred from Pauli's book [202], the nonrelativistic effect of material motion of the dielectric, $\boldsymbol{\beta}(\mathbf{r})$, is given by the so-called Fresnel drag term,

$$E' = \int (d\mathbf{r}) \frac{\epsilon\mu - 1}{\epsilon} \boldsymbol{\beta} \cdot (\mathbf{D} \times \mathbf{H}) = \int (d\mathbf{r})(\epsilon\mu - 1)\boldsymbol{\beta} \cdot (\mathbf{E} \times \mathbf{H}). \tag{5.38}$$

To preserve spherical symmetry (of course, this is likely not to be a realistic motion) we consider purely radial velocities,

$$\boldsymbol{\beta} = \beta\hat{\mathbf{r}}. \tag{5.39}$$

Then, what we seek is the asymmetrical structure

$$\hat{\mathbf{r}} \cdot \langle \mathbf{E}(\mathbf{r}) \times \mathbf{H}(\mathbf{r}') \rangle = -\hat{\mathbf{r}} \cdot \langle \mathbf{H}(\mathbf{r}') \times \mathbf{E}(\mathbf{r}) \rangle = -\frac{1}{i\mu} \epsilon_{ijk}\hat{r}_i \cdot \Phi_{jk}(\mathbf{r}',\mathbf{r})$$

$$= \omega\hat{\mathbf{r}} \cdot \sum_{lm} \left\{ \mathbf{X}_{lm}(\Omega') \times [\nabla \times G_l(r',r)\mathbf{X}_{lm}^*(\Omega)] \right.$$

$$+ [\boldsymbol{\nabla}' \times F_l(r',r)\mathbf{X}_{lm}(\Omega')] \times \mathbf{X}_{lm}^*(\Omega)\Big\}. \qquad (5.40)$$

This is easily seen to reduce to

$$\hat{\mathbf{r}} \cdot \langle \mathbf{E} \times \mathbf{H} \rangle = \omega \frac{1}{r}\frac{\partial}{\partial r} r \sum_{lm} G_l(r',r)\mathbf{X}_{lm}(\Omega') \cdot \mathbf{X}_{lm}^*(\Omega)$$

$$- \omega \frac{1}{r'}\frac{\partial}{\partial r'} r' \sum_{lm} F_l(r',r)\mathbf{X}_{lm}(\Omega') \cdot \mathbf{X}_{lm}^*(\Omega), \qquad (5.41)$$

so when Ω and Ω' are identified, and the angular integral is carried out, we obtain the corresponding energy for a *slow, adiabatic, radially symmetric motion*,

$$E' = \beta \int_0^\infty r^2 dr \, (\epsilon\mu - 1) \int_{-\infty}^\infty \frac{d\omega}{2\pi} \omega \, e^{-i\omega\tau} \sum_{l=1}^\infty (2l+1)$$

$$\times \frac{1}{r}\frac{\partial}{\partial r} r \, [G_l(r,r') - F_l(r,r')]\Big|_{r'=r}. \qquad (5.42)$$

It is clear, immediately, that if the cutoff τ is set equal to zero, this vanishes because the integrand is odd in ω; compare to (5.28). Since the sign of τ is certainly irrelevant, we therefore claim that in this quasistatic approximation Fresnel drag is absent.

If we were dealing with statics, of course $\langle \mathbf{E} \times \mathbf{H} \rangle$ would be zero by time-reversal invariance. Our argument extends that result to the quasistatic regime. Our point in presenting the result (5.42) is that it will make it possible to extend the calculation to the dynamical regime, where Fresnel drag is nonzero.

Related is the Abraham value of the field momentum [203, 204],

$$\mathbf{G} = \mathbf{E} \times \mathbf{H}, \qquad (5.43)$$

which gives then an extra contribution to the force density,

$$\mathbf{f}' = (\epsilon - 1)\frac{\partial}{\partial t}(\mathbf{E} \times \mathbf{H}). \qquad (5.44)$$

However, as Brevik noted [204], the expectation value of this is also zero, because, in the Fourier transform, successive action of the time derivative brings down ω and $-\omega$. So the continuing controversy about which field momentum to use is without consequence here. This should be already obvious, because the energy is well-defined, and we have already seen that the force is related to the energy by (5.32). Further implications of the results of this section will be given in Sec. 12.3.

5.5   Electrostriction

When a dielectric medium is deformed, there is an additional contribution to the force density, that of electrostriction [184, 203],

$$\mathbf{f}_{\text{ES}} = \frac{1}{2}\nabla\left(E^2\rho\frac{\partial\epsilon}{\partial\rho}\right), \tag{5.45}$$

where ρ is the density of the medium. This term is without effect for computation of the *force* on the dielectric, because it is a total derivative, yet here, where we are calculating the *stress* on the surface, it can be significant. The simplest model for describing the density dependence of the dielectric constant is that given by the Clausius-Mossotti equation, (for example, see [99], p. 58),

$$\frac{\epsilon - 1}{\epsilon + 2} = K\rho, \tag{5.46}$$

where K is a constant. Consequently the logarithmic derivative appearing in (5.45) is

$$\rho\frac{\partial\epsilon}{\partial\rho} = \frac{1}{3}(\epsilon - 1)(\epsilon + 2). \tag{5.47}$$

The calculation of the electrostrictive Casimir effect for a dielectric ball is given by Brevik [204]. We have confirmed his result, and generalized it to the situation at hand [37]. Again, the contribution if either medium fills all space has been subtracted. The result for the integrated stress on the spherical cavity, after the Euclidean transformation is performed, is

$$
\begin{aligned}
\mathcal{S}_{\text{ES}} = -\frac{1}{12a^2}\sum_{l=1}^{\infty}(2l+1)\int_{-\infty}^{\infty}dy\, e^{iy\delta} \\
\times\left\{\frac{(\epsilon'-1)(\epsilon'+2)}{\epsilon'}\left[\frac{A_G}{2x'}\left(x'^2 I_{l+1/2}^2(x')\right)'\right.\right. \\
\left.\left. - x'(A_F+A_G)\int_0^{x'}d\xi\, I_{l+1/2}^2(\xi) + x'A_G\int_0^{x'}\frac{d\xi}{\xi}I_{l+1/2}^2(\xi)\right]\right.
\end{aligned}
$$

$$+ \frac{(\epsilon - 1)(\epsilon + 2)}{\epsilon} \left(\frac{2}{\pi}\right)^2 \left[-\frac{B_G}{2x} \left(x^2 K_{l+1/2}^2(x)\right)' \right.$$

$$\left. - x(B_F + B_G) \int_x^\infty d\xi \, K_{l+1/2}^2(\xi) + x B_G \int_x^\infty \frac{d\xi}{\xi} K_{l+1/2}^2(\xi) \right] \Big\}.$$

$$(5.48)$$

It is rather difficult to extract numerical results from this formula. Indeed, Brevik [204] considered only two special cases, $\epsilon \gg 1$, appropriate to a perfect conductor, and $|\epsilon - 1| \ll 1$. In fact in the latter case he was able to consider only the $l = 1$ term in the sum. This is highly unreliable, as such a term may be completely unrepresentative (such as having the wrong sign, as we saw in the previous chapter). Because this electrostrictive stess presents divergences that are somewhat difficult to understand, we will not consider it further here. We will only remark that it is highly likely to contribute a term comparable to the finite Casimir estimate presented in Sec. 5.7, and urge that efforts be made to extract a value from the above formula.

5.6 Dilute Dielectric-Diamagnetic Sphere

We first discuss the special case $\sqrt{\epsilon \mu} = \sqrt{\epsilon' \mu'}$, that is, when the speed of light is the same in both media. Then $x = x'$ and the Casimir energy (5.24), (5.32) reduces to

$$E = -\frac{1}{4\pi a} \int_{-\infty}^\infty dy \, e^{iy\delta} \sum_{l=1}^\infty (2l + 1) x \frac{d}{dx} \ln[1 - \xi^2((s_l e_l)')^2], \qquad (5.49)$$

where

$$\xi = \frac{\mu - \mu'}{\mu + \mu'} = -\frac{\epsilon - \epsilon'}{\epsilon + \epsilon'}. \qquad (5.50)$$

If $\xi = 1$ we recover the case of a perfectly conducting spherical shell, treated in Chapter 4 [cf. (4.23)], for which E is finite. In fact (5.49) is finite for all ξ.

The evaluation for small ξ was considered first by Brevik and Kolbenstvedt [205, 206, 207, 208]. They applied only the leading uniform asymptotic approximation for the Bessel functions and obtained

$$E \sim \frac{3}{64a}\xi^2, \quad \xi^2 \ll 1. \tag{5.51}$$

This is just the leading term found in the case of a conducting spherical shell, where $\xi = 1$—see (4.38). Much more recently, Klich [209] showed how the evaluation may be carried out exactly in this case. The calculation hinges upon the identity

$$\sum_{l=0}^{\infty}(2l+1)s_l(x)e_l(y)P_l(\cos\theta) = \frac{xy}{\rho}e^{-\rho}, \quad \rho = \sqrt{x^2 + y^2 - 2xy\cos\theta}. \tag{5.52}$$

That fact, together with the orthogonality condition for the Legendre polynomials,

$$\int_{-1}^{1} d\cos\theta\, P_l(\cos\theta)P_{l'}(\cos\theta) = \frac{2}{2l+1}\delta_{ll'}, \tag{5.53}$$

allows us to perform the sum over Bessel functions occurring in the small-ξ expansion of (5.49),

$$E \approx \frac{\xi^2}{4\pi a}\int_{-\infty}^{\infty} dy e^{iy\delta}\sum_{l=1}^{\infty}(2l+1)x\frac{d}{dx}[(s_l e_l)']^2. \tag{5.54}$$

The sum required is then

$$\sum_{l=0}^{\infty}(2l+1)\left([s_l(x)e_l(x)]'\right)^2 = \frac{1}{2}\int_{-1}^{1} d\cos\theta\left[\frac{\partial}{\partial x}\frac{xe^{-x\sqrt{2-2\cos\theta}}}{\sqrt{2-2\cos\theta}}\right]^2$$

$$= \frac{1}{2}\int_{0}^{4x}\frac{dt}{t}e^{-t}\left(1 - t + \frac{t^2}{4}\right). \tag{5.55}$$

Here we have made the change of variable $t = 2x\sqrt{2-2\cos\theta}$. Since the sum in (5.54) starts at $l = 1$, we must subtract from (5.55) the $l = 0$ term, constructed from

$$s_0(x) = \sinh x, \quad e_0(x) = e^{-x}, \tag{5.56}$$

and then apply the operator $x\frac{d}{dx}$:

$$\sum_{l=1}^{\infty}(2l+1)x\frac{d}{dx}[(s_l e_l)']^2 = \frac{1}{2}e^{-4x}(1 + 4x + 4x^2). \tag{5.57}$$

The evaluation of the Casimir energy for this dilute dielectric-diamagnetic sphere is now immediate:

$$E = \frac{\xi^2}{16\pi a} \int_0^\infty du\, e^{-u} \left(1 + u + \frac{u^2}{4}\right)$$
$$= \frac{5\xi^2}{32\pi a} = \frac{0.0994718\xi^2}{2a}, \quad \xi \ll 1. \tag{5.58}$$

It is interesting to note that the Brevik and Kolbenstvedt approximation (5.51) is only 6% too low.[†] It is further remarkable that the value for a spherical conducting shell (4.40), for which $\xi = 1$, is only 7% lower, which as Klich remarks, is accounted for nearly entirely by the next term in the small ξ expansion.

Brevik and Einevoll [212] argue that dispersion should be included, on physical grounds. They assume the speed of light is uniform, $\epsilon(\omega)\mu(\omega) = 1$, but assume a dispersion relation

$$\mu(\omega) - 1 = \frac{\mu_0 - 1}{1 - \omega^2/\omega_0^2}. \tag{5.59}$$

The Casimir stress found is *attractive*, but very sensitive to the values of μ_0 and ω_0. In the leading uniform asymptotic approximation they find the stress on the sphere to be

$$\mathcal{F} = -\frac{1}{8\pi a^4} \left(\frac{\mu_0 - 1}{\mu_0 + 1}\right)^2 \sum_{l=1}^\infty J_l, \tag{5.60}$$

where, for $x_0 = \omega_0 a \to \infty$,

$$\sum_{l=1}^\infty J_l = -\frac{3}{32} + \frac{x_0}{8}\sqrt{\frac{\mu_0 + 1}{2}}. \tag{5.61}$$

If dispersion were neglected, the leading term $-3/32$ alone would be obtained [see (5.51)]. Brevik and Einevoll point out their results are qualitatively similar to those of Candelas [213], although he obtains a constant divergent dispersion energy, and therefore no corresponding stress contribution, while they find a dispersive term logarithmically dependent on a,

---

[†]Subsequently, by including two more terms in the uniform asymptotic expansion, Brevik, Nesterenko, and Pirozhenko [210] obtained an approximation to this result accurate to about 0.1%. See also Ref. [211].

and the corresponding stress/area having an inverse cube dependence:

$$E_{\text{disp}} \propto \omega_0 \ln a, \quad \mathcal{F}_{\text{disp}} \propto \frac{\omega_0}{a^3}. \tag{5.62}$$

Given the nonuniformity of the limits here, however, it seems fair to conclude that the nature of the divergences encountered in this calculation are poorly understood, and to maintain that, most likely, only the finite part of (5.61) is observable.

5.6.1   *Temperature Dependence*

As Klich et al. have observed [214] it is easy to work out the temperature in the leading order in ξ^2. We may simply make the replacement (2.58) in the zero temperature expressions for either the energy or the stress. For the former, we have from (5.54), with $x_n = 2\pi na/\beta$,

$$
\begin{aligned}
E^T &= \frac{\xi^2}{\beta} \sum_{n=0}^{\infty}{}' \sum_{l=1}^{\infty} (2l+1) x_n \frac{d}{dx_n} [(e_l s_l)']^2 (x_n) \\
&= \frac{\xi^2}{2\beta} \sum_{n=0}^{\infty}{}' e^{-4x_n} (1 + 4x_n + 4x_n^2) \\
&= \frac{\xi^2}{2\beta} \left(1 - \frac{\partial}{\partial\lambda} + \frac{1}{4} \frac{\partial^2}{\partial\lambda^2} \right) \sum_{n=0}^{\infty}{}' e^{-4x_n\lambda} \bigg|_{\lambda=1}.
\end{aligned}
\tag{5.63}
$$

Here we have used the evaluation (5.57). Thus the temperature dependence of the Casimir energy for a dilute dielectric-diamagnetic ball is exactly that given in Ref. [214]:

$$E^T = \xi^2 \frac{kT}{4} \left\{ \coth t + \frac{t}{\sinh^2 t} + \frac{t^2 \coth t}{2 \sinh^2 t} \right\}, \tag{5.64}$$

where $t = 4\pi akT$, as in (3.28). The high-temperature limit is similar to that seen in (3.30):

$$E^T \sim \xi^2 \frac{kT}{4} + \xi^2 e^{-8\pi kTa} \left[\frac{kT}{2} + 4\pi a(kT)^2 + 8\pi^2 a^2 (kT)^3 \right], \quad kTa \gg 1, \tag{5.65}$$

while the low-temperature limit

$$E^T \sim \frac{5\xi^2}{32\pi a} + \frac{8\xi^2}{45} a^3 \pi^3 (kT)^4, \quad kTa \ll 1, \tag{5.66}$$

resembles (3.36) for parallel plates. These results are very similar to those for a conducting sphere, where $\xi = 1$; note that "black-body" term in (5.66) differs from that in (4.44) by only a factor of 8/9. The formulas given in Secs. 2.5 or 3.2 for the temperature dependence referred to the force/area. However, because at zero temperature the stress S is given by

$$S = 4\pi a^2 \mathcal{F} = -\frac{\partial}{\partial a} E = \frac{\xi^2}{4\pi a^2} \int_{-\infty}^{\infty} dy \sum_{l=1}^{\infty} (2l+1) x \frac{d}{dx} (s_l e_l)'^2, \qquad (5.67)$$

the substitution (2.58) may be made for the stress, with the result that

$$\mathcal{S}^T = \frac{E^T}{a}. \qquad (5.68)$$

At finite temperature, the stress is obtained by differentiating the free energy, not the energy [see (2.63)], but nevertheless the simple connection above holds true. Indeed, after a somewhat elaborate calculation Klich et al. [214] find that (5.68) holds with E^T given by (5.64). [They seem not to have made this simple observation.]

Brevik and Yousef [215] again consider dispersion, and obtain a logarithmically divergent result for the free energy in the high temperature limit. Only the a dependent part can be observable, however, so that can be interpreted as

$$T \to \infty: \quad F \sim -\frac{\xi^2 kT}{4} \ln a, \qquad (5.69)$$

which gives a stress of $\xi^2 kT/4a$, coinciding with the leading term in (5.65).

5.7 Dilute Dielectric Ball

The general expression (5.24) is rather opaque. Therefore, we consider a dilute dielectric ball, which was already considered in Ref. [16]. (That is, we consider $\mu = 1$ everywhere, $\epsilon = 1$ outside of the ball, and $|\epsilon - 1| \ll 1$ inside the ball.) The formula, which still admits of dispersion, corresponds in that case to the energy

$$E \approx -\frac{1}{8\pi a} \sum_{l=1}^{\infty} (2l+1) \frac{1}{2} \int_{-\infty}^{\infty} dy \, e^{iy\delta} (\epsilon(y) - 1)^2 x \frac{d}{dx} F_l(x), \qquad (5.70)$$

where

$$F_l(x) = x^2 \left(1 + \frac{l(l+1)}{x^2}\right) - \frac{1}{4}\left(\frac{d}{dx} e_l s_l\right)^2$$

$$- x^2 \left[2\left(1 + \frac{l(l+1)}{x^2}\right) e_l s_l - \frac{1}{2}\frac{d^2}{dx^2} e_l s_l\right]^2. \qquad (5.71)$$

(The same result evidently holds if we consider a dielectric bubble, the general dilute effect being proportional to $(\epsilon - \epsilon')^2$.) The integrand here may be approximated by the uniform asymptotic approximation (4.29a), (4.29b) [111]:

$$e_l(x) s_l(x) \sim \frac{1}{2} zt \left(1 + \frac{a_1(t)}{\nu^2} + \frac{a_2(t)}{\nu^4} + \cdots\right), \qquad (5.72)$$

where $\nu = l + 1/2$, $x = \nu z$, and $t = (1 + z^2)^{-1/2}$. The coefficients $a_k(t)$ are polynomials in t of degree $3k$. If we ignore dispersion, and set the time-splitting parameter $\delta = 0$, we obtain [216] the leading uniform asymptotic approximation to (5.70),

$$E \sim \frac{(\epsilon - 1)^2}{64a} \sum_{l=1}^{\infty} \left\{\nu^2 - \frac{65}{128} + \frac{927}{16384\nu^2} + \mathcal{O}(\nu^{-4})\right\}. \qquad (5.73)$$

The first two terms are formally divergent, but may be evaluated by the zeta-function definition (4.39). (That is, we may replace the overall $2l + 1$ factor in Eq. (5.70) by $(2l + 1)^{1-\eta}$, and continue from $\text{Re}\,\eta > 3$ to $\eta = 0$.) Note that if only the leading term were kept, the result given in Ref. [16, 37] would be obtained, $E_1 = -(\epsilon - 1)^2/(256a)$, while including two terms reverses the sign and hardly changes the magnitude [216]: $E_2 = +33(\epsilon - 1)^2/(8192a)$. It is important to recognize that the same finite result is achieved if the point-split regularization is retained, as detailed in Ref. [37].‡

‡The leading ν term is

$$E \sim -\frac{(\epsilon' - \epsilon)^2}{16\pi a} \sum_{l=1}^{\infty} \nu^2 \frac{1}{2} \int_{-\infty}^{\infty} dz\, e^{i\nu z\delta} z \frac{d}{dz} \frac{1}{(1+z^2)^2}$$

$$= -\frac{(\epsilon' - \epsilon)^2}{64a}\left(\frac{16}{\delta^3} + \frac{1}{4}\right) \to -\frac{(\epsilon' - \epsilon)^2}{256a}. \qquad (5.74)$$

Here, the last arguable step is made plausible by noting that since $\delta = \tau/a$ the divergent term represents a contribution to the surface tension on the bubble, which should be cancelled by a suitably chosen counter term (contact term).

There seems to be no ambiguity in the procedure.[§]

Indeed, let us do the result exactly. We simply add and subtract the two leading asymptotic terms from the integrand in (5.70), so that $E = E_2 + E_R$, where the remainder is

$$E_R = \frac{(\epsilon-1)^2}{4\pi a} \sum_{l=1}^{\infty} \nu^2 \int_0^{\infty} dz \left[F_l(\nu z) - \frac{t^4}{4} + \frac{t^{10}}{8\nu^2}(1 + 8z^2 - 5z^4 + z^6) \right],$$

(5.75)

According to the third term in (5.73), the z integral here is asymptotic to $927\pi/262144\nu^4$; we evaluate the l sum by doing the integral numerically for the first ten terms, and using the asymptotic approximant thereafter. The result[¶] is, as first given in Ref. [39],

$$E = (\epsilon - 1)^2 \frac{0.004767}{a}.$$

(5.76)

(The approximation E_2 is 15% too low, whereas if the first three terms in (5.73) are kept, the estimate is 1.8% high.)

This result may be obtained analytically by use of the identity (5.52). Here, the calculation is not exactly straightforward [217], so we sketch it here. We write the energy expression (5.70) as the sum of three terms,

$$E = E^{(0)} + E^{(2)} + E^{(4)},$$

(5.77)

where the number in the superscript counts the number of Bessel-function products, $e_l s_l$, in each term. The first term is just a polynomial in x, so it may be absorbed by a contact term; effectively,

$$E^{(0)} = 0.$$

(5.78)

The second term is proportional to that encountered in (5.54), so the result

[§]As we have seen, zeta function regularization is a simple and effective method of capturing the finite part of the Casimir energy. It yields the same result as isolating the divergent part with a physical cutoff, such as the time-splitting parameter $\delta \neq 0$, and removing that term through the process of renormalization. (For more details see, for example, Ref. [41].) If we directly evaluate (5.70) as $\delta = \tau/a \to 0$ we obtain $E = A_{1/2}(a^2/\tau^3) + A_{3/2}(1/\tau) + \tilde{E}$, where, for example, $A_{1/2}$ is exhibited in (5.74), and $\tilde{E}$ is given by (5.76), below.

[¶]It is interesting that this result is more than two times smaller that the (unambiguously finite) result for a ball with the speed of light the same inside and outside, that is, with $\epsilon_1\mu_1 = \epsilon_2\mu_2$, as given in (5.58).

may be immediately written down from (5.58):

$$E^{(2)} = \frac{(\epsilon - 1)^2}{16} \frac{5}{32\pi a}. \tag{5.79}$$

The third term is more complicated:

$$
\begin{aligned}
E^{(4)} &= -\frac{(\epsilon - 1)^2}{32\pi a} \sum_{l=1}^{\infty} (2l + 1) \int_0^\infty dx \, x^2 (e_l'' s_l + s_l'' e_l - 2e_l' s_l')^2 \\
&= -\frac{(\epsilon - 1)^2}{256\pi a} \int_0^\infty dx \int_0^{4x} dt \, t \, e^{-t} \left(1 - \frac{2}{t} - \frac{16x^2}{t^2} - \frac{32x^2}{t^3} \right)^2,
\end{aligned}
\tag{5.80}
$$

where we have set the time-splitting parameter equal to zero, integrated by parts in x, and used the identity (5.52). (The $l = 0$ contribution here is merely a contact term.) Now let $t = 4xu$, and integrate first on x, which gives the result

$$E^{(4)} = -\frac{(\epsilon - 1)^2}{256\pi a} \int_0^1 du \left(\frac{1}{2u^2} + \frac{1}{u^4} + \frac{5}{2u^6} \right). \tag{5.81}$$

This last integral diverges at $u = 0$; however, if we retain only the evaluation at $u = 1$ we obtain the correct finite result:

$$E^{(4)} = \frac{(\epsilon - 1)^2}{192\pi a}. \tag{5.82}$$

Combining these two parts, we obtain an analytic result coincindent with (5.76),

$$E = \frac{23}{1536\pi} \frac{(\epsilon - 1)^2}{a}. \tag{5.83}$$

We will see in Sec. 5.9 that this result may be reproduced from the sum of van der Waals interactions.

5.7.1 *Temperature Dependence*

Nesterenko, Lambiase, and Scarpetta [218] use a mode summation technique to compute the temperature dependence of the Casimir effect for a dilute dielectric ball. The result is, for low temperature,

$$E^T = \frac{23}{1536} \frac{(\epsilon - 1)^2}{\pi a} - \frac{7}{90} (\epsilon - 1)^2 (\pi a)^3 T^4 + O(T^6), \tag{5.84}$$

so from the relation between the energy and the stress, or the derivative of the free energy, $E = -a\partial F/\partial a$, we have the following low temperature expansion for the free energy

$$F = \frac{23}{1536}\frac{(\epsilon - 1)^2}{\pi a} + \frac{7}{270}(\epsilon - 1)^2(\pi a)^3 T^4. \qquad (5.85)$$

They note that the negative sign of the T^4 term in the energy (or the stress) implies that the temperature correction tends to counteract the zero-temperature repulsion, unlike the case exemplified in Sec. 5.6.1.

5.8 Conducting Ball

In general, the Casimir effect for a dielectric ball is cubically divergent [16], as seen in the special case considered in Sec. 5.7. However, there is one case in which the divergence is softened, which is when

$$\xi \to 1, \quad \text{or} \quad \epsilon \to \infty. \qquad (5.86)$$

In this limit, the "inside" functions become

$$s_l(x') \to \frac{1}{2}e^{x'}, \quad s_l'(x') \to \frac{1}{2}e^{x'}, \qquad (5.87a)$$

while

$$x'(e_l's_l' - e_ls_l'')(x') \to -x'. \qquad (5.87b)$$

The expression (5.24) for the force per unit area on the surface of a perfectly conducting ball therefore becomes

$$\mathcal{F} = -\frac{1}{8\pi^2 a^4}\sum_{l=1}^{\infty}\nu\int_{-\infty}^{\infty}dy$$

$$\times e^{iy\delta}x\left\{\frac{d}{dx}\ln(-e_le_l')(x) - 2[e_l'(x)s_l'(x) - e_l(x)s_l''(x)]\right\}, \qquad (5.88)$$

since the effect of the exponentials (5.87a) is cancelled by the contact term (5.87b). Note that, apart from the remaining volume term, this is exactly the same as the "outside" part of the expression for the stress on the spherical shell obtained from (4.23).

Again we approximately evaluate (5.88) by using the uniform asymptotic behaviors of the Bessel functions. The $d\eta/dz$ term cancels between the logarithm term and the volume term,

$$\ln(-e_l e_l') \sim \frac{1}{\nu}\frac{1}{12}(3t + t^3) - 2\nu\eta, \qquad (5.89a)$$

$$2(e_l' s_l' - e_l s_l'') \sim -\frac{1}{\nu^2 z}\frac{1}{4}(t + t^5) + 2\frac{d\eta}{dz}, \qquad (5.89b)$$

leaving us with the leading term

$$\mathcal{F} \sim \frac{1}{8\pi^2 a^4} \sum_{l=1}^{\infty} \nu \frac{1}{2} \int_{-\infty}^{\infty} dz\, t^5\, e^{i\nu z\delta}. \qquad (5.90)$$

The z integral is

$$\int_0^{\infty} dz \cos \nu z\delta \frac{1}{(1 + z^2)^{5/2}} = \frac{1}{3}(\nu\delta)^2 K_2(\nu\delta) \to \frac{2}{3} \quad \text{as} \quad \nu\delta \to 0. \qquad (5.91)$$

Using the Euler-Maclaurin summation formula (2.84) we find the sum over l to be[||]

$$\sum_{l=1}^{L} \nu\frac{2}{3} + \frac{2}{3}\frac{1}{2}\left(L + \frac{3}{2}\right) - \frac{1}{12}\frac{2}{3} + \int_{L+3/2}^{\infty} d\nu\, \nu\frac{1}{3}(\nu\delta)^2 K_2(\nu\delta)$$

$$= \frac{8}{3\delta^2} - \frac{11}{36}. \qquad (5.92)$$

If we include the next-to-leading term in the uniform asymptotic expansion, which is easily seen to be just half the result for the spherical shell result (4.38), we obtain the approximate result for the Casimir energy for a conducting ball:

$$E_{\text{ball}} \sim \frac{1}{\pi a}\left(\frac{4}{3\delta^2} - \frac{11}{72}\right) + \frac{3}{128a}. \qquad (5.93)$$

Again it is plausible that the divergent term is an unobservable quantity: it gives rise to an term in the energy proportional to the radius of the sphere, and so corresponds to a constant stress. It would not appear if a zeta-function regularization were adopted.[**] If that term in (5.93) were

---

[||]There is an error in Ref. [16]: The derivative term in this evaluation [the third term on the first line of (5.92)] was inadvertently omitted so that $-1/4$ was obtained for the result instead of $-11/36$.

[**]Then the leading term becomes

$$E = -\frac{1}{2\pi a}\frac{1}{3}(1 + \zeta(-1)) = -\frac{11}{72\pi a}. \qquad (5.94)$$

simply omitted, we would be left with a constant *attractive* stress, or a negative energy

$$E \sim -\frac{0.025}{a},$$ (5.95)

so that it would be possible to resurrect Casimir's electron model (4.2), with a predicted value of the fine structure constant only an order of magnitude too large. However, as we will see in the next Chapter, it seems impossible to obtain a complete evaluation of the Casimir effect in this case, because an irreducible divergence still remains.

5.9   Van der Waals Self-Stress for a Dilute Dielectric Sphere

Here, our intention is to carry out the same simple calculation, summing the van der Waals interactions between the molecules that make up the material, that we performed on parallel slabs in Sec. 3.2.3, for a dilute dielectric ball. The first two steps are unambiguous, following from (3.51) (θ is the angle between $\mathbf{r}$ and $\mathbf{r}'$, and $\mathcal{N}$ is the number density of molecules):

$$
\begin{aligned}
E &= -\frac{1}{2}B\mathcal{N}^2 \int (d\mathbf{r})(d\mathbf{r}') \frac{1}{(r^2 + r'^2 - 2rr'\cos\theta)^{\gamma/2}} \\
&= -\frac{4\pi^2 B\mathcal{N}^2}{2-\gamma} \int_0^a dr \int_0^a dr'\, rr' \left[\frac{1}{(r+r')^{\gamma-2}} - \frac{1}{|r-r'|^{\gamma-2}} \right].
\end{aligned}
$$ (5.96)

Now, however, there are divergences of two types, "volume" ($r' \to r$) and "surface" ($r \to a$). The former is of a universal character. If we regulate it by a naive point separation, $r' \to r + \delta$, $\delta \to 0$, we find the most divergent part to be

$$E_{\text{vol}} = -\frac{\pi B\mathcal{N}^2}{10}\frac{1}{\delta^4}V, \quad V = \frac{4\pi a^3}{3},$$ (5.97)

which is identical to the corresponding (omitted) divergent term in the parallel dielectric calculation, where $V = aA$. This is obviously the self-energy divergence that would be present if the medium filled all space, and makes no reference to the interface, and is therefore quite unobservable. This is the analogue (although the ϵ dependence is different) of the volume divergence in the Casimir effect, (1.40), which we will discuss more fully in Chapter 12.

If, once again, the divergent terms are simply omitted, as may be weakly justified by continuing in the exponent γ from $\gamma < 3$, we obtain a positive energy,

$$E_{\text{vdW}} = \frac{23}{1536\pi a}(\epsilon - 1)^2. \tag{5.98}$$

This may be more rigorously justified by continuing in dimension, a procedure which has proved useful and illuminating in Casimir calculations [32]—See Chapter 9. Thus we replace the previous expression for the energy by

$$E = -\frac{1}{2}B\mathcal{N}^2 \int d^D r \, d^D r' \frac{1}{|\mathbf{r} - \mathbf{r}'|^\gamma} \tag{5.99}$$

where, in terms of the last angle in D-dimensional polar coordinates,

$$\int d^D r = \frac{2\pi^{(D-1)/2}}{\Gamma\left(\frac{D-1}{2}\right)} \int_0^a dr \, r^{D-1} \int_0^\pi d\theta \sin^{D-2}\theta. \tag{5.100}$$

If we take, say, $\mathbf{r}'$ to lie along the z axis, so that θ is again the angle between $\mathbf{r}$ and $\mathbf{r}'$, we find

$$E = -\frac{1}{2}B\mathcal{N}^2 \frac{2\pi^{D/2}}{\Gamma\left(\frac{D}{2}\right)} \frac{2\pi^{(D-1)/2}}{\Gamma\left(\frac{D-1}{2}\right)} \int_0^a dr' \, r'^{D-1} \int_0^a dr \, r^{D-1} \tag{5.101}$$

$$\times \int_{-1}^1 d\cos\theta \, (1 - \cos^2\theta)^{(D-3)/2}(r^2 + r'^2 - 2rr'\cos\theta)^{-\gamma/2}.$$

The angular integration can be given in terms of an associated Legendre function $P_b^a(z)$,

$$\int_{-1}^1 dt \, (1 - t^2)^{(D-3)/2}(r^2 + r'^2 - 2rr't)^{-\gamma/2}$$

$$= \sqrt{\pi}\Gamma\left(\frac{D-1}{2}\right)(rr')^{1-D/2}|r^2 - r'^2|^{(D-\gamma-2)/2}P_{(\gamma-D)/2}^{1-D/2}\left(\frac{r^2 + r'^2}{|r^2 - r'^2|}\right). \tag{5.102}$$

Now let us substitute this into the expression for the energy, and change variables from r, r' to

$$x = r^2 + r'^2, \quad y = \frac{r^2 + r'^2}{|r^2 - r'^2|}. \tag{5.103}$$

The x integral is then trivially done, leaving us with

$$E = -\frac{B\mathcal{N}^2\pi^D}{2^{D/2}\Gamma(D/2)} \frac{1}{D - \gamma/2} \int_1^\infty dy \left(\frac{2a^2}{y+1}\right)^{D-\gamma/2}$$
$$\times (y^2 - 1)^{(D-2)/4} P_{(\gamma-D)/2}^{1-D/2}(y), \tag{5.104}$$

valid for $D > \gamma/2$. Integrals of this type are given in [110]:

$$\int_1^\infty dy \, (y-1)^{-a/2}(y+1)^{b+a/2-1} P_b^a(y) = 2^b \frac{\Gamma(-2b)}{\Gamma(1-b-a)\Gamma(1-b)}, \tag{5.105}$$

valid for Re $a < 1$, Re $b < 0$. Then we have, using the duplication formula for the Γ function (2.33),

$$E = -B\mathcal{N}^2 \frac{\pi^{D-1/2}2^{D-\gamma}\Gamma\left(\frac{D-\gamma+1}{2}\right)}{\Gamma(D/2)\Gamma(D-\gamma/2+1)(D-\gamma)} a^{2D-\gamma}. \tag{5.106}$$

The resulting formula is regular when D and γ are both odd integers, so we can analytically continue from $D > \gamma$ to $D = 3$ for $\gamma = 7$. Doing so gives us, using Eq. (3.55),

$$E = B\mathcal{N}^2 \frac{\pi^2}{24} \frac{1}{a} = \frac{23}{24} \frac{(\epsilon-1)^2}{64\pi a}, \tag{5.107}$$

exactly the same as the naive result (5.98). This calculation[tt] was first presented in Ref. [38].

This precisely agrees with the Casimir result (5.83). The same identity between the van der Waals and Casimir forces has now been noted by several authors [39, 41, 42]. For example, Barton [40] used an elementary method of summing zero-point energies directly in powers of $(\epsilon - 1)$, using ordinary perturbation theory. Of course, the approach given in Sec. 5.7 is in principle more general, in that it allows for arbitrary ϵ and permits inclusion of dispersion.

[tt]We offer as evidence for the validity of our methodology the fact that formula (5.106) gives the correct Coulomb energy for a uniform ball of charge, for which $\gamma = 1$.

5.10   Discussion

There has recently been considerable controversy concerning the possible relevance of the Casimir effect to sonoluminescence [34]. The idea that the "dynamical Casimir effect" might be relevant to sonoluminescence originated in the work of Schwinger [35]. We will discuss these ideas, and subsequent work, in detail in a later chapter. However, now that we clearly see that the Casimir energy may be identified with sum of van der Waals interactions, it seems perfectly plain that the volume effect they consider, proportional to $\epsilon - 1$, simply cannot be present, because such cannot arise from pairwise interactions. (This point was already made in Ref. [16].) Our interpretation stands vindicated: an effect proportional to the volume represents a contribution to the mass density of the material, and cannot give rise to observable effects.

More subtle is the role of surface divergences [16, 38]. The zeta function regularization calculation we presented above simply discards such terms; but they appear in more physical regularization schemes. For example, if the time splitting parameter in (5.24) is retained, we get from the leading asymptotic expansion [see (5.74)]

$$E_{\text{div}} = -\frac{(\epsilon - 1)^2}{4a}\frac{1}{\delta^3};\qquad(5.108)$$

and if a simple model for dispersion is used, with characteristic frequency ω_0, the same result is obtained with $1/\delta \to \omega_0 a/4$ [37]. (A very similar result is given in Ref. [216].) We believe these terms are probably also unobservable, for they modify the surface tension of the liquid, which, like the bulk energy, is already phenomenologically described. (That surface tension has its origin in the Casimir effect was proposed in Ref. [11].)

We note that Barton in his perturbative work [40] seems to concur with our assessment: The terms "proportional to V [the volume] and to S [the surface] would be combined with other contributions to the bulk and to the surface energies of the material, and play no further role if one uses the measured values." However, he seems to give more credence to the physical observability of these divergent terms in his most recent analysis [219].

It is truly remarkable that however the (true) divergences in the theory are regulated, and subsequently discarded, the finite result is unchanged. That is, in the van der Waals energy, we can simply omit the point-split divergences, or proceed through dimensional continuation, where no divergences are explicit; in either case, the same result (5.98) is obtained. Likewise, the same result is obtained for the Casimir energy using either a temporal point-splitting, or an exponential wavenumber cutoff [40], and omitting the divergent terms; or through the formal trick of zeta-function regularization [39]. The finite parts are uniquely obtained by quite distinct methods [41, 42]. It is worth re-emphasizing that we are not claiming that the Casimir effect for a dielectric ball is finite, unlike the classic case of a spherical conducting shell described in Sec. 4.1. It is merely that those divergent terms serve to renormalize phenomenological parameters in the condensed matter system.

Chapter 6

Application to Hadronic Physics: Zero-Point Energy in the Bag Model

Quantum chromodynamics (QCD) is nearly universally believed to be the underlying theory of hadronic matter. Yet, the theory remains poorly understood. The phenomenon of color confinement has yet to be derived from QCD, but it may be roughly approximated by the phenomenologically successful bag model [84, 85, 86, 87, 88, 89, 220]. In this model, the normal vacuum is a perfect color magnetic conductor, that is, the color magnetic permeability μ is infinite, while the vacuum in the interior of the bag is characterized by $\mu = 1$. This implies that the color electric and magnetic fields are confined to the interior of the bag, and that they satisfy the following boundary conditions on its surface S:

$$\mathbf{n} \cdot \mathbf{E}\bigg|_S = 0, \quad \mathbf{n} \times \mathbf{B}\bigg|_S = 0, \tag{6.1}$$

where $\mathbf{n}$ is a unit normal to S. Now, even in an "empty" bag (i.e., one containing no quarks) there will be nonzero fields present because of quantum fluctuations. This gives rise to a zero-point or Casimir energy, as we have seen. It would be anticipated that this energy would have the form $-Z/a$, where a is the radius of a (spherical) bag and Z is some pure number. Indeed, such a term has been put in bag model calculations, and a good fit to hadronic masses has been obtained for $Z = 1.84$ [84, 85, 86, 87, 220]. However, in principle Z should be computable from the underlying dynamics of QCD.* In this Chapter we will calculate Z in the approxima-

---

*In fact, in Ref. [86], it is noted that "a real calculation of Z is needed, and eventually will be provided." Apparently, the authors were unaware of Boyer's calculation [13], which suggests $Z < 0$.

tion that the gluons are free inside the bag, which is roughly justified by asymptotic freedom. The result is a value of Z that appears quite incompatible with the phenomenological value. However, some twenty years after these considerations were first published [18, 19] the results still remain not clearcut because, as we observed in Chapter 5, when there is a discontinuity in the speed of light, the Casimir energy is not finite. Nevertheless, the results of that chapter give us some confidence that unambiguous finite observable Casimir energies can be extracted in such a case.

A related but somewhat different motivation for this work came from Johnson's model for the QCD ground-state wavefunction [17]. Effectively, he supposed that space was filled with bags, the boundaries of which confine color to small, asymptotically-free regions.[†] He used the classic result for the Casimir effect of a conducting spherical shell, as described in Chapter 4, together with various guesses for the higher-order effects, to estimate the parameters of the bag model. But those QED calculations cannot be properly extrapolated to this situation, for they refer to a single shell in otherwise empty space. As we have seen, there is a delicate balance between interior and exterior contributions, so that only the sum is cutoff independent. The closely packed bags in Johnson's model present a quite different situation. In fact the energy density required in Johnson's model will be provided by the result of the calculation presented here, since the energy of space filled with contiguous bags is simply the sum of the field energies contained within each bag.

We should remark on another simple application of zero-point energy considerations to hadronic physics. Fishbane, Gasiorowicz, and Kaus [222] calculated the zero-point energy in the flux-tube connecting a heavy quark-antiquark system. They found, not surprisingly, that it coincides with the Lüscher potential [83],

$$V = -\frac{\pi}{12(2a)}, \tag{6.2}$$

where $2a$ is the quark separation, which we have seen as the $d = 0$ case of (2.8) for parallel plates [see (1.35)], and we will see again as the result for a $D = 1$ hypersphere in (9.21).

[†]He later replaced this model with one in which real gluons were condensed in the ground state. See Ref. [221]. At that point he argued that surface energies and shape instabilities led to the vacuum bag model being inconsistent. Given our continuing poor understanding of the nature of these effects, that dismissal seems premature.

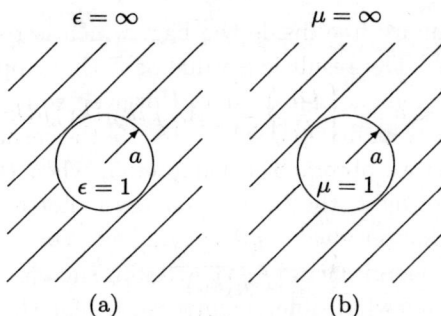

Fig. 6.1 (a) The geometry of a spherical cavity imbedded in a perfect conductor. (b) The dual geometry of the bag model.

6.1 Zero-point Energy of Confined Gluons

Our discussion follows closely on the formalism presented in Chapter 4, as extended to dielectric and conducting balls in Chapter 5. For electrodynamics the situation we consider is as shown in Fig. 6.1(a). Duality ($\mathbf{E} \to \mathbf{H}$, $\mathbf{H} \to -\mathbf{E}$) then allows us to extend the result to the QCD case, Fig. 6.1(b), where one must also allow for the fact that there are 8 gluon fields.

There are, as we have seen, several methods of proceeding. One can compute the energy (or the stress on the surface) when the dielectric constant ϵ is finite in the exterior region, letting $\epsilon \to \infty$ at the end of the calculation. This is the procedure followed in Sec. 5.8. Alternatively, one can calculate the result directly for a spherical cavity in an infinite conductor. Since all methods agree, we simply derive here the expression for the zero-point energy in the latter case. It may be obtained from the interior contribution of (4.15), as exhibited in terms of scalar Green's functions:

$$E = \frac{1}{2i} \int_{-\infty}^{\infty} \frac{d\omega}{2\pi} e^{-i\omega(t-t')} \sum_{l=1}^{\infty} (2l+1) \int_0^a r^2 \, dr$$

$$\times \left\{ 2k^2 [\tilde{F}_l(r,r) + \tilde{G}_l(r,r)] + \frac{1}{r^2} \frac{d}{dr} r \left(\frac{\partial}{\partial r'} r' [\tilde{F}_l + \tilde{G}_l](r,r') \Big|_{r'=r} \right) \right\},$$

$$(6.3)$$

We should emphasize that the cutoff $\tau \to 0$ emerges naturally from the overlap of field points, with no reference to the properties of the boundary. Here $k = |\omega|$ and $\tilde{F}_l$ and $\tilde{G}_l$ are the transverse electric and magnetic Green's functions from which the vacuum parts have been removed [see (4.13a),

(4.14a), and (4.14b)]:

$$r, r' < a : \quad \left\{ \begin{matrix} \tilde{G}_l \\ \tilde{F}_l \end{matrix} \right\} = -A_{G,F} i k j_l(kr) j_l(kr'), \qquad (6.4)$$

where

$$A_F = \frac{h_l^{(1)}(ka)}{j_l(ka)}, \qquad (6.5a)$$

$$A_G = \frac{[kah_l^{(1)}(ka)]'}{[kaj_l(ka)]'}, \qquad (6.5b)$$

In the spherical-shell calculation given in Chapter 4 there were both interior and exterior contributions to the energy, and as a consequence the surface term [the second term in (6.3)] vanished. In fact, the surface term cancels a portion of the first term in (6.3), leaving us with

$$\begin{aligned} E &= -\frac{1}{i} \sum_{l=1}^{\infty} (2l+1) \int_{-\infty}^{\infty} \frac{d\omega}{2\pi} e^{-i\omega\tau} (A_F + A_G) ka \{ ([kaj_l(ka)]')^2 \\ &\quad + [(ka)^2 - l(l+1)][j_l(ka)]^2 \} \\ &= -\frac{1}{2\pi a} \sum_{l=1}^{\infty} (2l+1) \int_0^{\infty} dx\, x \left\{ \frac{s_l'}{s_l} + \frac{s_l''}{s_l'} + 2(e_l'e_l' - e_l e_l'') \right\} \cos x\delta, \end{aligned}$$

$$(6.6)$$

Here, we have performed the Euclidean rotation as summarized in (4.27), and have introduced the spherical Bessel functions of imaginary argument (4.25), which are, more explicitly given by

$$s_l(x) = x^{l+1} \left(\frac{1}{x} \frac{d}{dx} \right)^l \frac{\sinh x}{x}, \quad e_l(x) = (-1)^l x^{l+1} \left(\frac{1}{x} \frac{d}{dx} \right)^l \frac{e^{-x}}{x}. \quad (6.7)$$

Expression (6.6) is nearly, but not quite, the same as that found by Bender and Hays [223]. Apart from an overall sign, their formula has an extra term which arises precisely because of the neglect of the surface term in (6.3).

The result (6.6) is exactly what one would anticipate from the earlier work described in Chapters 4 and 5. The term proportional to

$$\frac{s_l'}{s_l} + \frac{s_l''}{s_l'} = \frac{d}{dx} \ln s_l s_l', \qquad (6.8)$$

is just the inside part of the spherical shell result (4.23), while the remaining term is just the corresponding "volume term" [cf. (5.23)], which cancels for a shell.[‡] It is the negative of what one would obtain from the use of the free-space Green's function in (6.3).

To attempt to evaluate this expression, we employ the first two terms in the uniform asymptotic expansion for large l for the spherical Bessel functions, Eqs. (4.29a) and (4.29b), and approximate (6.6) by

$$E \sim -\frac{1}{2\pi a}\left(\sum_{l=1}^{\infty}\nu\int_{0}^{\infty}dz\, t^5 \cos\nu z\delta + \frac{1}{4}\sum_{l=1}^{\infty}\int_{0}^{\infty}dz\, t^6 \cos\nu z\delta\right), \qquad (6.9)$$

where

$$\nu = l + \frac{1}{2}, \quad x = \nu z, \quad t = (1 + z^2)^{-1/2}. \qquad (6.10)$$

If we rather cavalierly set $\delta = 0$, and define the sum in terms of the Riemann zeta function as in (4.39):

$$\sum_{l=0}^{\infty}\nu^s = (2^{-s} - 1)\zeta(-\zeta), \qquad (6.11)$$

we obtain a finite result, the leading terms of which are found in (5.93),

$$\tilde{E} = \left(\frac{11}{72\pi} + \frac{3}{128}\right)\frac{1}{a} = \frac{0.072}{a}. \qquad (6.12)$$

As noted in Sec. 5.8, this disagrees with the result reported in Ref. [18] because of a simple error in evaluating the Euler-Maclaurin summation formula.[§]

[‡]The exterior energy corresponding to (5.88) plus the interior energy (6.6) exactly equals the spherical shell result (4.23).

[§]It is of interest to compare this with the formal result for the interior Casimir energy of a cube of side L, as reported by Lukosz [77] and by Ambjørn and Wolfram [82]:

$$E_v^{\text{cube,int}} = \frac{0.0916}{L}. \qquad (6.13)$$

(See also Ref. [80, 81].) It is doubtful that there is any significance to this very approximate agreement. The fact that this number is within 1% of the spherical shell result (4.40) originally led some to attribute some significance to this answer. However, the fact that the corresponding result of a cylindrical shell differs by almost a factor of two (Ref. [24] and Sec. 7.1) seems strong evidence that any agreement is fortuitous.

It will be rightly objected that this procedure has not removed the quadratic divergence found in Refs [16, 18], where if δ is retained in the same leading asymptotic approximation [cf. (5.93)]

$$E \sim -\frac{4}{3\pi a\delta^2} + \tilde{E},$$
(6.14)

but merely defined it away. So let us, following Ref. [22], proceed in a more careful manner, keeping the convergence factor $\cos \nu z\delta$ until the end. Because of that factor it is better to rewrite the integrals in terms of x:

$$E \sim -\frac{1}{2\pi a} \sum_{l=1}^{\infty} \int_0^{\infty} dx \, \cos x\delta \left[(1 + x^2/\nu^2)^{-5/2} + \frac{1}{4\nu}(1 + x^2/\nu^2)^{-3} \right].$$
(6.15)

We carry out the l sum first using the Euler-Maclaurin summation formula (2.84),

$$\sum_{l=1}^{\infty} f(l) = \sum_{l=1}^{L} f(l) + \int_{L+1}^{\infty} dl \, f(l) + \frac{1}{2}[f(\infty) + f(L+1)]$$
$$+ \sum_{k=1}^{\infty} \frac{1}{(2k)!} B_{2k}[f^{(2k-1)}(\infty) - f^{(2k-1)}(L+1)].$$
(6.16)

Here we choose L so that $L\delta \ll 1$; consequently, for the first term in (6.16) we can carry out the x integration with $\cos x\delta \to 1$. Thus we can write the first term in (6.15) as

$$E^{(2)} = \frac{1}{2\pi a} \left[\int_0^{\infty} dz \cos \tilde{\nu}z\delta \left\{ \tilde{\nu}^2 \left[\frac{8}{3t} - \frac{8}{3}z - \frac{4}{3}t - \frac{1}{3}t^3 \right] \right. \right.$$
$$\left. \left. - \frac{1}{2}\tilde{\nu}t^5 + \frac{5}{12}\left(t^5 - t^7\right) \right\} - \frac{1}{3}\left(\tilde{\nu}^2 - \tilde{\nu} - \frac{3}{4}\right) \right].$$
(6.17)

Here $\tilde{\nu} = L + 3/2$ and consequently $z = x/\tilde{\nu}$. In obtaining this result we have inserted two contact terms: a term independent of x to cancel the "infinite" upper limit of the angular momentum integration, and, explicitly, a terms linear in x to make the z integral finite. These insertions are necessary because they correspond to δ functions in the time separation (or derivatives thereof) which must be consistently omitted. [Indeed such terms were omitted in the derivation of the basic formula (6.6).] The L

dependence of (6.17) of course cancels and we are left with

$$E^{(2)} = \frac{11}{72\pi a},$$ (6.18)

which is just the first term in (6.12).

We proceed in just the same way to evaluate the second term in (6.15). This time when the l sum is carried out we encounter not only a constant but a $\ln x$ term. This is spurious because it is canceled by an identical term which arises when the contour of integration is rotated from the real to the imaginary frequency axis. (See Chapter 7.) We are left with

$$E^{(1)} = -\frac{1}{8\pi a} \left[\int_0^\infty dz \left(\frac{1}{2}\tilde{\nu} \left\{ \ln(z^2 t^2) + t^2 + \frac{1}{2}t^4 \right\} + \frac{1}{2}t^6 \right) \nu \right.$$
$$\left. + \frac{3\pi}{16} \left(\tilde{\nu} - \frac{3}{2} \right) \right] = \frac{3}{128a},$$ (6.19)

exactly the second term in (6.12) [and one-half the leading (accurate) approximation for the spherical shell result (4.38)]. This last evaluation could be carried out much more simply by using in (6.9) [cf. (6.11)]

$$\sum_{l=0}^\infty \cos \nu z \delta = 0.$$ (6.20)

We do not advocate this latter approach, however, because it does not lead to a well defined integral¶ for $E^{(2)}$, and, more fundamentally, fails to reveal the divergence structure as being associated with contact terms.

The hopeful reader might now suspect that a finite Casimir energy for virtual gluon modes interior to a sphere has now been achieved. Alas, such is not the case. For if we write the full vector result (6.6) as

$$E = \tilde{E} - \frac{1}{2\pi a} \sum_{l=1}^\infty R_l,$$ (6.21)

with the lth remainder term being

$$R_l = (2l+1) \int_0^\infty dx \, x \left[\frac{d}{dx} \ln s_l s_l' + 2(e_l' s_l' - e_l s_l'') \right.$$

¶The pole in z can be interpreted, however. If we use the sum $\sum_{l=0}^\infty \sin \nu z \delta = 1/(2 \sin z\delta/2)$, we find a divergent term proportional to $1/\delta^2$, and a finite term equal to (6.18).

$$-\frac{1}{2\nu^2}\frac{t^5}{z} - \frac{1}{8\nu^3}\frac{t^6}{z}\Bigg] \cos x\delta, \tag{6.22}$$

it is easily seen that a logarithmic divergence remains, since

$$R_l \sim \frac{1}{8\nu} \int_0^\infty dz (4t^5 - 53t^7 + 120t^9 - 71t^{11}) \cos \nu z\delta. \tag{6.23}$$

The procedure described above applied to this expression leads directly to

$$E^{(0)} = \frac{1}{315\pi a}\left[8\ln(\delta/8) + \frac{211}{12}\right]. \tag{6.24}$$

Alternatively, we can use

$$\sum_{l=0}^\infty \frac{1}{\nu}\cos \nu z\delta = -\ln(z\delta/4), \quad \delta \to +0, \tag{6.25}$$

to obtain the same result. The point is, $1/x$ in the frequency integrand cannot be canceled by a contact term.

6.2   Zero-point Energy of Confined Virtual Quarks

Similar considerations can be given for massless quark fields confined within a spherical bag of radius a, which imposes the linear boundary condition (4.53) on the surface S,

$$(1 + i\mathbf{n}\cdot\boldsymbol{\gamma})G\Big|_S = 0. \tag{6.26}$$

It is straightforward to derive the Casimir energy expression in this case, following the formalism given in Sec. 4.2 (see Ref. [19]):

$$E_f = \frac{1}{\pi a}\sum_{l=0}^\infty (l+1)\int_0^\infty dx\, x \cos x\delta\left[\frac{d}{dx}\ln(s_l^2 + s_{l+1}^2) - 2(s'_{l+1}e_l + s'_l e_{l+1})\right], \tag{6.27}$$

where, once again, we have removed the "vacuum" or volume energy by removing from the fermionic Green's functions the vacuum part

$$G_l^0(r, r') = ikj_l(kr_<)h_l(kr_>). \tag{6.28}$$

Equation (6.27) is immediately recognized as the "inside" part of the complete fermionic Casimir energy (4.62).

6.2.1 *Numerical Evaluation*

6.2.1.1 *J = 1/2 Contribution*

In addition to the linear boundary condition imposed above, the bag model possesses an additional nonlinear boundary condition, which expresses momentum conservation: For the fermions,

$$-\frac{\partial}{\partial r}\sum_i \psi_i \gamma^0 \psi_i \bigg|_S = 2B, \tag{6.29}$$

where the sum ranges over the various species of quarks, and B is the bag constant. This restricts valence quark states to only those characterized by total angular momentum $J = 1/2$. However, it imposes no condition on the fermionic Green's function (how could it, since G is already uniquely determined) but rather expresses the zero-point, quantum fluctuation contribution to the bag constant. This will be discussed further below in Sec. 6.4.

Nevertheless, it is interesting to compute the $J = 1/2$ contribution to (6.27), since the lowest mode might be thought to be the most important, and because it provides a check on the accuracy of the approximations to be made subsequently. Since

$$s_0(x) = \sinh x, \quad s_1(x) = \cosh x - \frac{1}{x}\sinh x,$$

$$e_0(x) = e^{-x}, \quad e_1(x) = \left(1 + \frac{1}{x}\right)e^{-x}, \tag{6.30}$$

we have for the $J = 1/2$ contribution to the fermionic zero-point energy

$$E_f^{1/2} = \frac{2}{\pi a}\int_0^\infty dx\, x\, f(x), \tag{6.31}$$

with

$$f(x) = \left[\left(1 + \frac{1}{2x^2}\right)\coth 2x - \frac{1}{2x^2}\operatorname{csch} 2x - \frac{1}{x}\right]^{-1}$$
$$- \frac{1}{x} - \left[\left(1 + \frac{1}{x^2}\right)\sinh x + \cosh x\right]e^{-x}. \tag{6.32}$$

Numerical integration yields

$$E_f^{1/2} = -\frac{0.183414}{\pi a} = -\frac{0.0583826}{a}. \tag{6.33}$$

Note that this contribution is attractive. It is instructive to compare this with the lowest-mode contribution of a confined vector field. From (6.6) the latter is

$$
\begin{aligned}
E_v^1 &= -\frac{3}{2\pi a} \int_0^\infty dx\, x \left[\frac{s_1'}{s_1} + \frac{s_1''}{s_1'} + e(s_1'e_1' - e_1 s_1'') \right] \\
&= \frac{3}{2\pi a} \int_0^\infty dx\, x\, g(x) + E_{v,\text{shell}}^1,
\end{aligned}
\tag{6.34}
$$

where $E_{v,\text{shell}}^1$ is the result of interior and exterior mode contributions [15],

$$
E_{v,\text{shell}}^1 = -\frac{0.0472709}{a} \approx -\frac{3}{64a},
\tag{6.35}
$$

and the integral cancels off exterior modes. Explicitly,

$$
\begin{aligned}
g(x) = &-\left[\frac{1 + 1/x + 1/x^2}{1 + 1/x} + \frac{(1 + 2/x^2)(1 + 1/x)}{1 + 1/x + 1/x^2} \right] \\
&+ 2 + \frac{2}{x^2} - \frac{1}{x^4} + \left(\frac{2}{x^3} + \frac{1}{x^4} \right) e^{-2x},
\end{aligned}
\tag{6.36}
$$

and numerical integration yields

$$
\int_0^\infty dx\, x\, g(x) = -0.273074.
\tag{6.37}
$$

As a result, the lowest-mode vector contribution is

$$
E_v^1 = -\frac{0.177654}{a},
\tag{6.38}
$$

which is also attractive, but more than 3 times larger than the lowest-mode spinor contribution (6.33).

6.2.1.2 *Sum Over All Modes*

Again we use the uniform asymptotic approximants to carry out the sum over all modes. The leading approximation to (6.27) is

$$
E_f \sim -\frac{1}{2\pi a} \sum_{l=0}^\infty (l+1) \int_0^\infty dz\, z^2 t^5 \cos z\nu\delta.
\tag{6.39}
$$

Again setting $\delta = 0$ and using (6.11) we find (see Ref. [20]—again an error was present in Ref. [19])

$$\tilde{E}_f = -\frac{1}{144\pi a}.\tag{6.40}$$

Of course, we have again used the zeta-function magic to sweep away the infinity. If, instead, we do the integral over z in (6.39), we obtain

$$\int_0^\infty dz\, z^2 t^5 \cos \nu z \delta = \nu \delta K_1(\nu\delta) - \frac{1}{3}(\nu\delta)^2 K_2(\nu\delta),\tag{6.41}$$

making it simple to do the sum over l using $(n \geq m)$

$$\int_0^\infty dx\, x^n K_m(x) = 2^{n-1}\Gamma\left(\frac{n+m+1}{2}\right)\Gamma\left(\frac{n-m+1}{2}\right).\tag{6.42}$$

The result is again quadratically divergent, with the finite part given above,

$$E_f \sim \frac{1}{3\pi a \delta^2} + \tilde{E}_f.\tag{6.43}$$

Again, let us extract the finite part by the careful contact term method used in the gluon case. Here, the Euler-Maclaurin sum formula gives

$$E_f^{(2)} = -\frac{1}{2\pi a}\left[\int_0^\infty dz\left\{\tilde{\nu}^2\left[\frac{2}{3t} - \frac{2}{3}z - \frac{1}{3}t - \frac{1}{3}t^3\right] + \frac{1}{2}\tilde{\nu}\left[\frac{4}{3}t^3 - t^5\right]\right.\right.$$
$$\left.\left. -\frac{1}{2}t^3 + \frac{11}{12}t^5 - \frac{5}{12}t^7\right\} + \frac{1}{6}\left(\tilde{\nu}^2 - 2\tilde{\nu} + \frac{3}{4}\right)\right],\tag{6.44}$$

where now $\tilde{\nu} = L + 5/2$, and appropriate contact terms have been inserted. Evaluation of this yields precisely (6.40). [The $\mathcal{O}(\nu^1)$ term for fermions vanishes by virtue of the identity (6.20).]

6.2.1.3 *Asymptotic Evaluation of Lowest J Contributions*

It is of some interest to note how accurate the uniform asymptotic approximation is even for the lowest angular momentum contributions. Thus the $J = 1/2$ contribution is approximated by

$$E_f^{1/2} \sim -\frac{1}{6\pi a} = -\frac{0.0531}{a},\tag{6.45}$$

which is within 10% of the exact value (6.33). The similar approximation to the $l = 1$ contribution for the vector field is

$$E_v^1 \sim -\frac{1}{2\pi a} = -\frac{0.159}{a}, \qquad (6.46)$$

which is not far from the correct value given by (6.38). Including the next-to-leading approximation here changes this to

$$E_v^1 \sim \frac{1}{\pi a} \left(\frac{1}{2} + \frac{3\pi}{128} \right) = -\frac{0.183}{a}, \qquad (6.47)$$

which is only off by 3%.

6.3   Discussion and Applications

To summarize, we have not yet achieved a finite theory of the zero-point energy of fields confined within finite geometries. But, as in the case of a dielectric ball, we see that the dominant quadratic divergences are illusory, being merely contact terms. The residual logarithmic divergence can only be removed by a refinement of the bag model. Presumably, since the divergences are associated with the surface (see Chapter 11) a softening of the boundary conditions would yield a totally finite result. There are also additional divergences that occur with massive quarks, which were discussed by Baacke and Igarashi [224].

But already we can make some definite qualitative and semiquantitative conclusions which should have significant phenomenological implications. Putting together Eqs. (6.12) and (6.24) we have[‖]

$$E_v \sim \frac{1}{a}(0.0898 + 0.00808 \ln \frac{\delta}{8}). \qquad (6.49)$$

Plausibly $\ln \delta/8 \sim 1$ when realistic boundary conditions are imposed (the surface region should have a "skin depth" of at least 10% of the whole radius), so we can drop the logarithm term for a first crude estimate. Multiplying by 8 for the eight gluons gives then the following estimate for the

---

[‖]A very similar result has been found by Romeo [225] for a scalar field in a spherical bag:

$$E_s = \frac{1}{2a} \left(0.0089 + \frac{3}{315\pi} \ln \mu a \right), \qquad (6.48)$$

where μ is a mass parameter introduced in his zeta-function technique.

zero-point energy:

$$E_{ZP} \approx +0.7/a. \tag{6.50}$$

(It is certainly very hard to doubt the sign of the effect.) Numerically this is the whole story since the fermion contribution is far smaller (the leading approximation per degree of freedom is down more than a factor of 20).

It is not that the quadratic divergences are incorrect, but rather they are seen to correspond to contact terms, and thus can be removed *unambiguously*. These are precisely analogous to the volume divergence which must always be removed from the Casimir stress. These contact terms may be thought of as renormalizations of phenomenological parameters of the effective theory, just as we discussed in Chapter 5. Thus we expect there should be the following phenomenological terms in the energy describing the bag,

$$H' = BV + \sigma A + Fa, \tag{6.51}$$

where B is the bag constant, σ is a constant surface tension, and F is a constant force. The necessity of introducing B and σ has been previously recognized [84, 85, 86, 87, 220]; however, equally well there is no reason to exclude F. In fact, all such terms may be regarded as contact terms, since they are polynomials in the bag radius. Since $\delta = i\tau/a$, we appreciate that the divergent parts of (6.14) and (6.43) are of the form of Fa, and so merely renormalize that phenomenological parameter. (Note that the volume energy subtraction can be thought of as a renormalization of B, but no term of the form σA occurs in zero-point energy calculations [131, 226, 227]. Eq. (6.51) would seem to represent the correct phenomenology, not the usual

$$H' = BV + \sigma A - Z/a, \tag{6.52}$$

since the last, nonlocal term, is a calculable effect. As we have noted in Chapter 4, similar considerations are made in Elizalde, Bordag, and Kirsten [199], for quarks of mass μ. Blau, Visser, and Wipf [228] had earlier argued that the bag energy should have the form

$$H' = BV + \sigma A + Fa + k + \frac{Z}{a} - \frac{\lambda}{a} \ln \mu a, \tag{6.53}$$

and that only the coefficient λ of the logarithmic term is computable. Clearly, work remains to be done!

6.3.1 *Fits to Hadron Masses*

The bag-model Lagrangian is

$$\mathcal{L}_{\text{bag}} = (\mathcal{L}_{\text{D}} + \mathcal{L}_{\text{YM}} - B)\eta(\overline{\psi}\psi), \tag{6.54}$$

where $\mathcal{L}_{\text{D,YM}}$ are the Dirac and Yang-Mills Lagrangians, respectively, B is the bag constant, and η is the unit step function. Variation of (6.54) leads to the following linear and quadratic boundary conditions (n^μ is the outwardly directed normal)

$$-i\gamma n\psi = \psi, \tag{6.55}$$

equivalent to (6.26),

$$n_\alpha G^{\alpha\beta} = 0, \tag{6.56}$$

equivalent to (6.1) and

$$-\frac{1}{4}G^2 - \frac{1}{2}(n\partial)\overline{\psi}\psi = B, \tag{6.57}$$

equivalent to (6.29) in terms of its quark contribution. Here n_α is the unit normal to the bag boundary.

The MIT bag model consists in solving the Dirac equation subject to the boundary conditions (6.55), (6.56), and (6.57), which in the simplest realization refer to a *spherical cavity approximation*. The interactions, through the gauge fields, are treated perturbatively in the approximation that the strong coupling is regarded as small, $\alpha_s \ll 1$. The radial wavefunctions are then expressed in terms of spherical Bessel functions. The mass content of the model is consequently reduced to the following mass function (a is the bag radius):

$$M(a) = \frac{4}{3}\pi Ba^3 + \sum_i \frac{\omega_i}{a} - \frac{Z}{a} + E_{\text{mag}} + E_{\text{el}}. \tag{6.58}$$

The first term is the volume energy required to create the "bubble" in the normal vacuum, ω_i/a is the kinetic energy of the ith quark, Z is the so-called zero-point energy parameter, and E_{mag} and E_{el} are the chromomagnetic and chromoelectric energies, respectively. The latter appears small, and there are uncertainties in its calculation, so we shall henceforth ignore E_{el}.

The linear boundary conditions are implemented through

(1) the transcendental equation expressing $x = pa$ in terms of $u = \mu a$ (p and μ are the quark momentum and mass, respectively)

$$x \cot x = 1 - u - \sqrt{u^2 + x^2}, \tag{6.59}$$

and

(2) the expressions for E_{mag} and E_{el}.

The quadratic boundary conditions are implemented by minimizing $M(a)$ with respect to a,

$$\frac{dM}{da} = 0. \tag{6.60}$$

Particularly simple is the case of N massless quarks:

$$M(a) = \frac{4}{3} \pi B a^3 + \frac{1}{a}(N x_0 - Z + \frac{2}{3} \alpha_s a_T h_{00}), \tag{6.61}$$

where the coefficient of $1/a$ in the second term is independent of a. Here $x_0 = 2.0428$, $h_{00} = 0.177$, and $a_T = -6, -3, +2$, and $+3$ for spin states $S = 0, 1/2, 1$, and $3/2$, respectively. From (6.61) and (6.60) we can obtain explicit linear relations between bag-model parameters $C = (4/3)^{4/3}(4\pi B)^{1/3}$, Z, and α_s, and $4/3$-powers of input masses, for example

$$C x_0 = \frac{5}{6} M_\Delta^{4/3} + \frac{1}{6} M_N^{4/3} - M_\omega^{4/3},$$
$$C Z = 2 M_\Delta^{4/3} - 3 M_\omega^{4/3}, \tag{6.62}$$
$$C h_{00} \alpha_s = \frac{1}{4} M_\Delta^{4/3} - \frac{1}{4} M_N^{4/3}.$$

The experimental masses for Δ, N, and ω then determine the bag-model parameters:

$$B^{1/4} = 0.14535 \text{ GeV}, \tag{6.63a}$$
$$Z = 1.8327, \tag{6.63b}$$
$$\alpha_s = 2.1736. \tag{6.63c}$$

The other fit masses are shown in Table 6.1. The results are essentially identical with those found in Ref. [86].

After this review of the standard bag-model fits, we recognize that the solution (6.63b), (6.63c) for the parameters Z and α_s is inconsistent. We can say rather little about the difficulty associated with the largeness of α_s

and the consequent failure of first-order perturbation theory. However, we can do something about the incorrect sign and magnitude of Z.

The relatively reliable calculations of the zero-point energy (Casimir effect) of gluons confined within a spherical cavity described above (recall that the effect of zero-point fluctuations in the quark fields is negligible) suggests that an appropriate value of Z arising from this phenomenon is

$$Z^{\text{Casimir}} \sim -1. \tag{6.64}$$

However, other phenomena, particularly center-of-mass corrections [229, 230, 231] contribute to a term of this form in $M(a)$. Estimates suggest $Z^{\text{CM}} \sim +1$. Thus, while early guesses for Z^{Casimir} were that it was it was positive, and of a value consistent with the magnitude in (6.63b), it now appears that Z is small, possibly negative. Can such a value be accommodated? A possible scheme for doing so was suggested above—one could add to $M(a)$ a constant force term, Fa. (As we have noted, such a term appears as a contact term in the calculation of Z.) This is in the spirit of the Budapest group [220] who add a surface tension term, $4\pi\sigma a^2$. The question is, can we fix Z at a reasonable value, and still obtain a good fit to masses by allowing F to be a free parameter? The answer is yes, as the calculations summarized in Table 6.1 show.

In conclusion, we see that all the fits are fair; none of the bag-model fits agree well with the pseudoscalars (for that we need chiral bag models [232, 233]), whereas the baryons and vectors mesons are adequately represented. Noteworthy is the fact that all the fits agree far more closely with each other than with the data; this probably reflects the common simplicity of the models, and the common fit masses. This agreement is striking given the large variation in the parameters Z and F. In any case, the challenge raised in Ref. [18, 19] has been answered: a correct treatment of zero-point energy effects appears compatible with the bag model, provided an enlarged parameterization of the model is allowed.

6.4 Calculation of the Bag Constant

As we noted above, the nonlinear boundary condition (6.29) or (6.57) is not used in finding the zero-point energy, but rather we regard it as an expression of the fermionic (and gluonic) contribution to the bag constant

Parameters and Masses	Experimental Values	Model A	Model D	Model E
Z		-1.0	1.0	1.8327
$B^{1/4}$(GeV)		0.19526	0.16515	0.14535
F (GeV^2)		-0.21923	-0.064729	0
α_s		2.16066	2.17932	2.19897
m_s (GeV)		0.2807	0.2802	0.2797
R_N (GeV^{-1})		5.06	5.03	5.01
δm (GeV)		0.1325	0.1343	0.1408
m_π	0.138	0.3610	0.3229	0.2833
m_η	0.549	0.645	0.6105	0.5583
$m_{\eta'}$	0.958	0.5031	0.4702	0.4209
m_K	0.4957	0.5827	0.5489	0.4982
m_N	$\underline{0.9389}$	$\underline{0.9389}$	$\underline{0.9389}$	$\underline{0.9389}$
m_Λ	1.1156	1.1022	1.1017	1.1012
m_Σ	1.1931	1.1426	1.1426	1.1422
m_Ξ	1.13181	1.2885	1.2876	1.1867
m_ρ	0.769	0.7826	0.7826	0.7826
m_ω	$\underline{0.7826}$	$\underline{0.7826}$	$\underline{0.7826}$	$\underline{0.7826}$
m_ϕ	1.0195	1.0733	1.0711	1.0688
m_{K^*}	0.8921	0.9251	0.9246	0.9229
m_Δ	$\underline{1.232}$	$\underline{1.232}$	$\underline{1.232}$	$\underline{1.232}$
m_{Σ^*}	1.3839	1.3760	1.3761	1.3762
m_{Ξ^*}	1.5334	1.5229	1.5229	1.5230
m_Ω	$\underline{1.6725}$	$\underline{1.6725}$	$\underline{1.6725}$	$\underline{1.6725}$

Table 6.1 Fits to bag model parameters and hadron masses for various values of the "zero–point" energy parameter Z. The underlined values are fitted. The nucleon radius is R_N, while m_s is the strange–quark mass. δm is the variance in the predicted masses. (All masses are in GeV.) These calculations were performed some years ago by Gary Köhler (unpublished).

itself. We do this by replacing in (6.29)

$$i\psi(x)\psi(x')\gamma^0 \to G(x, x'), \tag{6.65}$$

so we find

$$i\frac{\partial}{\partial r}\,\mathrm{tr}\,G(x,x)\bigg|_{r=a} = \frac{2B}{N}, \tag{6.66}$$

where N is the number of quarks. From the explicit construction of the Green's function as given in Sec. 4.2, we obtain

$$\frac{2B}{N} = i \int \frac{d\omega}{2\pi} e^{-i\omega\tau} \sum_{JM} a_J \frac{\partial}{\partial a} \left[j^2_{J+1/2}(ka) \operatorname{tr} Z^{J+1/2}_{JM}(\Omega) Z^{J+1/2}_{JM}(\Omega)^* \right.$$
$$\left. - j^2_{J-1/2}(ka) \operatorname{tr} Z^{J-1/2}_{JM}(\Omega) Z^{J-1/2}_{JM}(\Omega)^* \right]$$
$$= -\frac{i}{a^2} \int \frac{d\omega}{2\pi} e^{-i\omega\tau} \sum_{l=0}^{\infty} \frac{(l+1)}{4\pi} \frac{\partial}{\partial a} \ln \left[j^2_{l+1}(ka) - j^2_l(ka) \right], \qquad (6.67)$$

or

$$B = \frac{N}{8\pi^2 a^4} \sum_{l=0}^{\infty} (l+1) \int_0^{\infty} x \, dx \cos x\delta \frac{d}{dx} \ln \left(\frac{s^2_{l+1}(x) + s^2_l(x)}{x^2} \right). \qquad (6.68)$$

This formula, which is very similar to (6.27) for the energy, is the expression of the quark-field fluctuation component of the bag constant. It opens up the possibility of computing the bag constant from first principles.

In fact, if we believe that only the finite part of (6.68) is meaningful, perhaps a dubious proposition, we might be led to the following connection between the fermionic Casimir energy (6.40) and the bag constant (in the following, because we are making only a heuristic argument, involving orders of magnitude, we ignore the sign discrepancy):

$$B = \frac{N}{4\pi a^3} |\tilde{E}_f|. \qquad (6.69)$$

If we assume there are three light flavors of quarks (of three colors), we have $N = 9$, and the bag constant is given by

$$B^{1/4} = \left(\frac{9}{4\pi} \frac{1}{144} \right)^{1/4} (0.2 \, \text{GeV}) = 0.053 \, \text{GeV}, \qquad (6.70)$$

which is about three times too small.

In view of the dominance of the gluonic contribution to the Casimir energy, it may not be surprising to suggest that this discrepancy may be attributed to the gluonic effect. In fact, according to (6.57), the gluonic contribution to the bag constant is proportional to the so-called gluon condensate, but now appropriate to a larger bag containing quarks,

$$B_g = \frac{1}{4} \langle G^2 \rangle \Big|_{r=a}. \qquad (6.71)$$

A value for the latter, for the case of empty bags populating the vacuum and arising from vacuum fluctuations, will be given in Sec. 11.4, which in fact is consistent with the estimates for the gluon condensates given in terms of QCD sum rules [234, 235]. Although a direct calculation shows a local surface divergence (see Sec. 11.4), this we believe is spurious, and we crudely estimate the condensate by taking the value at the center of the bag. Using the value given in (11.48a), multiplying by 8 for the number of gluons, and dividing by a bag radius of order 5 GeV^{-1}, we find for the gluonic contribution

$$B_g \sim \frac{8}{4} \frac{0.8}{(5\,\mathrm{GeV}^{-1})^4}, \tag{6.72}$$

or

$$B_g^{1/4} \sim 0.2\,\mathrm{GeV}, \tag{6.73}$$

which quite overwhelms (6.70), and is quite consistent with the estimates appearing in Table 6.1.

This speculative but provocative idea obviously warrants serious analysis.

6.5 Recent Work

We close this chapter by citing some recent references on the application of quantum zero-point energy to hadronic physics. This includes the work of Fahri, Graham, Jaffe, and Weigel, on the quantum stabilization of $1+1$-dimensional static solitons [236], of Graham, Jaffe, Quandt, and Weigel on quantum energies of interfaces [237], of Hofmann, Gutsche, Schumann, and Violler [238, 239, 240] on cavity quantum chromodynamics to order α_s^2, and of Cherednikov and collaborators on hybrid bag models [241, 242]. It is fair to conclude that the development of hadronic applications of zero-point energy is still, after all these years, in its infancy.

Chapter 7

Casimir Effect in Cylindrical Geometries

7.1 Conducting Circular Cylinder

Since parallel plates yield an attractive Casimir force, while a conducting sphere experiences a repulsive stress, one might guess that for a conducting cylinder a zero stress results. In fact, Balian and Duplantier [14] obtained the result in the two-scattering approximation (which was quite accurate for a spherical shell) that the Casimir energy vanished for a long cylinder. The actual situation is not so simple. The first calculation was carried out in 1981 by DeRaad and Milton [24]. The electrodynamic result turns out to be attractive but with rather small magnitude.

Consider a right circular perfectly conducting cylinder of infinite length and radius a. We compute the Casimir energy using the above Green's dyadic formalism, adapted to this cylindrical basis. The necessary information about vector spherical harmonics in this case is given in Stratton [184]. The results for the Green's dyadics are

$$\boldsymbol{\Gamma}(\mathbf{r}, \mathbf{r}') = \sum_{m=-\infty}^{\infty} \int \frac{dk}{2\pi} \left[-\frac{1}{\omega^2} \mathbf{M} \mathbf{M}'^* (d_m - k^2) F_m(r, r') \right.$$
$$\left. + \frac{1}{\omega^2} \mathbf{N} \mathbf{N}'^* G_m(r, r') \right] \chi_{mk}(\theta, z) \chi_{mk}^*(\theta', z'), \qquad (7.1a)$$

$$\boldsymbol{\Phi}(\mathbf{r}, \mathbf{r}') = \sum_{m=-\infty}^{\infty} \int \frac{dk}{2\pi} \left[-i\mathbf{M} \mathbf{N}'^* G_m(r, r') \right.$$
$$\left. - \frac{i}{\omega} \mathbf{N} \mathbf{M}'^* F_m(r, r') \right] \chi_{mk}(\theta, z) \chi_{mk}^*(\theta', z'), \qquad (7.1b)$$

where the nonradial eigenfunctions are

$$\chi_{mk}(\theta, z) = \frac{1}{\sqrt{2\pi}} e^{im\theta} e^{ikz}, \tag{7.2}$$

the vector differential operators $\mathbf{M}$ and $\mathbf{N}$ are

$$\mathbf{M} = \hat{\mathbf{r}}\frac{im}{r} - \hat{\boldsymbol{\theta}}\frac{\partial}{\partial r}, \quad \mathbf{N} = \hat{\mathbf{r}}ik\frac{\partial}{\partial r} - \hat{\boldsymbol{\theta}}\frac{mk}{r} - \hat{\mathbf{z}}d_m, \tag{7.3}$$

and the scalar cylinder differential operator d_m is

$$d_m = \frac{1}{r}\frac{\partial}{\partial r}r\frac{\partial}{\partial r} - \frac{m^2}{r^2}. \tag{7.4}$$

Maxwell's equations (2.96a)–(2.96d) for the Green's dyadic are to be solved subject to the perfect conductor boundary conditions

$$\hat{\boldsymbol{\theta}} \cdot \boldsymbol{\Gamma}'\big|_{r=a} = \hat{\mathbf{z}} \cdot \boldsymbol{\Gamma}'\big|_{r=a} = \hat{\mathbf{r}} \cdot \boldsymbol{\Phi}\big|_{r=a} = \mathbf{0}, \tag{7.5}$$

which means in terms of the scalar Green's functions

$$G_m(a, r') = 0, \quad m \neq 0, \tag{7.6a}$$

$$\frac{\partial}{\partial r}F_m(r, r')\bigg|_{r=a} = 0, \tag{7.6b}$$

$$\frac{k^2}{\omega^2}G_0(a, r') + \mathcal{G}_0^G(a, r') = 0. \tag{7.6c}$$

The solutions for the scalar Green's functions in the interior and exterior regions are

$$r, r' < a:$$

$$\frac{1}{\omega^2}F_m(r, r') = \frac{i\pi}{2\lambda^2}\left[J_m(\lambda r_<)H_m(\lambda r_>) - \frac{H'_m(\lambda a)}{J'_m(\lambda a)}J_m(\lambda r)J_m(\lambda r')\right]$$

$$+ \frac{1}{\lambda^2}\mathcal{G}_m^F(r, r'), \tag{7.7a}$$

$$\frac{1}{\omega}G_m(r, r') = \frac{i\pi}{2\lambda^2}\left[J_m(\lambda r_<)H_m(\lambda r_>) - \frac{H_m(\lambda a)}{J_m(\lambda a)}J_m(\lambda r)J_m(\lambda r')\right]$$

$$+ \frac{1}{\lambda^2}\mathcal{G}_m^G(r, r'), \tag{7.7b}$$

$$\mathcal{G}_m^{G,F}(r, r') = -\frac{1}{2|m|}\left(\frac{r_<}{r_>}\right)^{|m|} \pm \frac{1}{2|m|}\frac{r^{|m|}r'^{|m|}}{a^{2|m|}}, \quad m \neq 0, \tag{7.7c}$$

$r, r' > a$:

$$\frac{1}{\omega^2} F_m(r,r') = \frac{i\pi}{2\lambda^2} \left[J_m(\lambda r_<) H_m(\lambda r_>) - \frac{J'_m(\lambda a)}{H'_m(\lambda a)} H_m(\lambda r) H_m(\lambda r') \right]$$

$$+ \frac{1}{\lambda^2} \mathcal{G}^F_m(r,r'), \tag{7.8a}$$

$$\frac{1}{\omega} G_m(r,r') = \frac{i\pi}{2\lambda^2} \left[J_m(\lambda r_<) H_m(\lambda r_>) - \frac{J_m(\lambda a)}{H_m(\lambda a)} H_m(\lambda r) H_m(\lambda r') \right]$$

$$+ \frac{1}{\lambda^2} \mathcal{G}^G_m(r,r'), \tag{7.8b}$$

$$\mathcal{G}^{G,F}_m(r,r') = -\frac{1}{2|m|} \left(\frac{r_<}{r_>} \right)^{|m|} \pm \frac{1}{2|m|} \frac{a^{2|m|}}{r^{|m|} r'^{|m|}}, \quad m \neq 0, \tag{7.8c}$$

where $H_m = H_m^{(1)}$ is the Hankel function of the first kind, and $\lambda^2 = \omega^2 - k^2$. Although $\mathcal{G}^{G,F}_0$ are not determined, they do not contribute to physical quantities.

From the above result we can compute T_{rr} on the cylindrical surface, where, making use of the boundary conditions appropriate to a perfect conductor, we have

$$T_{rr} = \frac{1}{2}(B_\perp^2 - E_r^2). \tag{7.9}$$

In terms of the Green's dyadic given by (7.1a) and (7.7a)–(7.8c), we have

$$i\langle E_r(x) E_{r'}(x') \rangle = \int \frac{d\omega}{2\pi} e^{-i\omega(t-t')} \hat{\mathbf{r}} \cdot \boldsymbol{\Gamma}(\mathbf{r}, \mathbf{r}'; \omega) \cdot \hat{\mathbf{r}}', \tag{7.10}$$

which implies for the square of the electric field on the cylinder's surface

$$i\langle E_r^2(a+) \rangle = \mathcal{L} \left[\frac{1}{z} \left(k^2 \frac{H'_m(z)}{H_m(z)} - \frac{m^2 \omega^2}{z^2} \frac{H_m(z)}{H'_m(z)} \right) \right], \tag{7.11a}$$

$$i\langle E_r^2(a-) \rangle = \mathcal{L} \left[-\frac{1}{z} \left(k^2 \frac{J'_m(z)}{J_m(z)} - \frac{m^2 \omega^2}{z^2} \frac{J_m(z)}{J'_m(z)} \right) \right]. \tag{7.11b}$$

Here we have set $z = \lambda a$ and used the notation

$$\mathcal{L}[f_m(z; k, \omega)] = \int_{-\infty}^{\infty} \frac{d\omega}{2\pi} \psi(\omega) \sum_{m=-\infty}^{\infty} \int_{-\infty}^{\infty} \frac{dk}{2\pi} f_m(z; k, \omega). \tag{7.12}$$

Here we have introduced a high-frequency cutoff function $\psi(\omega)$ which will be discussed in detail below.

Similarly, the product of magnetic fields is given by

$$i\langle B_z(x)B_z(x')\rangle = \int \frac{d\omega}{2\pi}e^{-i\omega(t-t')}\frac{1}{i\omega}\hat{\mathbf{z}}\cdot\mathbf{\Phi}(\mathbf{r},\mathbf{r}';\omega)\cdot\overset{\leftarrow}{\mathbf{\nabla}}'\times\hat{\mathbf{z}}, \quad (7.13)$$

so as a consequence

$$i\langle B_z^2(a+)\rangle = \mathcal{L}\left[-\frac{\lambda^2}{2\pi z}\frac{H_m(z)}{H_m'(z)}\right], \quad (7.14a)$$

$$i\langle B_z^2(a-)\rangle = \mathcal{L}\left[\frac{\lambda^2}{2\pi z}\frac{J_m(z)}{J_m'(z)}\right], \quad (7.14b)$$

while

$$i\langle B_\theta(x)B_\theta(x')\rangle = \int \frac{d\omega}{2\pi}e^{-i\omega(t-t')}\frac{1}{i\omega}\hat{\boldsymbol{\theta}}\cdot\mathbf{\Phi}(\mathbf{r},\mathbf{r}';\omega)\cdot\overset{\leftarrow}{\mathbf{\nabla}}'\times\hat{\boldsymbol{\theta}}, \quad (7.15)$$

so

$$i\langle B_\theta^2(a+)\rangle = \mathcal{L}\left[\frac{1}{2\pi z}\left(\omega^2\frac{H_m'(z)}{H_m(z)} - \frac{m^2k^2}{z^2}\frac{H_m(z)}{H_m'(z)}\right)\right], \quad (7.16a)$$

$$i\langle B_\theta^2(a-)\rangle = \mathcal{L}\left[-\frac{1}{2\pi z}\left(\omega^2\frac{J_m'(z)}{J_m(z)} - \frac{m^2k^2}{z^2}\frac{J_m(z)}{J_m'(z)}\right)\right], \quad (7.16b)$$

In deriving these results, we have ignored δ-function terms, since the coincidence of field points is to be understood in the limiting sense. We have also used the Wronskian

$$J_m(z)H_m'(z) - H_m(z)J_m'(z) = \frac{2i}{\pi z}. \quad (7.17)$$

Inserting these field products into the stress tensor expression (7.9), we obtain the result for the force per unit area is*

$$\begin{aligned}
\mathcal{F} &= \langle T_{rr}\rangle(a_-) - \langle T_{rr}\rangle(a_+) \\
&= -\frac{1}{4\pi i}\mathcal{L}\left[\frac{\lambda^2}{z^2}\left(\frac{H_m'}{H_m} + \frac{H_m''}{H_m'} + \frac{J_m'}{J_m} + \frac{J_m''}{J_m'} + \frac{2}{z}\right)\right] \\
&= -\frac{1}{2ia^2}\int_{-\infty}^{\infty}\frac{d\omega}{2\pi}\psi(\omega)\frac{1}{2\pi}\sum_{m=-\infty}^{\infty}\int_{-\infty}^{\infty}\frac{dk}{2\pi}z\frac{\partial}{\partial z}\ln\left(1+\lambda_m^2\right), \quad (7.19)
\end{aligned}$$

*If we drop the regularization factor, perform the naive substitution $\omega \to i\zeta$, and introduce polar coordinates $\zeta^2a^2 + k^2a^2 = r^2$, $d\zeta a\,dka = r\,dr\,d\theta$, (7.19) reads

$$\mathcal{F} = \frac{1}{a^4}\frac{1}{(2\pi)^2}\sum_{m=-\infty}^{\infty}\int_0^{\infty}r\,dr\,\ln(1 - [r(I_m(r)K_m(r))']^2). \quad (7.18)$$

where

$$\lambda_m = \frac{z\pi}{2} \left[J_m(z) H_m(z) \right]'.$$

(7.20)

We are interested only in the real part of this expression. The path of ω integration in (7.19) is understood to be just above the real axis for $\omega > 0$, and just below for $\omega < 0$. (See Sec. 4.1.)

Here, because convergence is more subtle than in the previous cases, we have, following Balian and Duplantier [14], inserted a frequency cutoff function $\psi(\omega)$ that has the properties

$$\psi(0) = 1,$$
$$\psi(\omega) \sim \frac{1}{|\omega|^4}, \quad |\omega| \to \infty,$$

(7.21)

$$\psi(\omega) \quad \text{real for } \omega \text{ real or imaginary.}$$

A way of satisfying these conditions is to take

$$\psi(\omega) = \sum_i \left(\frac{a_i}{a^2\omega^2 - \mu_i^2} + \frac{a_i^*}{a^2\omega^2 - \mu_i^*} \right),$$

(7.22)

with the following conditions on the residues and poles:

$$\sum_i \operatorname{Re} a_i = 0,$$

(7.23a)

$$2 \sum_i \operatorname{Re} \frac{a_i}{\mu_i^2} = -1.$$

(7.23b)

Of course, the poles are to recede to infinity. When we rotate the contour to imaginary frequencies, the contribution of these poles is necessary to achieve a real, finite result.

Once again we use the uniform asymptotic expansion to extract the leading behavior. We evaluate the m sum in (7.19) for the leading term,

$$\ln \sim -\frac{z^4}{4(m^2 - z^2)^3},$$

(7.24)

which gives

$$z \frac{d}{dz} \left(-\frac{z^2}{4(m^2 - z^2)^3} \right) = \left(\frac{1}{2} + \frac{5}{2} \frac{d}{d\rho} + \frac{7}{4} \frac{d^2}{d\rho^2} + \frac{1}{4} \frac{d^3}{d\rho^3} \right) \frac{1}{m^2\rho^2 - z^2} \Big|_{\rho=1},$$

(7.25)

by using the identity (2.45), or

$$2 \sum_{m=1}^{\infty} \frac{1}{m^2 \rho - z^2} = -\frac{\pi}{\sqrt{-z^2 \rho}} \left(1 - \coth \pi \sqrt{\frac{-z^2}{\rho}} \right)$$
$$+ \frac{1}{z^2} + \frac{\pi}{\sqrt{-z^2 \rho}}. \tag{7.26}$$

Only the terms here of order z^{-1} and z^{-2} yield cutoff dependence. The last term in (7.26) for $\rho = 1$ has a corresponding k integral ($\tilde{y} = ka$, $\tilde{x} = \omega a$)

$$\int_{-\infty}^{\infty} d\tilde{y} \, \frac{1}{2} \frac{1}{\sqrt{\tilde{y}^2 - \tilde{x}^2}} = -\ln \tilde{x} + c, \tag{7.27}$$

where a cutoff in $\tilde{y}$ is incorporated in the complex constant c. This constant, corresponding to a δ-function in time, is to be ignored. The remaining frequency integral

$$\int_0^{\infty} d\tilde{x} \, \psi(\tilde{x}) \ln \tilde{x} \tag{7.28}$$

is evaluated by rotating the path of integration to lie along the imaginary axis. In so doing, we pick up contributions from first quandrant poles of $\psi(x)$,

$$2\pi i \sum_i \frac{a_i}{2\mu_i} \ln \mu_i, \tag{7.29}$$

which cancels the imaginary part of the integral along the imaginary axis,[†]

$$-i2 \mathrm{Re} \, \frac{\pi}{2} \sum_i \frac{a_i}{\mu_i} \ln \mu_i. \tag{7.30}$$

The z^{-2} term in (7.26) gives

$$-\frac{1}{2} \int_{-\infty}^{\infty} d\tilde{x} \, \psi(\tilde{x}) \int_{-\infty}^{\infty} d\tilde{y} \frac{1}{\tilde{y}^2 - \tilde{x}^2}, \tag{7.31}$$

which is identically cancelled by the cutoff-dependent term coming from the $m = 0$ contribution in (7.19). The behavior for large argument there is

---

[†]We take μ_i, without loss of generality, to lie in the first quadrant, so $-\mu_i^*$, $-\mu_i$, and μ_i^* lie in the second, third, and fourth quadrants, respectively.

just the $m = 0$ limit of (7.24)

$$\ln \sim \frac{1}{4z^2}, \tag{7.32}$$

which when $z\frac{d}{dz}$ is applied gives just the negative of the integral (7.31). All that remains is finite, for it is easy to see that terms of order $z^{-3/2}$ have vanishing pole contributions as the poles in ψ recede to infinity, and finite integrals along the imaginary frequency axis.

The energy per unit length $\mathcal{E}$ is expressed in terms of integrals over polar coordinates,

$$\tilde{x} \to ix, \quad \tilde{y} \to y, \quad r^2 = x^2 + y^2, \quad dx\,dy = 2\pi r\,dr, \tag{7.33}$$

of functions constructed from modified Bessel functions,

$$I_\nu(x) = e^{-\frac{1}{2}\nu\pi i} J_\nu(xe^{\frac{1}{2}\pi i}), \tag{7.34a}$$

$$K_\nu(x) = \frac{\pi}{2} i e^{\frac{1}{2}\nu\pi i} H_\nu(xe^{\frac{1}{2}\pi i}). \tag{7.34b}$$

The result is

$$\mathcal{E} = \pi a^2 \mathcal{F} = -\frac{1}{8\pi a^2}(S + R + R_0), \tag{7.35}$$

where, from the asymptotic behavior,

$$S = \frac{1}{2}\int_{\epsilon \to 0}^{\infty} dr\,\pi(\coth \pi r - 1)$$
$$+ \frac{1}{2}\left(5\frac{d}{d\rho} + \frac{7}{2}\frac{d^2}{d\rho^2} + \frac{1}{2}\frac{d^3}{d\rho^3}\right)\int_0^\infty dr\left[\frac{\pi}{\sqrt{\rho}}\left(\coth\frac{\pi r}{\sqrt{\rho}} - 1\right) - \frac{1}{r}\right]\Bigg|_{\rho=1}$$
$$= -\frac{1}{2}\ln 2\pi\epsilon + \frac{5}{8}, \tag{7.36a}$$

while the remainders are integrated by parts,

$$R = -4\sum_{m=1}^{\infty}\int_0^\infty r\,dr\left\{\ln\left[1 - (r(I_m(r)K_m(r))')^2\right] + \frac{r^4}{4(m^2 + r^2)^3}\right\}$$
$$= -0.0437, \tag{7.36b}$$

$$R_0 = -2\int_{\epsilon \to 0}^{\infty} r\,dr\left\{\ln\left[1 - (r(I_0(r)K_0(r))')^2\right] + \frac{1}{4r^2}\right\} - \frac{1}{4}$$
$$= \frac{1}{2}\ln\epsilon + 0.6785. \tag{7.36c}$$

Adding these numbers, we see that the infrared singularity $\ln \epsilon$ cancels, and we obtain an attractive result,

$$\mathcal{E} = -\frac{0.01356}{a^2}, \qquad (7.37)$$

equivalent to (1.33).

Recently, this result has been confirmed by two independent calculations. First, Gosdzinsky and Romeo obtained the same answer, to eight significant figures, using a zeta-function technique [243]. Shortly thereafter, an earlier calculation by Nesterenko was corrected to yield the same answer [43]. This latter method is based not on the Green's function, but rather an analytic evaluation of the sum of zero-point energies through the formal formula for the energy per unit length

$$\mathcal{E} = \frac{1}{2} \int_{-\infty}^{\infty} \frac{dk}{2\pi} \sum_{m=-\infty}^{\infty} \frac{1}{2\pi i} \frac{1}{2} \oint_C \omega \, d_\omega \ln f_n(k, \omega, a), \qquad (7.38)$$

where f_n is a function, the zeroes of which are the mode frequencies of the system, and where C is a contour initially chosen to encircle those zeroes. The divergences there are regulated by a zeta function technique; the results, both analytically and numerical, are identical to those reported here and derived in Ref. [24]. This calculation will be described in Sec. 7.2 below.

7.1.1   Related Work

Nesterenko and Pirozhenko [244] have worked out the Casimir energy for a massless scalar field with Dirichlet boundary conditions on a cylinder. The result is repulsive, but with a very small magnitude,

$$\mathcal{E}_s = \frac{0.000606}{a^2}. \qquad (7.39)$$

Scandurra [245] considered a cylindrical δ-function shell potential for a massive scalar field. The result is divergent, but if the divergent terms are subtracted by requiring the vacuum fluctuations to vanish for infinite mass (the method advocated by the Leipzig group), a finite energy results, which goes like $\ln \mu a / a^2$ as $\mu a \to 0$, and like $-1/\mu a^3$ for $\mu a \to \infty$.

Related is the Casimir energy of a field in the presence of a magnetic fluxon. Leseduarte and Romeo [246] considered the influence of a magnetic fluxon on a massless scalar field confined by a circular, and by a spherical bag, and also a massless fermion in the former case. For a sphere, the result (9.34) was recovered. Sitenko and Babansky [247] considered the Casimir-Aharonov-Bohm effect, that is, the vacuum energy of a massive scalar field in the presence of a magnetic vortex line, where the scalar field vanishes on the line. If the magnetic flux vanishes, the Casimir energy diverges.

7.1.2 *Parallelepipeds*

It may be of interest to compare the result (7.37) with the calculation given by Lukosz [77, 78, 79] for the force/area due to interior electromagnetic zero-point fluctuations for a cylinder with a square cross section:

$$\mathcal{F}_{\text{square cylinder}} = -\frac{0.0382}{a^4}. \tag{7.40}$$

(See also Refs. [80, 81] and [82].) If we compare this with the corresponding force per area of a circular cylinder of the same diameter, with both interior and exterior electromagnetic fluctuations, which from (7.37) is

$$\mathcal{F}_{\text{circular cylinder}} = -\frac{0.0691}{(2a)^4}, \tag{7.41}$$

we see nearly a factor of two discrepancy. Actually, we have no reason to trust the result (7.40) because when the exterior modes are excluded the result is terribly divergent; Lukosz obtains a finite result through the magic of zeta-function regularization. Thus the numerical coincidence of Lukosz' result for (the interior modes of) a cube with that for (the interior and exterior modes of) a spherical shell—see (6.13)—is surely fortuitous.

For later work on the interior modes in parallelepipedal geometries, see Refs. [248, 249, 250, 251, 252]. The temperature dependence for the interior Casimir effect for hypercuboids in n dimensions was considered by Kirsten [253].

7.1.3 *Wedge-Shaped Regions*

The formalism developed above may be applied readily to calculate the Casimir effect in wedge-shaped geometries, say space divided by two conducting planes making an angle α with each other. Such a situation was examined by Brevik and Lygren [254], see also [255], which was closely based on our work [24] and on Ref. [256]. Physically, this is of interest because it can describe the geometry in the neighborhood of a cosmic string [257, 258]. We will discuss some of the local results in Chapter 11. Other relevant citations are Refs. [102, 259, 260, 261, 262].[‡]

7.2 Dielectric-Diamagnetic Cylinder—Uniform Speed of Light

In this section we consider the Casimir energy of an infinite solid cylinder surrounded by an uniform medium. The permittivity and permeability of the cylinder material (ϵ_1, μ_1) and those of the surroundings (ϵ_2, μ_2) are considered to be arbitrary. In principle they may depend on the frequency of the electromagnetic oscillations (dispersive media), but we will ignore this dependence at present. In Sec. 7.2.1 we derive the general integral representation for the Casimir energy, based on (7.38), and then in Sec. 7.2.2 we specialize to the circumstance when the speed of light is the same inside and outside the cylinder, $\epsilon_1\mu_1 = \epsilon_2\mu_2 = c^{-2}$. When this condition is satisfied, all the divergences cancel between interior and exterior modes. Here, we regulate those divergences by employing the zeta function technique. In Sec. 7.2.3 the cases when $\xi^2 \ll 1$ and $\xi^2 = 1$ are considered numerically, ξ^2 being $(\epsilon_1 - \epsilon_2)^2/(\epsilon_1 + \epsilon_2)^2$. The first case gives the Casimir energy of a dilute dielectric-diamagnetic cylinder, while the second case corresponds to a perfectly conducting infinitely thin cylindrical shell. Remarkably, the Casimir energy obtained for a tenuous medium vanishes, as it does for a tenuous dielectric cylinder. The result obtained in the second case is identical to that obtained by the Green's function method of calculating the energy and regulating the divergences by use of an ultraviolet regulator as obtained in the previous section. We conclude this section by discussing the significance of the results obtained. The calculations presented in this section were first published in Ref. [43]. The Green's function formulation of this problem was given in Ref. [256], a straightforward generalization of the procedure described in Sec. 7.1.

[‡]Aliev [263] calculated the Casimir force between two parallel cosmic strings (there is no classical force between the strings). If the (dimensionless) mass densities of the strings are μ_1, μ_2 and the separation is a, the force per unit length is

$$\mathcal{F} = -\frac{16}{15\pi}\frac{\mu_1\mu_2}{a^3}. \tag{7.42}$$

As Aliev remarks, this may be regarded as a gravitational analog of an Aharonov-Bohm interaction between two thin solenoids.

7.2.1 *Integral Representation for the Casimir Energy*

We shall consider the following configuration. An infinite circular cylinder of radius a is placed in a uniform unbounded medium. The permittivity and the permeability of the material making up the cylinder are ϵ_1 and μ_1, respectively, and those for surrounding medium are ϵ_2 and μ_2. It is assumed that the conductivity in both the media is zero. We will compute the Casimir energy per unit length of the cylinder.

In the mode summation method, the Casimir energy is defined by

$$E = \frac{1}{2} \sum_{\{p\}} (\omega_p - \bar{\omega}_p), \qquad (7.43)$$

where ω_p are the classical eigenfrequencies of the electromagnetic oscillations in the system under consideration, and $\bar{\omega}_p$ are those in the absence of any boundary, that is, when either medium fills all space. (When the precise meaning is not required, we denote this by the formal limit $a \to \infty$.) The set $\{p\}$ stands for a complete set of quantum numbers (discrete and continuous) which is determined by the symmetry of the problem. Either sum in (7.43) diverges, therefore a preliminary regularization is required.

In order to find the eigenfrequencies one needs to solve Maxwell's equations for the given configuration with the appropriate boundary conditions on the lateral surface of the cylinder. As is well known, it is sufficient to require the continuity of the tangential components of the electric field $\mathbf{E}$ and of the magnetic field $\mathbf{H}$ [184]. In terms of the cylindrical coordinates (r, θ, z) the eigenfunctions for this boundary value problem contain the factor

$$\exp(-i\omega t + ikz + im\theta), \qquad (7.44)$$

and their dependence on r is described by cylindrical Bessel functions J_m for $r < a$ and by Hankel functions of the first kind $H_m \equiv H_m^{(1)}$ for $r > a$. The eigenfrequencies are the roots of the equation (Ref. [184], p. 526)

$$f_m(k, \omega, a) = 0, \qquad (7.45)$$

where

$$\begin{aligned} f_m \equiv {}& \lambda_1^2 \lambda_2^2 \Delta_m^{\mathrm{TE}}(\lambda_1 a, \lambda_2 a)\, \Delta_m^{\mathrm{TM}}(\lambda_1 a, \lambda_2 a) \\ & - m^2 \omega^2 k^2 (\epsilon_1 \mu_1 - \epsilon_2 \mu_2)^2 \left(J_m(\lambda_1 a)\, H_m(\lambda_2 a) \right)^2, \end{aligned} \qquad (7.46)$$

with

$$\Delta_m^{\text{TE}}(\lambda_1 a, \lambda_2 a) = a\mu_1\lambda_2 J_m'(\lambda_1 a)\, H_m(\lambda_2 a) - a\mu_2\lambda_1 J_m(\lambda_1 a)\, H_m'(\lambda_2 a),$$
$$\Delta_m^{\text{TM}}(\lambda_1 a, \lambda_2 a) = a\epsilon_1\lambda_2 J_m'(\lambda_1 a)\, H_m(\lambda_2 a) - a\epsilon_2\lambda_1 J_m(\lambda_1 a)\, H_m'(\lambda_2 a),$$

$$(7.47)$$

$$\lambda_i^2 = k_i^2 - k^2, \quad k_i^2 = \epsilon_i\mu_i\omega^2, \quad i = 1, 2, \quad m = 0, \pm 1, \pm 2, \ldots. \qquad (7.48)$$

The indices TE and TM refer to transverse electric and transverse magnetic modes, and will be explained further below. For given k and m, (7.45) has an infinite sequence of roots $\omega_{mn}(k)$, $n = 1, 2, \ldots$, these frequencies being the same inside and outside the cylinder. In view of this the Casimir energy per unit length (7.43) can be rewritten as

$$\mathcal{E} = \frac{1}{2}\int_{-\infty}^{\infty}\frac{dk}{2\pi}\sum_{m=-\infty}^{\infty}\sum_{n=1}^{\infty}[\omega_{mn}(k) - \bar{\omega}_{mn}(k)], \qquad (7.49)$$

where $\bar{\omega}_{mn}(k)$ stands for the uniform medium subtraction referred to above.

We next represent the formal sum in (7.49) in terms of the contour integral [264]§

$$\mathcal{E} = \frac{1}{2}\int_{-\infty}^{\infty}\frac{dk}{2\pi}\sum_{m=-\infty}^{\infty}\frac{1}{2\pi i}\frac{1}{2}\oint_C \omega\, d_\omega \ln\frac{f_m(k,\omega,a)}{f_m(k,\omega,\infty)}. \qquad (7.50)$$

The integration in (7.50) is carried out along a closed path C in the complex ω plane which consists of two parts: C_+ which encloses the positive roots of Eq. (7.45) in a counterclockwise sense, and C_- which encircles the negative roots in a clockwise sense. Therefore, we face the task of investigating the analytic properties of the function $f_m(k,\omega,a)$, which specifies the frequency eigenvalues. Generally this is a problem of extreme difficulty. Therefore, in the next two sections we shall consider specific simple cases.

The method of calculation of the Casimir energy proposed above can be straightforwardly generalized to dispersive media. To this end, it is sufficient to treat the parameters ϵ_i and μ_i, $i = 1, 2$ in the frequency equation (7.45) as given functions of the frequency ω. However, we will not address this issue in the present section.

§The result (7.50) may be easily shown to be equivalent to the corresponding Green's function formulation. See Appendix A.

7.2.2 *Casimir Energy of an Infinite Cylinder when* $\epsilon_1\mu_1 = \epsilon_2\mu_2$

We now assume that the permittivity and permeability of the cylinder material (ϵ, μ_1) and of the surroundings (ϵ_2, μ_2) are not arbitrary but satisfy the condition

$$\epsilon_1\mu_1 = \epsilon_2\mu_2 = c^{-2}, \tag{7.51}$$

where c is the light speed in either medium (in units of the speed of light in vacuum). The physical implications of this condition can be found in Refs. [265, 266, 267, 205, 268, 269]. When (7.51) holds, we have $\lambda_1 = \lambda_2 = \lambda$, and the frequency equation (7.45) is simplified considerably. It breaks up into two equations: for the transverse-electric (TE) oscillations

$$\Delta_m^{\text{TE}}(\lambda, a) \equiv \lambda a \left[\mu_1 J_m'(\lambda a) H_m(\lambda a) - \mu_2 J_m(\lambda a) H_m'(\lambda a)\right] = 0 \tag{7.52a}$$

and for the transverse-magnetic (TM) oscillations

$$\Delta_m^{\text{TM}}(\lambda, a) \equiv \lambda a \left[\epsilon_1 J_m'(\lambda a) H_m(\lambda a) - \epsilon_2 J_m(\lambda a) H_m'(\lambda a)\right] = 0. \tag{7.52b}$$

In the general case (7.46) such a decomposition occurs only for oscillations with $m = 0$. In (7.52a) and (7.52b) λ is the eigenvalue of the corresponding transverse [membrane-like] boundary value problem [270]

$$\lambda^2 = \frac{\omega^2}{c^2} - k^2. \tag{7.53}$$

Classification of the solutions of Maxwell's equations without sources in terms of the TE- and TM-modes originates in waveguide theory [184, 270, 99]. The main distinction of the propagation of electromagnetic waves in waveguides, in contrast to the same process in unbounded space, is that a purely transverse wave cannot propagate in a waveguide. The wave in a waveguide must necessarily contain either longitudinal electric or magnetic fields. The first case is referred to as the waves of electric type [transverse-magnetic (TM) waves] and in the second case one is dealing with waves of magnetic type [or transverse-electric (TE) waves]. As we have seen, this classification proves to be convenient in studies of electromagnetic oscillations in closed resonators as well.

Replacing the function $f_m(k, \omega, a)$ in (7.50) by the left hand sides of (7.52a) and (7.52b) and changing the integration variable to λ we arrive at

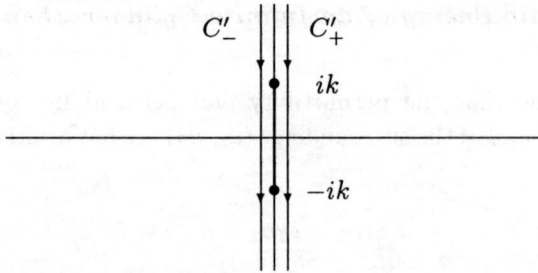

Fig. 7.1   Contour of integration $C' = C'_+ + C'_-$ in (7.54).

the following representation for the Casimir energy per unit length

$$\mathcal{E} = -\frac{c}{2} \int_{-\infty}^{\infty} \frac{dk}{2\pi} \sum_{m=-\infty}^{\infty} \frac{1}{2\pi i} \frac{1}{2} \oint_{C'} \sqrt{\lambda^2 + k^2}\, d\lambda \ln \frac{\Delta_m^{\mathrm{TE}}(\lambda a)\Delta_m^{\mathrm{TM}}(\lambda a)}{\Delta_m^{\mathrm{TE}}(\infty)\Delta_m^{\mathrm{TM}}(\infty)}.$$

(7.54)

Here we have distorted the contour of integration to $C' = C'_+ + C'_-$, as shown in Fig. 7.1. We take C'_+ to consist of a straight line parallel to, and just to the right of, the imaginary axis $(-i\infty, +i\infty)$ closed by a semicircle of an infinitely large radius in the right half-plane. C'_- similarly is a line parallel to, and just to the left of, the imaginary axis, closed by an infinite semicircle in the left hand plane. On both semicircles the argument of the logarithm function in (7.54) tends to 1. As a result these parts of the contour C' do not give any contribution to the Casimir energy $\mathcal{E}$. When integrating along the imaginary axis we chose the branch line of the function $\varphi(\lambda) = \sqrt{\lambda^2 + k^2}$ to run between $-ik$ and ik, where $k = +\sqrt{k^2} > 0$. In terms of $y = \mathrm{Im}\,\lambda$ we have

$$\varphi(iy) = \begin{cases} i\sqrt{y^2 - k^2}, & y > k, \\ \pm\sqrt{k^2 - y^2}, & |y| < k, \\ -i\sqrt{y^2 - k^2}, & y < -k, \end{cases}$$

(7.55)

where the sign on the middle form depends on whether we are to the right or the left of the cut. Thus contributions to (7.54) due to the integration along the segment of the imaginary axis $(-ik, ik)$ cancel between C'_+ and C'_-, and (7.54) acquires the form

$$\mathcal{E} = -\frac{c}{2\pi^2} \sum_{m=-\infty}^{\infty} \int_0^{\infty} dk \int_k^{\infty} \sqrt{y^2 - k^2}\, d_y \ln \frac{\Delta_m^{\mathrm{TE}}(iay)\Delta_m^{\mathrm{TM}}(iay)}{\Delta_m^{\mathrm{TE}}(i\infty)\Delta_m^{\mathrm{TM}}(i\infty)}.$$

(7.56)

Changing the order of integration of k and y and using the value of the integral

$$\int_0^y dk \sqrt{y^2 - k^2} = \frac{\pi}{4} y^2 , \qquad (7.57)$$

we obtain after the substitution $ay \to y$

$$\mathcal{E} = -\frac{c}{8\pi a^2} \sum_{m=-\infty}^{\infty} \int_0^{\infty} y^2 \, dy \, \ln \frac{\Delta_m^{\mathrm{TE}}(iy) \Delta_m^{\mathrm{TM}}(iy)}{\Delta_m^{\mathrm{TE}}(i\infty) \Delta_m^{\mathrm{TM}}(i\infty)} . \qquad (7.58)$$

Now we encounter the modified Bessel functions $I_m(y)$ and $K_m(y)$ (7.34a), (7.34b) and their asymptotic behavior for fixed m and $y \to \infty$

$$I_m(y) \sim \frac{e^y}{\sqrt{2\pi y}}, \qquad I_m'(y) \sim \frac{e^y}{\sqrt{2\pi y}}, \qquad (7.59a)$$

$$K_m(y) \sim \sqrt{\frac{\pi}{2y}} e^{-y}, \qquad K_m'(y) \sim -\sqrt{\frac{\pi}{2y}} e^{-y}. \qquad (7.59b)$$

With the help of these we derive from (7.52a) and (7.52b)

$$\frac{\Delta_m^{\mathrm{TE}}(iy)}{\Delta_m^{\mathrm{TE}}(i\infty)} = \frac{2y}{\mu_1 + \mu_2} [\mu_1 I_m'(y) K_m(y) - \mu_2 I_m(y) K_m'(y)],$$

$$\frac{\Delta_m^{\mathrm{TM}}(iy)}{\Delta_m^{\mathrm{TM}}(i\infty)} = \frac{2y}{\epsilon_1 + \epsilon_2} [\epsilon_1 I_m'(y) K_m(y) - \epsilon_2 I_m(y) K_m'(y)] . \qquad (7.60)$$

Making use of all this, we can recast (7.58) into the form

$$\mathcal{E} = \frac{c}{4\pi a^2} \sum_{m=-\infty}^{\infty} \int_0^{\infty} y \, dy \, \ln \Bigg\{ \frac{4y^2}{\varepsilon + \varepsilon^{-1} + 2} \Big[(I_m'(y) K_m(y))^2$$

$$+ (I_m(y) K_m'(y))^2 - (\varepsilon + \varepsilon^{-1}) I_m(y) I_m'(y) K_m(y) K_m'(y) \Big] \Bigg\} . \qquad (7.61)$$

Here a new notation $\varepsilon = \epsilon_1/\epsilon_2$ has been introduced, μ has been eliminated by the condition (7.51), and when going from (7.58) to (7.61) an integration by parts has been done, the boundary terms being omitted. The argument of the logarithm in (7.61) is simplified considerably if one uses the value of the Wronskian of the modified Bessel functions $I_m(y)$ and $K_m(y)$ [111]

$$I_m(y) K_m'(y) - I_m'(y) K_m(y) = -\frac{1}{y}. \qquad (7.62)$$

Finally (7.61) acquires the form

$$\mathcal{E} = \sum_{m=-\infty}^{\infty} \mathcal{E}_m, \tag{7.63}$$

where

$$\mathcal{E}_m = \frac{c}{4\pi a^2} \int_0^\infty y \, dy \, \ln\left\{1 - \xi^2 \left[y(I_m(y)K_m(y))'\right]^2\right\} \tag{7.64}$$

with $\xi = (1 - \varepsilon)/(1 + \varepsilon)$. This is a simple unregulated generalization of Eq. (7.19) for a conducting cylindrical shell, where $\xi = 1$, [see (7.18)] and the cylindrical analog of the spherical form (5.49) for a dielectric-diamagnetic ball.

From the asymptotic behaviors given in (7.59a) and (7.59b) it follows that the integral in (7.64) diverges logarithmically when $y \to \infty$. At the same time the sum over m in (7.63) also diverges because for large m the uniform asymptotic expansion of the modified Bessel functions gives [111] [see (7.24)]

$$\mathcal{E}_m\Big|_{m\to\infty} \simeq -\frac{c\xi^2}{16\pi a^2} \int_0^\infty \frac{z^5 \, dz}{(1 + z^2)^3} \equiv \mathcal{E}^\infty. \tag{7.65}$$

Here the change of variables $y = mz$ has been performed. Disregarding for the moment that the integral in (7.65) is divergent, we employ here the Riemann zeta function technique [271, 272] for attributing a finite value to the sum in (7.63),

$$\mathcal{E} = \sum_{m=-\infty}^{\infty} (\mathcal{E}_m - \mathcal{E}^\infty + \mathcal{E}^\infty)$$

$$= \sum_{m=-\infty}^{\infty} (\mathcal{E}_m - \mathcal{E}^\infty) + \sum_{m=-\infty}^{\infty} \mathcal{E}^\infty$$

$$= \sum_{m=-\infty}^{\infty} \bar{\mathcal{E}}_m + \mathcal{E}^\infty \sum_{m=-\infty}^{\infty} m^0, \tag{7.66}$$

where $\bar{\mathcal{E}}_m$ stands for the "renormalized" partial Casimir energy per length

$$\bar{\mathcal{E}}_m = \mathcal{E}_m - \mathcal{E}^\infty, \qquad m = 0, \pm 1, \dots. \tag{7.67}$$

We now have to treat the product of two divergent expressions $\mathcal{E}^\infty \sum m^0$ more precisely, by presenting it in the following form

$$\mathcal{E}^\infty \sum_{m=-\infty}^{\infty} m^0 = -\frac{c\xi^2}{16\pi a^2} \lim_{s\to 0+} \int_0^\infty \frac{z^{5-s}dz}{(1+z^2)^3} [2\zeta(s) + 1]$$

$$= -\frac{c\xi^2}{16\pi a^2} \lim_{s\to 0+} \left(\frac{1}{s} - \frac{3}{4}\right) [2\zeta'(s)\, s]$$

$$= \frac{c\xi^2}{16\pi a^2} \ln(2\pi). \tag{7.68}$$

Thus, the Casimir energy can be written in the form

$$\mathcal{E} = \bar{\mathcal{E}}_0 + 2 \sum_{m=1}^{\infty} \bar{\mathcal{E}}_m + \frac{c\xi^2}{16\pi a^2} \ln(2\pi), \tag{7.69}$$

because

$$\bar{\mathcal{E}}_{-m} = \bar{\mathcal{E}}_m, \qquad m = 0, 1, 2, \ldots. \tag{7.70}$$

Now we deduce from (7.67), (7.64), and (7.65) that

$$\bar{\mathcal{E}}_m = \frac{c}{4\pi a^2} \int_0^\infty y\, dy \left\{ \ln\left[1 - \xi^2\sigma_m^2(y)\right] + \frac{\xi^2}{4} \frac{y^4}{(m^2 + y^2)^3} \right\}, \quad m = 1, 2, \ldots, \tag{7.71}$$

while for $m = 0$

$$\bar{\mathcal{E}}_0 = \frac{c}{4\pi a^2} \int_0^\infty y\, dy \left\{ \ln\left[1 - \xi^2\sigma_0^2(y)\right] + \frac{\xi^2}{4} \frac{y^4}{(1 + y^2)^3} \right\}, \tag{7.72}$$

where $\sigma_m(y) = y(I_m(y)K_m(y))'$. The integrals in these formulæ converge because for $y \to 0$ and $m \neq 0$ we have [111]

$$I_m(y)K_m(y) \to \frac{1}{2m}. \tag{7.73}$$

In this limit

$$\sigma_0^2(y) \to 1. \tag{7.74}$$

On the other hand, for large y

$$\sigma_m^2(y) \to \frac{1}{4y^2}. \tag{7.75}$$

By making use of the uniform asymptotics for the Bessel functions [111] we deduce from (7.71)

$$\bar{\mathcal{E}}_m\bigg|_{m\to\infty} \sim \frac{c\xi^2}{4\pi a^2}\left[\frac{10-3\xi^2}{960m^2} - \frac{28224-7344\xi^2+720\xi^4}{15482880m^4} + O\left(\frac{1}{m^6}\right)\right].$$
(7.76)

Thus the Casimir energy given in (7.69) is finite.¶ One can advance further only by considering special cases and applying numerical calculations.

7.2.3 Dilute Compact Cylinder and Perfectly Conducting Cylindrical Shell

We begin by addressing the case when $\xi^2 \ll 1$. Because we are assuming the condition $\epsilon_1\mu_1 = \epsilon_2\mu_1 = c^{-2}$, this is not the same situation as a dilute compact cylinder with $|\epsilon_1-\epsilon_2| \ll 1$ and $\mu_1 = \mu_2 = 1$, which we shall discuss in Sec. 7.3. Recall that

$$\xi^2 = \left(\frac{\epsilon_1-\epsilon_2}{\epsilon_1+\epsilon_2}\right)^2 = \frac{(\epsilon_1-\epsilon_2)^2}{4\epsilon^2},$$
(7.77)

where $\epsilon = (\epsilon_1+\epsilon_2)/2$. Retaining in (7.72) only the terms proportional to ξ^2 we obtain

$$\bar{\mathcal{E}}_0 \simeq \frac{c\xi^2}{4\pi a^2}\int_0^\infty y\,dy\left[\frac{y^4}{4(1+y^2)^3} - \sigma_0^2(y)\right]$$

$$\simeq \frac{c\xi^2}{4\pi a^2}\,(-0.490878).$$
(7.78)

To estimate $\bar{\mathcal{E}}_m$, $m > 0$, we can use the leading asymptotic behavior (7.76)

$$\bar{\mathcal{E}}_m \sim \frac{c\xi^2}{4\pi a^2}\left(\frac{1}{96m^2} - \frac{7}{3840m^4}\right), \quad m \to \infty.$$
(7.79)

¶As noted above, this problem has been considered in Ref. [256] using the Green's function technique. However, there Brevik and Nyland include dispersion, and supply an abrupt frequency cutoff in the dispersion relation, as well as an angular momentum cutoff. The results are divergent as the cutoffs tend to infinity. This is similar to what Brevik and Einevoll [212] found for a spherical ball with the same type of dispersion relation—see Sec. 5.6. There is here, as in the latter case, room for suspicion that these divergences are spurious. For example, keeping only a finite number of angular momentum modes in the conducting spherical shell problem gives even the incorrect sign of the energy. See Sec. 4.1.

To a precision of 10^{-6}, we evaluate (7.69) by substituting in the value of $\bar{\mathcal{E}}_0$, (7.78), integrating $\bar{\mathcal{E}}_m$ numerically for $m = 1, \dots 5$, and asymptotically using (7.79) for $n \geq 6$, with the result[||]

$$
\begin{aligned}
\mathcal{E} &\simeq \frac{c\xi^2}{4\pi a^2} \left[-0.490878 + 2\sum_{m=1}^{5} \bar{\mathcal{E}}_m + \sum_{m=6}^{\infty} \left(\frac{1}{48m^2} - \frac{7}{1920m^4} \right) + \frac{1}{4}\ln(2\pi) \right] \\
&= \frac{c\xi^2}{4\pi a^2} \left(-0.490878 + 0.027638 + 0.003778 - 0.000007 + 0.459469 \right) \\
&= \frac{c\xi^2}{4\pi a^2} (0.000000) .
\end{aligned}
\tag{7.80}
$$

Thus the Casimir energy of a dilute cylinder possessing the same speed of light inside and outside proves to be zero!

Following Klich and Romeo [273], we can obtain this result analytically. We use the addition theorem given there

$$
\sum_{m=-\infty}^{\infty} r^2 \left([I_m(r)K_m(r)]' \right)^2 = \int_0^{2\pi} \frac{d\phi}{2\pi} \left[R(r,\phi) K_1(R(r,\phi)) \right]^2 ,
\tag{7.81}
$$

where

$$
R(r,\phi) = 2r|\sin\phi/2| .
\tag{7.82}
$$

Thus the Casimir energy of a dilute cylinder is from (7.64)

$$
\mathcal{E} = \sum_{m=-\infty}^{\infty} \mathcal{E}_m = -\frac{2c\xi^2}{\pi^2 a^2} \int_0^{\infty} dy\, y^3 \int_0^1 du \frac{u^2}{\sqrt{1-u^2}} [K_1(2yu)]^2 .
\tag{7.83}
$$

Here we have made the substitution $u = |\sin\phi/2|$. This latter integral is obviously positive, and divergent. We may regulate it by replacing there $y^3 \to y^s$, which gives a finite integral if $\operatorname{Re} s < 2$. The Macdonald function may be given by the integral representation [99]

$$
K_1(x) = \int_0^{\infty} d\theta \cosh\theta\, e^{-x\cosh\theta} .
\tag{7.84}
$$

[||] The cancellations here are very severe. If the asymptotic approximation were used for all m, a positive result would be found, $E \sim (c\xi^2/4\pi a^2)(-0.00108)$. Unlike for the spherical case, doing the integral exactly is essential.

The y integral then becomes trivial, and the continued energy is given by

$$\mathcal{E}_s = -\frac{2c\xi^2\Gamma(s+1)}{\pi^2a^22^{s+1}} \int_0^\infty d\theta \int_0^\infty d\theta' \frac{\cosh\theta\cosh\theta'}{(\cosh\theta+\cosh\theta')^{s+1}}$$

$$\times \int_0^1 du\, u^{1-s}(1-u^2)^{-1/2}. \tag{7.85}$$

Clearly the integral over θ, θ' is finite for $s > 0$, but the u integral is

$$\int_0^1 du\, u^{1-s}(1-u^2)^{-1/2} = \frac{\sqrt{\pi}}{2}\frac{\Gamma(1-s/2)}{\Gamma(3/2-s/2)}, \quad \mathrm{Re}\, s < 2, \tag{7.86}$$

so if we take this to define the integral for all s, we see that it has a zero for $s = 3$. Thus this analytic method of regularization also gives a vanishing result (to first order in ξ^2) for the Casimir energy of a dilute dielectric-diamagnetic cylinder.

Of course, this vanishing does not persist beyond lowest order. In Ref. [274] Nesterenko and Pirozhenko show that in order ξ^4 an attractive energy results,

$$E \approx -\frac{0.007602\xi^4}{a}. \tag{7.87}$$

This zero is to be contrasted with the positive Casimir energy found for a dilute ball with the same property given in (5.58),

$$E_{\mathrm{ball}} = \frac{5}{32\pi a}\xi^2 = 0.0497359\frac{\xi^2}{a}. \tag{7.88}$$

It is further remarkable that the same zero result is found for a dilute *dielectric* cylinder, that is, one with $\mu = 1$ everywhere and $\epsilon > 1$ inside the cylinder, and $\epsilon = 1$ outside, a result which may be most easily confirmed by summing the intermolecular van der Waals energies. That calculation is given in Sec. 7.3. However, zero is not the universal value of the Casimir energy for cylinders, as we now remind the reader.

Of particular interest is the case when $\xi^2 = 1$. With $c = 1$ in our formulas it corresponds to infinitely thin, perfectly conducting cylindrical shell. Setting $\xi = 1$ and $c = 1$ in (7.72) we obtain by numerical integration[**]

$$\bar{\mathcal{E}}_0 = \frac{1}{4\pi a^2}(-0.6517) = -0.05186\frac{1}{a^2}. \tag{7.89}$$

[**]In the notation of Sec. 7.1 this is $-\frac{1}{8\pi a^2}(S + R_0 + \frac{1}{2}\ln 2\pi)$—see (7.36a) and (7.36c). The $\ln 2\pi$ term is cancelled by that in (7.69) here.

The sum $2 \sum_{n=1}^{\infty} \bar{\mathcal{E}}_n$ in (7.69) can be found approximately by making use of the two leading terms in the uniform asymptotic expansion (7.76)

$$
\begin{aligned}
2 \sum_{m=1}^{\infty} \bar{\mathcal{E}}_m &\simeq \frac{1}{4\pi a^2} \left(\frac{7}{480} \sum_{m=1}^{\infty} \frac{1}{m^2} - \frac{5}{1792} \sum_{m=1}^{\infty} \frac{1}{m^4} \right) \\
&= \frac{1}{4\pi a^2} \left(\frac{7}{480} \frac{\pi^2}{6} - \frac{5}{1792} \frac{\pi^4}{90} \right) \\
&= \frac{1}{4\pi a^2} 0.0210 = 0.0018 \frac{1}{a^2} .
\end{aligned}
\tag{7.90}
$$

With higher accuracy (up to 10^{-5}) this sum was calculated in Sec. 7.1 by integration of (7.71)[††]

$$
2 \sum_{m=1}^{\infty} \bar{\mathcal{E}}_m \simeq \frac{1}{4\pi a^2} \frac{1}{2} 0.0437 = \frac{1}{4\pi a^2} 0.0218 = 0.00174 \frac{1}{a^2} .
\tag{7.91}
$$

Substituting Eqs. (7.89) and (7.91) into (7.69) we obtain for the Casimir energy of a perfectly conducting cylindrical shell

$$
E_{\text{cyl. shell}} = \frac{1}{4\pi a^2} (-0.1704) = -0.01356 \frac{1}{a^2} .
\tag{7.92}
$$

This is exactly the result first obtained by DeRaad and Milton [24] and given in (7.37). It is worth noting here that unlike in that approach the use of the ζ function technique enables us to dispense with the introduction of a high-frequency cutoff function, although the latter is undoubtedly more physical.

Gosdzinsky and Romeo independently computed [243] the electromagnetic vacuum energy of space divided by an infinite perfectly conducting cylindrical surface, to much higher accuracy, but using a rather more elaborate zeta-function method.

Nesterenko, Lambiase, and Scarpetta [275] recently computed the Casimir energy of a semicircular conducting cylinder. They encountered divergences, which are not unexpected due to the sharp corners of the semicircular cross section [276].

[††]This is exactly the same as $-\frac{1}{8\pi a^2} R$ given by (7.36b).

7.3 Van der Waals Energy of a Dielectric Cylinder

It is now established that for tenuous media the Casimir effect and the sum of molecular van der Waals forces are identical [39]. The basis for this assertion is the equivalence between the Casimir energy for a dilute dilectric sphere (5.83) and the renormalized van der Waals energy for the same system (5.107). Here we calculate the latter for a dilute solid cylinder, with dielectric constant $\epsilon \neq 1$ in the interior, $\epsilon = 1$ in the exterior, and $\mu = 1$ everywhere. We follow the procedure given in Sec. 5.9. The van der Waals energy for this cylinder is

$$\mathcal{E}_{\mathrm{vdW}} = -\frac{1}{2}B\mathcal{N}^2 \int d^D r\, d^D r' [|\mathbf{r}_\perp - \mathbf{r}'_\perp|^2 + r^2 + r'^2 - 2rr'\cos\theta]^{-\gamma/2}, \quad (7.93)$$

where $B = (23/4\pi)\alpha^2$, $\alpha = (\epsilon - 1)/4\pi\mathcal{N}$ being the molecular polarizability, and $\mathcal{N}$ being the number density of molecules. We have regulated the integral by dimensional continuation, D being the number of spatial dimensions, and γ being the (inverse) power of the Casimir-Polder potential. The following calculation is valid providing $D > \gamma$; the final result will be obtained by violating this condition, by setting $D = 3$ and $\gamma = 7$.

We assume translational invariance in the $D - 2$ transverse directions, so the transverse integral is easy: (L is the length of the cylinder and $b^2 = r^2 + r'^2 - 2rr'\cos\theta$)

$$\int_{-\infty}^{\infty} d^{D-2}r_\perp\, d^{D-2}r'_\perp \, [|\mathbf{r}_\perp - \mathbf{r}'_\perp|^2 + b^2]^{-\gamma/2}$$

$$= L^{D-2} \int_{-\infty}^{\infty} d^{D-2}r_\perp \, [r_\perp^2 + b^2]^{-\gamma/2}$$

$$= \frac{L^{D-2}}{\Gamma(\gamma/2)} \int_{-\infty}^{\infty} d^{D-2}r_\perp \int_0^{\infty} \frac{dt}{t} t^{\gamma/2} e^{-t(r_\perp^2 + b^2)}$$

$$= \frac{L^{D-2}}{\Gamma(\gamma/2)} \int_0^{\infty} dt\, t^{\gamma/2 - 1} e^{-tb^2} \left[\int_{-\infty}^{\infty} dx\, e^{-tx^2} \right]^{D-2}$$

$$= (L\sqrt{\pi})^{D-2}(b^2)^{D/2 - \gamma/2 - 1} \frac{\Gamma(\gamma/2 - D/2 + 1)}{\Gamma(\gamma/2)}. \quad (7.94)$$

The remaining integral over r, r', θ, θ' is just that given in Sec. 5.9. In

(5.106) there, we merely set $D = 2$ and $\gamma = \gamma - D + 2$. The result is

$$\mathcal{E}_{\text{vdW}} = -B\mathcal{N}^2 \frac{L^{D-2}}{a^{\gamma - D - 2}} \frac{2^{D-\gamma} \pi^{D/2+1/2} \Gamma\left(\frac{\gamma}{2} - \frac{D}{2} + 1\right) \Gamma\left(\frac{D}{2} - \frac{\gamma}{2} + \frac{1}{2}\right)}{\Gamma\left(\frac{\gamma}{2}\right) \Gamma\left(\frac{D}{2} - \frac{\gamma}{2} + 2\right) (D - \gamma)}.$$

(7.95)

This is exactly the result found by Romeo [44]. Now when we set $D = 3$ and $\gamma = 7$ everything is finite except for the second gamma function in the denominator, which has a simple pole, and thus the van der Waals energy vanishes in this case.

It is very likely that the same zero value is obtained by a direct calculation of the Casimir energy, either through a zeta function, or through a Green's function, technique. Yet, because of the intrinsic difficulty of cylindrical calculations, that explicit demonstration has not yet been completed. However, Bordag and Pirozhenko [277] have very recently shown that the second heat-kernel coefficient (the residue of the pole of the zeta function at $s = -1/2$) is proportional to $(\epsilon - 1)^3$, i.e., it vanishes in the dilute approximation. In other words, the analog of (7.56) [with the argument of the logarithm replaced by f_m given in (7.46)] is finite in $(\epsilon - 1)^2$ order. This does not prove that the Casimir energy vanishes in this order, but rather that it may be uniquely calculated. Thus, the zero result found by summing van der Waals energies must coincide with the Casimir energy in order $(\epsilon - 1)^2$.

Chapter 8

Casimir Effect in Two Dimensions: The Maxwell-Chern-Simons Casimir Effect

8.1 Introduction

It has been known for two decades that one can construct gauge theories with interesting properties in odd-dimensional spaces. In addition to the usual Maxwell–Yang-Mills action for the gauge fields, one can put in a gauge-invariant mass term [278, 279, 280, 281, 282, 283, 284, 285]. In this Chapter we are interested in the $(2 + 1)$-dimensional Abelian theory for which the Lagrangian is

$$\mathcal{L} = -\frac{1}{4}F^{\mu\nu}F_{\mu\nu} + \frac{1}{4}\mu\epsilon^{\mu\alpha\beta}F_{\alpha\beta}A_{\mu}. \tag{8.1}$$

In terms of the dual tensor

$$F^{\mu} \equiv \epsilon^{\mu\alpha\beta}\frac{1}{2}F_{\alpha\beta} = \epsilon^{\mu\alpha\beta}\partial_{\alpha}A_{\beta}, \tag{8.2}$$

we can rewrite (8.1) as

$$\mathcal{L} = \frac{1}{2}F^{\mu}F_{\mu} + \frac{1}{2}\mu F^{\mu}A_{\mu}. \tag{8.3}$$

The equations of motion,

$$\epsilon_{\mu\alpha\beta}\partial^{\alpha}F^{\beta} + \mu F_{\mu} = 0, \tag{8.4}$$

yield the Bianchi identity

$$\partial_{\mu}F^{\mu} = 0, \tag{8.5}$$

consistent with (8.2). Further, we have the constraint

$$F^\mu F_\mu = 0, \tag{8.6}$$

which says that the theory possesses but one degree of freedom. Obviously the equations of motion are invariant under a gauge transformation

$$A_\mu \to A_\mu + \partial_\mu \vartheta, \tag{8.7}$$

while the Lagrangian changes only by a total derivative,

$$\mathcal{L} \to \mathcal{L} + \frac{1}{2}\mu\partial_\mu(F^\mu\vartheta). \tag{8.8}$$

We can use (8.4) to show that the gauge field is indeed massive:

$$(-\partial^2 + \mu^2)F^\lambda = 0. \tag{8.9}$$

The non-Abelian generalization of the mass term is the Chern-Simons secondary characteristic. Accordingly, we call the mass term in (8.1) topological.

Although the topological mass term is gauge invariant, due to the structure of the Levi-Civita symbol it is odd under a parity transformation and under time reversal. In contrast to a single, massless, spin-0 excitation in the Maxwell theory in $2+1$ dimensions, the Maxwell-Chern-Simons electrodynamics given by (8.1) possesses a single, massive, spin-1 excitation [282]. Also, one may compare it with the conventional massive electrodynamics given by the Lagrangian

$$\mathcal{L} = -\frac{1}{4}F^{\mu\nu}F_{\mu\nu} - \frac{1}{2}\mu^2 A^\mu A_\mu, \tag{8.10}$$

which has a pair of spin-1 degrees of freedom, and whose mass term, although invariant under P and T transformations, violates gauge invariance.

Subsequently, there was renewed interest in Chern-Simons electrodynamics. It has been proposed that such a $(2 + 1)$-dimensional Abelian theory may be relevant for the fractional quantum Hall effect [92, 93, 94, 95, 286, 287, 288, 289, 290, 291] in semiconductors and for high-T_c superconductivity [96, 97, 98, 292] in copper oxide crystals. The proposals involve anyons (with fractional statistics and fractional charges) and Chern-Simons topological gauge fields. However, in contrast with the true electromagnetic fields, the Chern-Simons fields are not given the conventional Maxwell kinetic energy;* consequently, they have no independent dynamics and their presence is merely to implement fractional statistics.

In our work, the Abelian gauge fields are allowed to have the conventional Maxwell action as well as the topological mass term given in (8.1). We proposed [30, 31] a new test of the gauge-field sector in such a theory, involving the Casimir effect. In Section 8.2, the Casimir force between two conducting parallel lines is calculated. We would like to think that our results provide a way to measure the topological mass of the "photon" in (2 + 1)-dimensional spacetime. [Another possibility would be to measure the rate of fall-off of the Yukawa force as given by (8.9).] The finite temperature effect is considered in Sec. 8.2.1. Section 8.2.2 contains a brief discussion, and a comparison with the case of a massive scalar field. Then in Sec. 8.3 we turn to the much more interesting case of the Casimir effect with a circular boundary. It is then most convenient to use a curved-space formalism. Results for both zero and finite temperature are obtained. Unfortunately, although finite results are obtained in the leading asymptotic approximation, it appears that the Casimir effect for a circle, for scalar or vector fields, is divergent, a subject we will explore more fully in the following Chapter.

Throughout we use the metric $(-1, 1, 1)$ and the antisymmetric symbol $\epsilon^{012} = +1$.

8.2 Casimir Effect in 2 + 1 Dimensions

We start with the Lagrangian given by (8.3) supplemented by a source term:

$$\mathcal{L} = \frac{1}{2} F^{\mu} F_{\mu} + \frac{1}{2} \mu F^{\mu} A_{\mu} + J^{\mu} A_{\mu}. \tag{8.11}$$

The energy-momentum tensor for the "photon,"

$$T^{\mu\nu} = F^{\mu} F^{\nu} - \frac{1}{2} g^{\mu\nu} F_{\lambda} F^{\lambda}, \tag{8.12}$$

is independent of μ since the mass term is topological. But, as we will see, the vacuum expectation values of $T^{\mu\nu}$ (which determine the Casimir

---

*Integrating out the anyon gives the Maxwell term in the effective action, however. See Refs. [293, 294].

attraction) do depend on μ. We will compute the Casimir energy from the Green's function $G^{\mu\nu}$ defined by

$$F^{\mu}(x) = \int (dx')\, G^{\mu\nu}(x, x') J_{\nu}(x'). \qquad (8.13)$$

The field equations for F^{μ} can be used to yield the equation satisfied by the Green's function,

$$\epsilon^{\mu}{}_{\alpha\beta}\partial^{\alpha} G^{\beta\nu}(x, x') + \mu G^{\mu\nu}(x, x') + g^{\mu\nu}\delta(x - x') = 0. \qquad (8.14)$$

In terms of the propagator $D_{\mu\nu}$ for the A_{μ} field given by the time-ordered vacuum expectation values

$$D_{\mu\nu}(x, x') = i\langle A_{\mu}(x) A_{\nu}(x')\rangle, \qquad (8.15)$$

the Green's function can be written as

$$G^{\mu\nu}(x, x') = \epsilon^{\mu}{}_{\alpha\beta}\partial^{\alpha} D^{\beta\nu}(x, x') = i\langle F^{\mu}(x) A^{\nu}(x')\rangle, \qquad (8.16)$$

where we have used (8.2). The expectation values we need to evaluate the Casimir energy are

$$\langle F_{\mu}(x) F_{\nu}(x')\rangle = \frac{1}{i}\epsilon_{\nu\lambda\sigma}\partial'^{\lambda} G_{\mu}{}^{\sigma}(x, x'). \qquad (8.17)$$

We can choose the coordinate system so that the two conducting lines are parallel to the second (y) spatial axis and lie, respectively, on $x = 0$ and $x = a$, where x denotes the first coordinate. We will use perfect conductor boundary conditions so the Bianchi identity (8.5) coupled with the statics requirement ($\partial_0 F^0 = 0$) give

$$F_1 = 0 \quad \text{(boundary condition)}. \qquad (8.18)$$

This is just the statement that the tangential electric field is zero. [See Sec. 8.3.1 for further discussion of this boundary condition, in particular (8.92).] For the present geometry, it is convenient to introduce a transverse spatial Fourier transform together with a Fourier transform in time:

$$G^{\mu\nu}(x, x') = \int \frac{d\omega}{2\pi} e^{-i\omega(t-t')} \int \frac{dk}{2\pi} e^{ik(y-y')} \mathcal{G}^{\mu\nu}(x, x'; k, \omega). \qquad (8.19)$$

The arguments x and x' in $G^{\mu\nu}$ refer to three-dimensional spacetime points, while those in $\mathcal{G}^{\mu\nu}$ refer only to the first spatial coordinates.

The expectation values in (8.17) now take the form

$$
\begin{aligned}
\langle F_0 F_0 \rangle &= k\mathcal{G}^{01} + \frac{1}{i}\frac{\partial}{\partial x'}\mathcal{G}^{02}, \\
\langle F_1 F_1 \rangle &= k\mathcal{G}^{10} + \omega\mathcal{G}^{12}, \\
\langle F_2 F_2 \rangle &= \frac{1}{i}\frac{\partial}{\partial x'}\mathcal{G}^{20} - \omega\mathcal{G}^{21},
\end{aligned}
\tag{8.20}
$$

where, for convenience, we have suppressed all the (obvious) arguments. It follows from (8.12) that the stress tensor component T^{11} is given by

$$
T^{11}(x) = \lim_{x' \to x}\left[\frac{1}{2}F^0(x)F^0(x') + \frac{1}{2}F^1(x)F^1(x') - \frac{1}{2}F^2(x)F^2(x')\right],
\tag{8.21}
$$

so the vacuum expectation value of the stress tensor for a given frequency ω and wavenumber k is

$$
\langle t^{11} \rangle = \lim_{x' \to x}\left[\frac{1}{2i}\frac{\partial}{\partial x'}(\mathcal{G}^{02} - \mathcal{G}^{20}) + \frac{k}{2}(\mathcal{G}^{01} + \mathcal{G}^{10}) + \frac{\omega}{2}(\mathcal{G}^{12} + \mathcal{G}^{21})\right].
\tag{8.22}
$$

where the limit $x' \to x$ is to be taken symmetrically.

Our task is to solve for the various Green's functions. With the aid of (8.14) and (8.19) we see that three of them satisfy the system of equations

$$
\begin{aligned}
ik\mathcal{G}^{01} + \mu\mathcal{G}^{11} - i\omega\mathcal{G}^{21} &= \delta(x - x'), \\
\mu\mathcal{G}^{01} + ik\mathcal{G}^{11} - \frac{\partial}{\partial x}\mathcal{G}^{21} &= 0, \\
-\frac{\partial}{\partial x}\mathcal{G}^{01} + i\omega\mathcal{G}^{11} + \mu\mathcal{G}^{21} &= 0.
\end{aligned}
\tag{8.23}
$$

It is not hard to combine these equations to find the equation for $\mathcal{G}^{11}$:

$$
\left(-\frac{\partial^2}{\partial x^2} + k^2 + \mu^2 - \omega^2\right)\mathcal{G}^{11} = \left(\mu - \frac{1}{\mu}\frac{\partial^2}{\partial x^2}\right)\delta(x - x'),
\tag{8.24}
$$

or, for $\tilde{\mathcal{G}}^{11} = \mathcal{G}^{11} - \frac{1}{\mu}\delta(x - x')$:

$$
\left(-\frac{\partial^2}{\partial x^2} + k^2 + \mu^2 - \omega^2\right)\tilde{\mathcal{G}}^{11} = -\frac{k^2 - \omega^2}{\mu}\delta(x - x').
\tag{8.25}
$$

Actually in the calculation of the Casimir energy, for which we take the *limit* $x' \to x$, the δ functions are irrelevant so that $\tilde{\mathcal{G}}^{11}$ and $\mathcal{G}^{11}$ are effectively the same. Let

$$-\lambda^2 = k^2 - \omega^2 + \mu^2 \equiv \kappa^2; \tag{8.26}$$

then a standard calculation gives

$$\widetilde{\mathcal{G}}^{11} = \frac{\kappa^2 - \mu^2}{\lambda\mu} \frac{(ss)}{\sin\lambda a}, \tag{8.27}$$

where we have used the perfect conductor boundary conditions, (8.18). Also we have made use of the first of the following notations, as introduced in Sec. 2.6:

$$\begin{aligned}
(ss) &= \sin\lambda x_< \sin\lambda(x_> - a), \\
(cs) &= \begin{cases} \cos\lambda x \sin\lambda(x' - a), & \text{if } x < x', \\ \sin\lambda x' \cos\lambda(x - a), & \text{if } x > x', \end{cases} \\
(sc) &= \begin{cases} \sin\lambda x \cos\lambda(x' - a), & \text{if } x < x', \\ \cos\lambda x' \sin\lambda(x - a), & \text{if } x > x', \end{cases} \\
(cc) &= \cos\lambda x_< \cos\lambda(x_> - a),
\end{aligned} \tag{8.28}$$

with $x_>$ $(x_<)$ being the larger (smaller) of x and x'.

With the aid of (8.23) we can calculate $\mathcal{G}^{21}$ from $\mathcal{G}^{11}$ using

$$\mathcal{G}^{11} = \frac{\mu}{k^2 + \mu^2}\left(i\omega - \frac{ik}{\mu}\frac{\partial}{\partial x}\right)\mathcal{G}^{21}, \tag{8.29}$$

where we have dropped irrelevant δ functions. Writing $\mathcal{G}^{21}$ as a linear combination of the terms in (8.28) we easily find

$$\mathcal{G}^{21} = \frac{i}{\sin\lambda a}\left[\frac{\omega}{\lambda}(ss) + \frac{k}{\mu}(cs)\right]. \tag{8.30}$$

In the same way, we can calculate $\mathcal{G}^{01}$ from

$$\left(-i\omega\frac{\partial}{\partial x} + ik\mu\right)\mathcal{G}^{01} = (\omega^2 - \mu^2)\mathcal{G}^{11}, \tag{8.31}$$

yielding

$$\mathcal{G}^{01} = \frac{i}{\sin\lambda a}\left[\frac{k}{\lambda}(ss) + \frac{\omega}{\mu}(cs)\right]. \tag{8.32}$$

Next, we deal with the equations satisfied by $\mathcal{G}^{20}$, $\mathcal{G}^{10}$, and $\mathcal{G}^{00}$:

$$\mu\mathcal{G}^{00} + ik\mathcal{G}^{10} - \frac{\partial}{\partial x}\mathcal{G}^{20} = -\delta(x - x'),$$
$$ik\mathcal{G}^{00} + \mu\mathcal{G}^{10} - i\omega\mathcal{G}^{20} = 0, \qquad (8.33)$$
$$-\frac{\partial}{\partial x}\mathcal{G}^{00} + i\omega\mathcal{G}^{10} + \mu\mathcal{G}^{20} = 0,$$

from which follows

$$\left(-\lambda^2 - \frac{\partial^2}{\partial x^2}\right)\mathcal{G}^{10} = \frac{1}{i}\left(\frac{\omega}{\mu}\frac{\partial}{\partial x} - k\right)\delta(x - x'). \qquad (8.34)$$

We find

$$\mathcal{G}^{10} = \frac{1}{i\sin\lambda a}\left[\frac{k}{\lambda}(ss) + \frac{\omega}{\mu}(sc)\right]. \qquad (8.35)$$

From this we can compute $\mathcal{G}^{20}$ using

$$\left(-\frac{\omega}{k}\frac{\partial}{\partial x} + \mu\right)\mathcal{G}^{20} = -\left(\frac{\mu}{ik}\frac{\partial}{\partial x} + i\omega\right)\mathcal{G}^{10}, \qquad (8.36)$$

which gives

$$\mathcal{G}^{20} = \frac{1}{(\kappa^2 - \mu^2)\sin\lambda a}\left[\frac{k\mu\omega}{\lambda}(ss) + k^2(cs) + \omega^2(sc) + \frac{\omega\lambda k}{\mu}(cc)\right]. \qquad (8.37)$$

Noting that $\mathcal{G}^{10} \leftrightarrow \mathcal{G}^{01}$ under complex conjugation plus $(sc) \leftrightarrow (cs)$, we are led to guess from $\mathcal{G}^{20}$ and $\mathcal{G}^{21}$, respectively,

$$\mathcal{G}^{02} = \frac{1}{(\kappa^2 - \mu^2)\sin\lambda a}\left[\frac{k\mu\omega}{\lambda}(ss) + k^2(sc) + \omega^2(cs) + \frac{\omega\lambda k}{\mu}(cc)\right] \qquad (8.38)$$

and

$$\mathcal{G}^{12} = \frac{1}{i\sin\lambda a}\left[\frac{\omega}{\lambda}(ss) + \frac{k}{\mu}(sc)\right]. \qquad (8.39)$$

We successfully confirm our conjecture by checking the required identities given by (8.14).

We can now insert the various Green's functions into (8.22) to find

$$\langle t^{11}(x)\rangle = \lim_{x'\to x}\left\{\frac{1}{2i}\frac{1}{\sin\lambda a}\frac{\partial}{\partial x'}[(sc) - (cs)] + \frac{k}{2}\frac{1}{i\sin\lambda a}\frac{\omega}{\mu}[(sc) - (cs)]\right.$$
$$\left. + \frac{\omega}{2}\frac{1}{i\sin\lambda a}\frac{k}{\mu}[(sc) - (cs)]\right\}. \qquad (8.40)$$

The second and third terms do not contribute because (8.28) implies

$$\lim_{x' \to x} ((sc) - (cs)) = 0. \tag{8.41}$$

For the first term we use

$$\lim_{x' \to x} \frac{\partial}{\partial x'} (sc) = -\lambda \sin \lambda x \sin \lambda (x - a),$$

$$\lim_{x' \to x} \frac{\partial}{\partial x'} (cs) = \lambda \cos \lambda x \cos \lambda (x - a), \tag{8.42}$$

to give

$$\langle t^{11} \rangle (0 \text{ or } a) = \frac{i\lambda}{2} \cot \lambda a. \tag{8.43}$$

This is exactly the same form as for a massless scalar (2.24), except that λ now incorporates the mass, according to (8.26). This is the flux of momentum incident on the conducting lines, in terms of which the Casimir force per unit length is given by

$$\text{Force/length} = \mathcal{F} = \int \frac{d\omega}{2\pi} \int \frac{dk}{2\pi} \langle t^{11} \rangle. \tag{8.44}$$

It is, as usual, convenient to change to Euclidean variables

$$\omega \to i\zeta, \quad \lambda \to i\kappa, \tag{8.45}$$

so that from (8.26)

$$\kappa^2 = k^2 + \zeta^2 + \mu^2, \tag{8.46}$$

and from (8.43)

$$\langle t^{11} \rangle (0 \text{ or } a) = \frac{i}{2}\kappa \left[1 + \frac{2}{e^{2\kappa a} - 1} \right]. \tag{8.47}$$

Omitting the constant term (as usual in Casimir calculations[†]) we get

$$\mathcal{F} = i \int \frac{d\zeta}{2\pi} \frac{dk}{2\pi} \frac{i}{2} \frac{2\kappa}{e^{2\kappa a} - 1} = -\frac{1}{2\pi} \int_\mu^\infty \kappa \, dk \frac{\kappa}{e^{2\kappa a} - 1} \tag{8.48}$$

[†]As usual, such a constant is omitted because 1) it may be regarded as a contact term, 2) it would be cancelled by the contribution from the exterior fields, and 3) it would be present even if no "plates" were present. For example, see Ref. [11] and Sec. 2.3.

or

$$\mathcal{F} = -\frac{1}{16\pi a^3} \int_{2\mu a}^{\infty} dy \, \frac{y^2}{e^y - 1}, \tag{8.49}$$

an attractive force (which, one recognizes, has the form of a Debye function). This is exactly what we obtained for a massive scalar in $d = 1$ transverse directions in Section 2.4, see (2.54). Of course, then the massless limit corresponds to the $d = 1$ version of (2.35):

$$\mathcal{F}(\mu = 0) = -\frac{1}{8\pi a^3}\zeta(3), \tag{8.50}$$

which follows from (2.34). The large mass ($\mu a \gg 1$) limit is given by

$$\mathcal{F} \approx -\frac{1}{8\pi a^3}(2\mu^2 a^2 + 2\mu a + 1)e^{-2\mu a}. \tag{8.51}$$

As a check, let us supply another derivation of this Casimir force, which employs the 00 component of the stress tensor, that is, the energy density,

$$T^{00}(x) = \lim_{x' \to x} \frac{1}{2}\left(F^0(x)F^0(x') + F^1(x)F^1(x') + F^2(x)F^2(x')\right). \tag{8.52}$$

The calculations are similar to those for t^{11}; we find the same form given in (2.39)

$$\int_0^a \langle t^{00}\rangle(x)\,dx = -\frac{1}{2i}a\frac{\omega^2}{\lambda}\cot\lambda a + (\text{independent of } a). \tag{8.53}$$

The energy per unit length is

$$\begin{aligned}\mathcal{E} &= i\int \frac{d\zeta}{2\pi}\frac{dk}{2\pi} \int_0^a \langle t^{00}\rangle(x)\,dx \\ &= -\frac{1}{32\pi a^2}\int_{2\mu a}^{\infty} dy \, \frac{y^2}{e^y - 1} + \frac{\mu^2}{8\pi}\int_{2\mu a}^{\infty} dy \, \frac{1}{e^y - 1},\end{aligned} \tag{8.54}$$

so that the force per unit length is

$$\mathcal{F} = -\frac{\partial \mathcal{E}}{\partial a} = -\frac{1}{16\pi a^3}\int_{2\mu a}^{\infty} dy \, \frac{y^2}{e^y - 1}, \tag{8.55}$$

which agrees with (8.49).

8.2.1   Temperature Effect

The temperature dependence can be included by the replacement ($\beta = 1/kT$)

$$\zeta \to \zeta_n = \frac{2\pi n}{\beta}, \tag{8.56}$$

just as in Sec. 2.5, so that

$$
\begin{aligned}
\mathcal{F}^T &= -\frac{2}{\pi\beta}\sum_{n=0}^{\infty}{}'\int_0^\infty dk\,\frac{\kappa_n}{e^{2\kappa_n a}-1} \\
&= -\frac{1}{2\pi\beta a^2}\sum_{n=0}^{\infty}{}'\int_{y_n}^\infty dy\,\frac{y^2}{e^y-1}\left(y^2-y_n^2\right)^{-1/2},
\end{aligned} \tag{8.57}
$$

where $y_n^2 = 4\mu^2 a^2 + 16\pi^2 n^2 a^2/\beta^2$ and where the prime in the summation means that the $n = 0$ term is counted with half weight. We can consider two simple limits of (8.57) just as in Sec. 2.5. We will content ourselves here by writing down the high-temperature limit ($T \to \infty$ or $\beta \to 0$), where the dominant term comes from $n = 0$ ($n \neq 0$ terms are exponentially small), yielding

$$\mathcal{F}^T \approx -\frac{1}{2}\frac{1}{2\pi\beta a^2}g(2\mu a), \quad \beta \ll 4\pi a, \tag{8.58}$$

where

$$g(z) = \int_z^\infty dy\,\frac{y^2}{e^y-1}\frac{1}{\sqrt{y^2-z^2}} \tag{8.59}$$

is a monotonically (and rapidly) decreasing function, with $g(0) = \zeta(2) = \frac{\pi^2}{6} \approx 2g(2)$.

8.2.2   Discussion

In this section we have made a thoroughly consistent calculation of the Casimir effect between parallel lines for the $(2 + 1)$-dimensional Maxwell-Chern-Simons Abelian gauge theory. We see the result is exactly the same as for a massive spin-0 particle. We can have a qualitative understanding of this agreement as follows. Recall that, like the massive scalar field, the topologically massive spin-1 field obeys the Klein-Gordon equation (8.9) and it has only one polarization degree of freedom. The boundary condition

$F_1 = 0$, (8.18), is the Dirichlet condition that completes the analogy with the massive scalar case. So the agreement of the respective Casimir forces is perhaps not surprising, but hardly obvious. This agreement does not persist for other geometries; for example, in $(3 + 1)$-dimensional electrodynamics, the Casimir force between perfectly conducting parallel plates is twice that of a scalar field, as we saw in Sec. 2.6, but such is not the case for a spherical shell, treated in Sec. 4.1.

What about the case of massive, gauge-noninvariant electrodynamics given by (8.10)? This case is also considered in Ref. [82]. However, the situation is not completely clear. For one thing, the question of boundary conditions is subtle [82, 295, 296]. And then there is the issue of polarization degrees of freedom [297]. Presumably the additional polarization state for (conventionally) massive vector fields contributes no Casimir energy because it is found to decouple in the $\mu \to 0$ limit. But we do not know for sure whether the two theories given by (8.1) and (8.10) give the same Casimir energy or not. In any case, one should have a certain bias against gauge-noninvariant theories.

8.2.3 *Casimir Force between Chern-Simons Surfaces*

Bordag and Vassilevich [298] consider $3 + 1$ dimensional electrodynamics with a Chern-Simons mass term on the 3-dimensional bounding surface. The latter gives rise to nontrivial boundary conditions and, in the approximation that the Chern-Simons term is fully described by this boundary condition, they calculate the Casimir energy for the case of parallel plates when the charges (Chern-Simons masses) are different on the two plates. They find the energy per area to be

$$\mathcal{E} = -\frac{\pi^2}{1440a^3} h(\delta), \tag{8.60}$$

where the deviation from the Dirichlet scalar result depends on a function h of the phase

$$\delta = \tan^{-1} \mu_1 - \tan^{-1} \mu_2. \tag{8.61}$$

$h(\delta)$ is a function with period π, with

$$h(0) = 1, \quad h(\pi/4) = -\frac{7}{128}, \quad h(\pi/2) = -\frac{7}{8}; \tag{8.62}$$

the latter case corresponds to the interaction of a Dirichlet plate with a Neumann plate (cf. Sec. 2.6.1).

8.3   Circular Boundary Conditions

Now we wish to consider the same system bounded by a circle. The original analysis of this case was carried out in Ref. [31]. It is most convenient for this purpose to reformulate the theory in curvilinear coordinates. The Lagrangian for the Maxwell-Chern-Simons theory written in curvilinear coordinates is

$$\mathcal{L} = -\sqrt{-g}\frac{1}{4}F^{\mu\nu}F_{\mu\nu} + \frac{1}{4}\mu\epsilon^{\mu\alpha\beta}F_{\alpha\beta}A_{\mu}, \tag{8.63}$$

where g is the determinant of the metric $g_{\mu\nu}$ and

$$\epsilon^{\mu\alpha\beta} = \sqrt{-g}e^{\mu\alpha\beta} \tag{8.64}$$

is a tensor density. In terms of the dual tensor

$$F^{\lambda} = \frac{1}{2}e^{\lambda\alpha\beta}F_{\alpha\beta} = \frac{1}{2\sqrt{-g}}\epsilon^{\lambda\alpha\beta}(\partial_{\alpha}A_{\beta} - \partial_{\beta}A_{\alpha}), \tag{8.65}$$

we can rewrite (8.63) as

$$\mathcal{L} = \frac{1}{2}\sqrt{-g}(F^{\lambda}F_{\lambda} + \mu F^{\lambda}A_{\lambda}). \tag{8.66}$$

Varying $\mathcal{L}$ with respect to A_{μ} we find the equations of motion

$$\epsilon^{\mu\alpha\beta}\partial_{\alpha}F_{\beta} + \mu\sqrt{-g}F^{\mu} = 0, \tag{8.67}$$

which satisfy the Bianchi identity

$$\partial_{\mu}(\sqrt{-g}F^{\mu}) = 0, \tag{8.68}$$

consistent with (8.65). We can identify μ as the mass of the gauge field by using (8.67) to show that in Cartesian coordinates

$$(-\partial^{\lambda}\partial_{\lambda} + \mu^2)F^{\nu} = 0. \tag{8.69}$$

The equations of motion (8.67) are obviously invariant under a gauge transformation, while the Lagrangian changes only by an irrelevant total derivative. All of this generalizes the flat-space analysis of Sec. 8.1.

In the previous section we considered the Casimir effect between parallel conducting lines in two spatial dimensions for the Maxwell-Chern-Simons theory defined by (8.67). At zero temperature, we found an attractive force per unit length given by (8.49), while the high temperature limit was given by (8.58). In this section we turn to the case of a circular boundary. In Sec. 8.3.1 we compute the Casimir self-stress starting with the reduced Green's functions, both inside and outside the conducting circle. The representation of the product of two fields at a point is obtained by allowing the two field points in the Green's function to become infinitesimally close. A formula for the force on the circle is obtained in terms of Bessel functions. In Sec. 8.3.2 uniform asymptotic expansions for the Bessel functions are used to attempt to obtain approximate numerical results for zero and nonzero values of the Chern-Simons mass μ. In Sec. 8.3.3 we examine this effect in the limit of high temperatures. Physically, to obtain a(n effectively) two-dimensional system, one needs to freeze the degrees of freedom in the third direction perpendicular to the two-dimensional plane. According to quantum mechanics, it takes a finite amount of energy to excite the motion in the third direction. So, if the temperature is low enough, all the particles will remain in the ground state for those degrees of freedom and the system behaves as if there were only two spatial directions. Implicit in our discussion of the high-temperature limit in Sec. 8.3.3 is the assumption that the temperature is sufficiently low so that the degrees of freedom in the perpendicular directions are not excited. However, the high-temperature result for the $(2+1)$ theory is of field-theoretic interest in its own right. In Sec. 8.3.4 we give a brief discussion, and a comparison with the massless $(3+1)$-dimensional Casimir effect for a cylinder, see Sec. 7.1. For that comparison, we also need the result of the (massless) scalar field case for the circle, the derivation of which is given in Sec. 8.4. The work described in this section was originally published in Ref. [31].

8.3.1 *Casimir Self-Stress on a Circle*

We can rewrite the Lagrangian in (8.66) in terms of the fundamental variable A_μ as

$$\mathcal{L} = \frac{1}{2} \frac{1}{\sqrt{-g}} g_{\lambda\mu} \epsilon^{\lambda\alpha\beta} \epsilon^{\mu\sigma\tau} \partial_\alpha A_\beta \partial_\sigma A_\tau + \frac{1}{2} \mu \epsilon^{\lambda\alpha\beta} \partial_\alpha A_\beta A_\lambda. \tag{8.70}$$

Note that the last term is independent of $g_{\mu\nu}$. Varying the Lagrangian (8.70) with respect to $g_{\mu\nu}$,

$$\delta\mathcal{L} = \frac{1}{2}\delta g_{\mu\nu}T^{\mu\nu}, \tag{8.71}$$

we find

$$T^{\mu\nu} = \sqrt{-g}(F^{\mu}F^{\nu} - \frac{1}{2}g^{\mu\nu}F_{\lambda}F^{\lambda}) \tag{8.72}$$

for the stress tensor density for the photon, where we have used $\delta\sqrt{-g} = \sqrt{-g}\frac{1}{2}g^{\mu\nu}\delta g_{\mu\nu}$.

Next, we introduce the propagator $D_{\mu\nu}$ for the A_{μ} field according to

$$A_{\mu}(x) = \int dx' \sqrt{-g(x')}D_{\mu}{}^{\nu}(x,x')J_{\nu}(x'), \tag{8.73}$$

where J_{ν} is the source for the A_{μ} field. Equivalently, $D_{\mu\nu}$ is given by the time-ordered vacuum expectation values as in (8.15). Similarly, we introduce the Green's function $G_{\mu\nu}$ according to

$$F_{\mu}(x) = \int dx' \sqrt{-g(x')}G_{\mu}{}^{\nu}(x,x')J_{\nu}(x'). \tag{8.74}$$

The equations of motion (8.67) imply that $G_{\mu\nu}$ satisfies the equation

$$e_{\mu}{}^{\nu\lambda}\partial_{\nu}G_{\lambda}{}^{\alpha} + \mu G_{\mu}{}^{\alpha} = -\frac{1}{\sqrt{-g}}g_{\mu}{}^{\alpha}\delta(x-x'), \tag{8.75}$$

where $e_{\mu}{}^{\nu\lambda} = g_{\mu\beta}\epsilon^{\beta\nu\lambda}/\sqrt{-g}$. Eqs. (8.65), (8.73), and (8.74) can be used to show that

$$G_{\mu\nu}(x,x') = \frac{1}{\sqrt{-g}}\epsilon_{\mu}{}^{\alpha\beta}\partial_{\alpha}D_{\beta\nu}(x,x') \tag{8.76}$$

or, with the help of (8.15),

$$G_{\mu\nu}(x,x') = i\langle F_{\mu}(x)A_{\nu}(x')\rangle. \tag{8.77}$$

The vacuum expectation value of the stress tensor can now be put in terms of the Green's function by using

$$i\langle F^{\mu}(x)F^{\nu}(x')\rangle = \frac{i}{\sqrt{-g'}}\epsilon^{\nu\alpha\beta}\partial_{\alpha}'\langle F^{\mu}(x)A_{\beta}(x')\rangle$$

$$= \frac{1}{\sqrt{-g'}} \epsilon^{\nu\alpha\beta} \partial'_\alpha G^\mu{}_\beta(x, x'), \tag{8.78}$$

where $g' = g(x')$. We then have

$$\langle T^{\alpha\beta} \rangle = \lim_{x' \to x} \frac{1}{i} \left(\epsilon^{\beta\gamma\sigma} \partial'_\gamma G^\alpha{}_\sigma - \frac{1}{2} g^{\alpha\beta} g_{\mu\nu} \epsilon^{\nu\gamma\sigma} \partial'_\gamma G^\mu{}_\sigma \right), \tag{8.79}$$

where the limit $x' \to x$ is to be taken symmetrically. For the problem at hand we use polar coordinates,

$$x^\mu = (t, r, \theta), \tag{8.80a}$$

so that the metric is given by

$$g_{\mu\nu} = (-1, 1, r^2), \quad \sqrt{-g} = r. \tag{8.80b}$$

Then the $\langle T^{11} \rangle$ component is given by

$$\langle T^{11} \rangle = \langle T^{rr} \rangle = \frac{1}{2i} \left(\frac{\partial}{\partial \theta'} G^1{}_0 - \frac{\partial}{\partial t'} G^1{}_2 \right) + \frac{1}{2i} \left(\frac{\partial}{\partial r'} G^0{}_2 - \frac{\partial}{\partial \theta'} G^0{}_1 \right)$$
$$- \frac{r^2}{2i} \left(\frac{\partial}{\partial t'} G^2{}_1 - \frac{\partial}{\partial r'} G^2{}_0 \right). \tag{8.81}$$

We further introduce the Fourier transform appropriate to the polar coordinates:

$$G_\mu{}^\nu(x, x') = \int \frac{d\omega}{2\pi} e^{-i\omega(t-t')} \sum_{m=-\infty}^\infty e^{im(\theta-\theta')} \mathcal{G}_\mu{}^\nu(r, r'), \tag{8.82}$$

where we have suppressed the dependence of the reduced Green's function $\mathcal{G}$ on m and ω. The Fourier transform of (8.81) is

$$\langle t^{11} \rangle = -\frac{m}{2}(\mathcal{G}^1{}_0 - \mathcal{G}^0{}_1) - \frac{\omega}{2}(\mathcal{G}^1{}_2 + r^2 \mathcal{G}^2{}_1) + \frac{1}{2i} \left(\frac{\partial}{\partial r'} \mathcal{G}^0{}_2 + r^2 \frac{\partial}{\partial r'} \mathcal{G}^2{}_0 \right)$$
$$= \frac{m}{2}(\mathcal{G}_1{}^0 - \mathcal{G}_0{}^1) - \frac{\omega}{2}(r^2 \mathcal{G}_1{}^2 + \mathcal{G}_2{}^1) - \frac{1}{2i} \left(\frac{\partial}{\partial r'} r'^2 \mathcal{G}_0{}^2 + \frac{\partial}{\partial r'} \mathcal{G}_2{}^0 \right), \tag{8.83}$$

where the limit $r' \to r$ is understood, and we have suppressed all the (obvious) arguments. Henceforth, unless stated otherwise, by $\mathcal{G}$ we mean $\mathcal{G}(r, r')$.

We now must solve the Green's function equation (8.75) for the various components which appear in (8.83). The corresponding equations for the reduced Green's functions fall into three groups. The first involves the $_0{}^0$, $_1{}^0$, and $_2{}^0$ components:

$$-\frac{1}{r}\frac{\partial}{\partial r}\mathcal{G}_2{}^0 + \frac{im}{r}\mathcal{G}_1{}^0 + \mu\mathcal{G}_0{}^0 = -\frac{1}{2\pi r}\delta(r - r'), \qquad (8.84\text{a})$$

$$\frac{im}{r}\mathcal{G}_0{}^0 + \frac{i\omega}{r}\mathcal{G}_2{}^0 + \mu\mathcal{G}_1{}^0 = 0, \qquad (8.84\text{b})$$

$$-i\omega r\mathcal{G}_1{}^0 - r\frac{\partial}{\partial r}\mathcal{G}_0{}^0 + \mu\mathcal{G}_2{}^0 = 0. \qquad (8.84\text{c})$$

We combine these equations to find the second-order equation satisfied by $\mathcal{G}_0{}^0$:

$$\left(\frac{\partial^2}{\partial r^2} + \frac{1}{r}\frac{\partial}{\partial r} - \frac{m^2}{r^2} + \omega^2 - \mu^2\right)\mathcal{G}_0{}^0 = -\frac{\omega^2 - \mu^2}{2\pi\mu r}\delta(r - r'). \qquad (8.85\text{a})$$

From $\mathcal{G}_0{}^0$ we can determine the two other Green's functions according to

$$\mathcal{G}_2{}^0 = -\frac{m\omega + \mu r\frac{\partial}{\partial r}}{\omega^2 - \mu^2}\mathcal{G}_0{}^0, \qquad (8.85\text{b})$$

$$\mathcal{G}_1{}^0 = -\frac{im}{\mu r}\mathcal{G}_0{}^0 - \frac{i\omega}{\mu r}\mathcal{G}_2{}^0. \qquad (8.85\text{c})$$

Similarly, the $_0{}^1$, $_1{}^1$, and $_2{}^1$ components of (8.75) are

$$-\frac{1}{r}\frac{\partial}{\partial r}\mathcal{G}_2{}^1 + \frac{im}{r}\mathcal{G}_1{}^1 + \mu\mathcal{G}_0{}^1 = 0, \qquad (8.86\text{a})$$

$$\frac{im}{r}\mathcal{G}_0{}^1 + \frac{i\omega}{r}\mathcal{G}_2{}^1 + \mu\mathcal{G}_1{}^1 = -\frac{1}{2\pi r}\delta(r - r'), \qquad (8.86\text{b})$$

$$-i\omega r\mathcal{G}_1{}^1 - r\frac{\partial}{\partial r}\mathcal{G}_0{}^1 + \mu\mathcal{G}_2{}^1 = 0, \qquad (8.86\text{c})$$

which can be combined to yield

$$\left(\frac{\partial^2}{\partial r^2} + \frac{1}{r}\frac{\partial}{\partial r} - \frac{m^2}{r^2} + \omega^2 - \mu^2\right)\mathcal{G}_0{}^1 = -\frac{i}{2\pi\mu r}\left(\frac{m\mu}{r} - \omega\frac{\partial}{\partial r}\right)\delta(r - r'), \qquad (8.87\text{a})$$

$$\mathcal{G}_1{}^1 = \frac{1}{\omega^2 - \mu^2}\left(\frac{im\mu}{r} + i\omega\frac{\partial}{\partial r}\right)\mathcal{G}_0{}^1 - \frac{\mu}{2\pi r}\frac{\delta(r - r')}{\mu^2 - \omega^2}, \qquad (8.87\text{b})$$

$$\mathcal{G}_2{}^1 = -\frac{m}{\omega}\mathcal{G}_0{}^1 + \frac{i\mu r}{\omega}\mathcal{G}_1{}^1 + \frac{i}{2\pi\omega}\delta(r - r'). \qquad (8.87\text{c})$$

Henceforth, we will ignore the δ functions in (8.87b) and (8.87c) because we are interested in the *limit* $r \to r'$.

Finally, the $_0{}^2$, $_1{}^2$, and $_2{}^2$ components satisfy

$$-\frac{1}{r}\frac{\partial}{\partial r}\mathcal{G}_2{}^2 + \frac{im}{r}\mathcal{G}_1{}^2 + \mu\mathcal{G}_0{}^2 = 0, \tag{8.88a}$$

$$\frac{im}{r}\mathcal{G}_0{}^2 + \frac{i\omega}{r}\mathcal{G}_2{}^2 + \mu\mathcal{G}_1{}^2 = 0, \tag{8.88b}$$

$$-i\omega r \mathcal{G}_1{}^2 - r\frac{\partial}{\partial r}\mathcal{G}_0{}^2 + \mu\mathcal{G}_2{}^2 = -\frac{1}{2\pi r}\delta(r - r'), \tag{8.88c}$$

which can be combined to yield

$$\left(\frac{\partial^2}{\partial r^2} + \frac{1}{r}\frac{\partial}{\partial r} - \frac{m^2}{r^2} + \omega^2 - \mu^2\right)\mathcal{G}_0{}^2 = -\left(\frac{m\omega}{\mu r^2} - \frac{1}{r}\frac{\partial}{\partial r}\right)\frac{1}{2\pi r}\delta(r - r'), \tag{8.89a}$$

$$\mathcal{G}_2{}^2 = -\frac{1}{\omega^2 - \mu^2}\left(m\omega + \mu r\frac{\partial}{\partial r}\right)\mathcal{G}_0{}^2 + \frac{\mu\delta(r - r')}{2\pi r(\omega^2 - \mu^2)}, \tag{8.89b}$$

$$\mathcal{G}_1{}^2 = -\frac{im}{\mu r}\mathcal{G}_0{}^2 - \frac{i\omega}{\mu r}\mathcal{G}_2{}^2. \tag{8.89c}$$

Again we ignore the δ function in (8.89b) in what follows.

We solve these equations for the reduced Green's functions subject to perfect conductor boundary conditions at $r = a$. That is, the tangential electric field must vanish on the circle, or in terms of the dual field,

$$F_1 = F_r = 0 \quad \text{at} \quad r = a. \tag{8.90}$$

It is interesting to note that this is precisely the condition necessary to ensure the gauge invariance of the Lagrangian (8.66). That is, the mass term

$$\frac{1}{2}\mu\int dx\,\sqrt{-g}F^\lambda A_\lambda \tag{8.91}$$

is gauge invariant only if we neglect the surface term [see (8.68)]

$$\frac{1}{2}\mu\int dx\,\sqrt{-g}F^\lambda\partial_\lambda\Lambda = \frac{1}{2}\mu\int dS_\lambda\sqrt{-g}F^\lambda\Lambda = 0, \tag{8.92}$$

which is true if the normal component of F^λ vanishes on the bounding surfaces.

We begin by solving the system of equations (8.85a)–(8.85c). The solution to (8.85a) is

$$\mathcal{G}_0{}^0 = -\frac{\lambda^2}{4i\mu} J_m(\lambda r_<) H_m(\lambda r_>) + A J_m(\lambda r) + B H_m(\lambda r), \tag{8.93}$$

where $\lambda^2 = \omega^2 - \mu^2$. The constants A and B are to be determined by the boundary condition (8.90). When we insert (8.93) into (8.85b) we find

$$\mathcal{G}_2{}^0 = \frac{m\omega}{4i\mu}(\widetilde{HJ})(r,r') - \frac{m\omega}{\lambda^2}[A\tilde{J}_m(\lambda r) + B\tilde{H}_m(\lambda r)], \tag{8.94}$$

where we have introduced the abbreviations

$$\tilde{J}_m(x) = J_m(x) + \frac{\mu x}{m\omega} J'_m(x), \tag{8.95a}$$

$$\tilde{H}_m(x) = H_m(x) + \frac{\mu x}{m\omega} H'_m(x), \tag{8.95b}$$

and

$$(\widetilde{HJ})(r,r') = \begin{cases} H_m(\lambda r')\tilde{J}_m(\lambda r), & r < r', \\ J_m(\lambda r')\tilde{H}_m(\lambda r), & r > r'. \end{cases} \tag{8.96}$$

When (8.93) and (8.94) are inserted into (8.85c) we obtain

$$\mathcal{G}_1{}^0 = \frac{m\lambda^2}{4\mu^2 r}[J_m(\lambda r_<)H_m(\lambda r_>) + \tilde{A}J_m(\lambda r) + \tilde{B}H_m(\lambda r)]$$

$$+ \frac{\omega^2 m}{4\mu^2 r}[-(\widetilde{HJ})(r,r') - \tilde{A}\tilde{J}_m(\lambda r) - \tilde{B}\tilde{H}_m(\lambda r)]$$

$$= -\frac{m}{4r}[(HJ)(r,r') + \tilde{A}\mathcal{J}_m(\lambda r) + \tilde{B}\mathcal{H}_m(\lambda r)], \tag{8.97}$$

where we have rescaled the constants,

$$\tilde{A} = -\frac{4i\mu}{\lambda^2}A, \quad \tilde{B} = -\frac{4i\mu}{\lambda^2}B, \tag{8.98}$$

and have defined

$$\mathcal{J}_m(x) = J_m(x) + \frac{\omega x}{m\mu} J'_m(x) \tag{8.99a}$$

$$= \frac{\omega^2}{\mu^2}\tilde{J}_m(x) - \frac{\lambda^2}{\mu^2}J_m(x), \tag{8.99b}$$

$$\mathcal{H}_m(x) = H_m(x) + \frac{\omega x}{m\mu} H'_m(x) \tag{8.99c}$$

$$= \frac{\omega^2}{\mu^2}\tilde{\mathcal{H}}_m(x) - \frac{\lambda^2}{\mu^2}H_m(x), \tag{8.99d}$$

$$(HJ)(r,r') = \begin{cases} H_m(\lambda r')J_m(\lambda r), & r < r', \\ J_m(\lambda r')H_m(\lambda r), & r > r'. \end{cases} \tag{8.100}$$

Now we are in a position to impose the boundary condition (8.90) on $\mathcal{G}_1{}^0$. First, we consider points inside the circle, $r, r' < a$. From (8.97) we see that $B = 0$ in order that the solution be finite at the origin. Then the boundary condition $\mathcal{G}_1{}^0 = 0$ at $r = a$ implies from (8.97) that

$$\tilde{A} = -\frac{\mathcal{H}_m(\lambda a)}{\mathcal{J}_m(\lambda a)}J_m(\lambda r'), \tag{8.101}$$

from which we deduce the explicit form for these components, for $r, r' < a$:

$$\mathcal{G}_0{}^0 = -\frac{\lambda^2}{4i\mu}\left[J_m(\lambda r_<)H_m(\lambda r_>) - \frac{\mathcal{H}_m(\lambda a)}{\mathcal{J}_m(\lambda a)}J_m(\lambda r)J_m(\lambda r')\right], \tag{8.102a}$$

$$\mathcal{G}_2{}^0 = -\frac{m\omega}{4i\mu}\left[-(\widetilde{HJ})(r,r') + \frac{\mathcal{H}_m(\lambda a)}{\mathcal{J}_m(\lambda a)}\tilde{J}_m(\lambda r)J_m(\lambda r')\right], \tag{8.102b}$$

$$\mathcal{G}_1{}^0 = -\frac{m}{4r}\left[(HJ)(r,r') - \frac{\mathcal{H}_m(\lambda a)}{\mathcal{J}_m(\lambda a)}J_m(\lambda r)J_m(\lambda r')\right]. \tag{8.102c}$$

Outside the circle, $r, r' > a$, we must have *outgoing* circular waves, so $A = 0$, and the boundary condition $\mathcal{G}_1{}^0 = 0$ at $r = a$ implies from (8.97) that

$$\tilde{B} = -\frac{\mathcal{J}_m(\lambda a)}{\mathcal{H}_m(\lambda a)}H_m(\lambda r'), \tag{8.103}$$

from which we deduce the explicit form for these components, for $r, r' > a$:

$$\mathcal{G}_0{}^0 = -\frac{\lambda^2}{4i\mu}\left[J_m(\lambda r_<)H_m(\lambda r_>) - \frac{\mathcal{J}_m(\lambda a)}{\mathcal{H}_m(\lambda a)}H_m(\lambda r)H_m(\lambda r')\right], \tag{8.104a}$$

$$\mathcal{G}_2{}^0 = -\frac{m\omega}{4i\mu}\left[-(\widetilde{HJ})(r,r') + \frac{\mathcal{J}_m(\lambda a)}{\mathcal{H}_m(\lambda a)}\tilde{\mathcal{H}}_m(\lambda r)H_m(\lambda r')\right], \tag{8.104b}$$

$$\mathcal{G}_1{}^0 = -\frac{m}{4r}\left[(HJ)(r,r') - \frac{\mathcal{J}_m(\lambda a)}{\mathcal{H}_m(\lambda a)}\mathcal{H}_m(\lambda r)H_m(\lambda r')\right]. \tag{8.104c}$$

Next, we solve the system (8.87a), (8.87b), (8.87c). It is slightly harder to solve (8.87a). We write

$$\mathcal{G}_0{}^1 = A_\pm J_m(\lambda r) + B_\pm H_m(\lambda r), \tag{8.105}$$

where the upper (lower) sign holds if $r > r'$ ($r < r'$). Equation (8.87a) implies that the derivative of $\mathcal{G}_0{}^1$ is discontinuous at $r = r'$:

$$-\frac{im}{2\pi r'} = (A_+ - A_-)\lambda r' J'_m(\lambda r') + (B_+ - B_-)\lambda r' H'_m(\lambda r'), \tag{8.106a}$$

while the function itself is discontinuous:

$$\frac{i\omega}{2\pi\mu r'} = (A_+ - A_-)J_m(\lambda r') + (B_+ - B_-)H_m(\lambda r'). \tag{8.106b}$$

These equations are solved by

$$A_+ - A_- = \frac{m}{4r'}H_m(\lambda r'), \tag{8.107a}$$

$$B_+ - B_- = -\frac{m}{4r'}J_m(\lambda r'). \tag{8.107b}$$

Inside the circle, $r, r' < a$, we have $B_- = 0$ and the boundary condition given through (8.87b) implies

$$B_+ H_m(\lambda a) + A_+ J_m(\lambda a) = 0. \tag{8.108}$$

Solving these equations gives, for $r, r' < a$:

$$\mathcal{G}_0{}^1 = \frac{m}{4r'}\left[-(JH)(r,r') + \frac{H_m(\lambda a)}{J_m(\lambda a)}J_m(\lambda r)J_m(\lambda r')\right], \tag{8.109a}$$

$$\mathcal{G}_1{}^1 = \frac{im^2\mu}{4\lambda^2 rr'}\left[-J_m(\lambda r_<)H_m(\lambda r_>) + \frac{H_m(\lambda a)}{J_m(\lambda a)}J_m(\lambda r)J_m(\lambda r')\right], \tag{8.109b}$$

$$\mathcal{G}_2{}^1 = -\frac{m^2\omega}{4\lambda^2 r'}\left[-[\widetilde{HJ}](r,r') + \frac{H_m(\lambda a)}{J_m(\lambda a)}\tilde{J}_m(\lambda r)J_m(\lambda r')\right]. \tag{8.109c}$$

Here we have introduced the abbreviations

$$(JH)(r,r') = \begin{cases} J_m(\lambda r)H_m(\lambda r'), & r < r' \\ H_m(\lambda r)J_m(\lambda r'), & r > r' \end{cases} = (HJ)(r',r) \tag{8.110}$$

and

$$[\widetilde{HJ}](r,r') = \begin{cases} H_m(\lambda r')\tilde{J}_m(\lambda r), & r < r', \\ J_m(\lambda r')\tilde{H}_m(\lambda r), & r > r'. \end{cases} \tag{8.111}$$

Outside the circle, $r, r' > a$, $A_+ = 0$ and (8.87b) implies

$$B_- \mathcal{H}_m(\lambda a) + A_- \mathcal{J}_m(\lambda a) = 0. \tag{8.112}$$

So now the solution to the system (8.87a)–(8.87c) is, for $r, r' > a$

$$\mathcal{G}_0{}^1 = \frac{m}{4r'} \left[-(JH)(r, r') + \frac{\mathcal{J}_m(\lambda a)}{\mathcal{H}_m(\lambda a)} \mathcal{H}_m(\lambda r)\mathcal{H}_m(\lambda r') \right], \tag{8.113a}$$

$$\mathcal{G}_1{}^1 = \frac{im^2\mu}{4\lambda^2 rr'} \left[-\mathcal{J}_m(\lambda r_<)\mathcal{H}_m(\lambda r_>) + \frac{\mathcal{J}_m(\lambda a)}{\mathcal{H}_m(\lambda a)} \mathcal{H}_m(\lambda r)\mathcal{H}_m(\lambda r') \right], \tag{8.113b}$$

$$\mathcal{G}_2{}^1 = -\frac{m^2\omega}{4\lambda^2 r'} \left[-[\widehat{HJ}](r, r') + \frac{\mathcal{J}_m(\lambda a)}{\mathcal{H}_m(\lambda a)} \tilde{\mathcal{H}}_m(\lambda r)\mathcal{H}_m(\lambda r') \right]. \tag{8.113c}$$

The system (8.89a), (8.89b), (8.89c) is solved in just the same way. The result is, inside the circle $(r, r' < a)$:

$$\mathcal{G}_0{}^2 = -\frac{im\omega}{4\mu r'^2} \left[-(\widetilde{JH})(r, r') + \frac{\mathcal{H}_m(\lambda a)}{\mathcal{J}_m(\lambda a)} \mathcal{J}_m(\lambda r)\tilde{\mathcal{J}}_m(\lambda r') \right], \tag{8.114a}$$

$$\mathcal{G}_2{}^2 = \frac{im^2\omega^2}{4\mu\lambda^2 r'^2} \left[-\tilde{\mathcal{J}}_m(\lambda r_<)\tilde{\mathcal{H}}_m(\lambda r_>) + \frac{\mathcal{H}_m(\lambda a)}{\mathcal{J}_m(\lambda a)} \tilde{\mathcal{J}}_m(\lambda r)\tilde{\mathcal{J}}_m(\lambda r') \right], \tag{8.114b}$$

$$\mathcal{G}_1{}^2 = \frac{m^2\omega}{4\lambda^2 rr'^2} \left[-[\widetilde{JH}](r, r') + \frac{\mathcal{H}_m(\lambda a)}{\mathcal{J}_m(\lambda a)} \mathcal{J}_m(\lambda r)\tilde{\mathcal{J}}_m(\lambda r') \right], \tag{8.114c}$$

and outside the circle $(r, r' > a)$:

$$\mathcal{G}_0{}^2 = -\frac{im\omega}{4\mu r'^2} \left[-(\widetilde{JH})(r, r') + \frac{\mathcal{J}_m(\lambda a)}{\mathcal{H}_m(\lambda a)} \mathcal{H}_m(\lambda r)\tilde{\mathcal{H}}_m(\lambda r') \right], \tag{8.115a}$$

$$\mathcal{G}_2{}^2 = \frac{im^2\omega^2}{4\mu\lambda^2 r'^2} \left[-\tilde{\mathcal{J}}_m(\lambda r_<)\tilde{\mathcal{H}}_m(\lambda r_>) + \frac{\mathcal{J}_m(\lambda a)}{\mathcal{H}_m(\lambda a)} \tilde{\mathcal{H}}_m(\lambda r)\tilde{\mathcal{H}}_m(\lambda r') \right], \tag{8.115b}$$

$$\mathcal{G}_1{}^2 = \frac{m^2\omega}{4\lambda^2 rr'^2} \left[-[\widetilde{JH}](r, r') + \frac{\mathcal{J}_m(\lambda a)}{\mathcal{H}_m(\lambda a)} \mathcal{H}_m(\lambda r)\tilde{\mathcal{H}}_m(\lambda r') \right], \tag{8.115c}$$

where

$$(\widetilde{JH})(r, r') = \begin{cases} \tilde{\mathcal{H}}_m(\lambda r')J_m(\lambda r), & r < r' \\ \tilde{\mathcal{J}}_m(\lambda r')H_m(\lambda r), & r > r' \end{cases} = (\widehat{HJ})(r', r) \tag{8.116}$$

and

$$[\widetilde{JH}](r,r') = \begin{cases} \tilde{\mathcal{H}}_m(\lambda r')\mathcal{J}_m(\lambda r), & r < r' \\ \tilde{\mathcal{J}}_m(\lambda r')\mathcal{H}_m(\lambda r), & r > r' \end{cases} = [\widetilde{HJ}](r',r). \qquad (8.117)$$

Note that there are only six independent Green's functions because of the following symmetry relations between them:

$$r'^2 \mathcal{G}_1{}^2(r,r') = -\mathcal{G}_2{}^1(r',r), \qquad (8.118\text{a})$$

$$r'^2 \mathcal{G}_0{}^2(r,r') = -\mathcal{G}_2{}^0(r',r), \qquad (8.118\text{b})$$

$$\mathcal{G}_0{}^1(r,r') = \mathcal{G}_1{}^0(r',r), \qquad (8.118\text{c})$$

$$\mathcal{G}_0{}^0(r,r') = \mathcal{G}_0{}^0(r',r), \qquad (8.118\text{d})$$

$$\mathcal{G}_1{}^1(r,r') = \mathcal{G}_1{}^1(r',r), \qquad (8.118\text{e})$$

$$r'^2 \mathcal{G}_2{}^2(r,r') = r^2 \mathcal{G}_2{}^2(r',r). \qquad (8.118\text{f})$$

Using the above symmetry relations we can write the expression for the vacuum expectation value of the rr component of the stress tensor (8.83) as (recall that the limit $r' \to r$ is understood)

$$\langle t^{11} \rangle = -\frac{1}{2i}\left(\frac{\partial}{\partial r'}r'^2 \mathcal{G}_0{}^2 + \frac{\partial}{\partial r'}\mathcal{G}_2{}^0\right), \qquad (8.119)$$

for a given m and ω. What we require, in fact, is the discontinuity across the circumference of the circle:

$$\Delta\langle t^{11} \rangle = \langle t^{11} \rangle\big|_{r=r'=a-} - \langle t^{11} \rangle\big|_{r=r'=a+} \qquad (8.120)$$

From (8.114a), (8.115a), (8.102b), and (8.104b) we find for this discontinuity

$$\Delta\langle t^{11} \rangle = -\frac{\lambda^2 a}{8}\left\{\frac{\mathcal{H}_m(\lambda a)}{\mathcal{J}_m(\lambda a)}\left[\left(1 - \frac{m^2}{(\lambda a)^2}\right)(J_m(\lambda a))^2 + (J'_m(\lambda a))^2\right]\right.$$
$$\left. - \frac{\mathcal{J}_m(\lambda a)}{\mathcal{H}_m(\lambda a)}\left[\left(1 - \frac{m^2}{(\lambda a)^2}\right)(H_m(\lambda a))^2 + (H'_m(\lambda a))^2\right]\right\}.$$
$$(8.121)$$

Here, we have used (8.95a) and (8.95b) as well as the Bessel equation

$$(zJ'_m(z))' = -z\left(1 - \frac{m^2}{z^2}\right)J_m(z). \qquad (8.122)$$

Equation (8.121), when integrated over the frequency ω and summed over m, is our general analytic expression for the Casimir stress on a conducting circle:

$$S = 2\pi \sum_{m=-\infty}^{\infty} \int_{-\infty}^{\infty} \frac{d\omega}{2\pi} \Delta \langle t^{11} \rangle \qquad (8.123)$$

is the total stress on the circle.[‡] [Recall T^{11} is the stress tensor density, so a factor of $a = \sqrt{-g}$ on the surface is already absorbed.]

8.3.2 Numerical Results at Zero Temperature

We now turn to the task of extracting a numerical result from the expression (8.123) and (8.121). To do so, it is convenient as usual to rotate the contour of frequency integration, $\omega \to i\zeta$, define the dimensionless real variable x by $x^2 = \zeta^2 a^2 + \mu^2 a^2$, and introduce the modified Bessel functions (7.34a), (7.34b). [For details of the contour rotation see Sec. 4.1.] When we explicitly symmetrize between positive and negative values of x (or m), we find the following result:

$$S = -\frac{1}{2\pi a^2} \frac{1}{\mu^2 a^2} \sum_{m=-\infty}^{\infty} \frac{1}{m^2} \int_{\mu a}^{\infty} \frac{dx\, x^4}{\sqrt{x^2 - \mu^2 a^2}} (x^2 - \mu^2 a^2 + m^2)$$

$$\times \frac{[I_m(x) K'_m(x) + K_m(x) I'_m(x)]}{[I_m^2(x) + x^2(x^2 - \mu^2 a^2) I_m'^2(x)/m^2 \mu^2 a^2]}$$

$$\times \frac{[K_m(x) I_m(x) + x^2(x^2 - \mu^2 a^2) K'_m(x) I'_m(x)/m^2 \mu^2 a^2]}{[K_m^2(x) + x^2(x^2 - \mu^2 a^2) K_m'^2(x)/m^2 \mu^2 a^2]}. \qquad (8.124)$$

In the massless limit, $\mu a \to 0$, (8.124) simplifies dramatically:

$$S = -\frac{1}{2\pi a^2} \sum_{m=-\infty}^{\infty} \int_0^{\infty} dx\, x \frac{d}{dx} \ln[x^2 I'_m(x) K'_m(x)]. \qquad (8.125)$$

We consider the $m = 0$ and the $m \neq 0$ terms in (8.125) separately. As usual in Casimir calculations, we ignore terms in the integrand of (8.125) which are powers of x (contact terms). (Note that a power of x corresponds to derivatives of δ functions in time.) In particular, for the $m = 0$ term,

---

[‡]We should also be able to derive the result (8.123), (8.121) from the vacuum energy. It is easy to do so for $\mu = 0$, but much more elaborate for $\mu \neq 0$, so we forgo further discussion of this point.

we add a contact term to the integrand to make it converge, so that the corresponding contribution to the stress becomes

$$S_0 = -\frac{1}{2\pi a^2} \int_0^\infty dx\, x \frac{d}{dx} \ln[2x I_1(x) K_1(x)]. \qquad (8.126)$$

Straightforward numerical integration gives the attractive result

$$S_0 = -\frac{1}{2\pi a^2} 1.5929 = -\frac{0.25352}{a^2}. \qquad (8.127)$$

For $m \neq 0$ we make use of the uniform asymptotic expansions for the modified Bessel functions (4.29a) and (4.29b), see Ref. [111]. The leading term gives

$$I'_m(x) K'_m(x) \sim -\frac{1}{2m}\frac{1}{z^2}(1 + z^2)^{1/2}, \qquad (8.128)$$

where $z = x/m$. The corresponding contribution to the stress is

$$S_{\text{LT}} \sim -\frac{1}{2\pi a^2} \int_0^\infty dx\, x^2\, 2 \sum_{m=1}^\infty \frac{1}{m^2 + x^2}, \qquad (8.129)$$

where the sum is performed according to (7.26), or

$$2 \sum_{m=1}^\infty \frac{1}{m^2 + x^2} = \frac{\pi}{x} \coth \pi x - \frac{1}{x^2}. \qquad (8.130)$$

Again, we supply appropriate contact terms, so that this leading $m \neq 0$ contribution is

$$S_{\text{LT}} = -\frac{1}{4\pi^2 a^2} \int_0^\infty \frac{dy\, y}{e^y - 1} = -\frac{1}{24a^2}, \qquad (8.131)$$

only 16% of (8.127). We should now correct (8.131) by including the next-to-leading corrections. However, it is not hard to see that these possess an infrared divergence, a phenomenon which is associated with the low dimensionality of the problem. One might think that this divergence is probably spurious: Each integrand in (8.125) is quite accurately represented by leading term given in (8.128). [Even at $m = 1$, the maximum value of $\ln(-2x^2 I'_m(x) K'_m(x)(m^2 + x^2)^{-1/2})$ is less than 7% of the value of $\ln(-x^2 I'_m(x) K'_m(x))$, and globally the fit is excellent.] This divergence, of course, is regulated by the mass μ, so we will discuss this point further below.

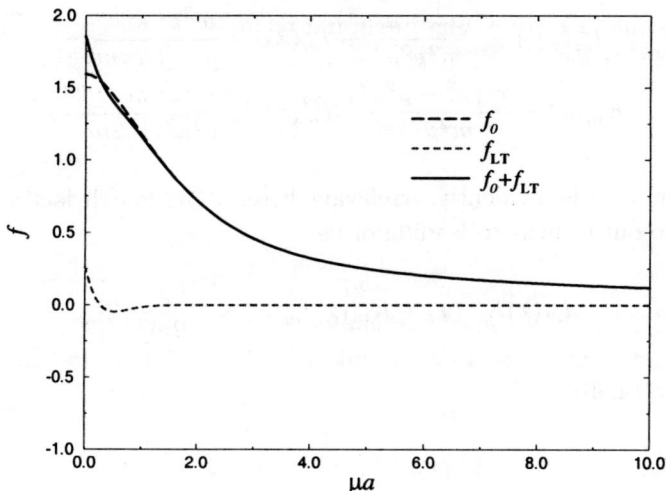

Fig. 8.1 The contribution of $m = 0$ to the Casimir stress at zero temperature, S_0, given in (8.132). Further graphed is the leading uniform asymptotic approximation to the $m \neq 0$ contributions to the Casimir stress at zero temperature, S_{LT}, given by (8.136). The sum of these contributions is also shown. In each case what is plotted is $f = -2\pi a^2 S$ as a function of μa.

When $\mu \neq 0$, the calculation proceeds similarly. We first treat the $m = 0$ term, which is easily seen from (8.124) to be the obvious generalization of (8.126):

$$S_0 = -\frac{1}{2\pi a^2} \int_{\mu a}^{\infty} dx \, \frac{x^2}{\sqrt{x^2 - \mu^2 a^2}} \frac{d}{dx} \ln[2xI_1(x)K_1(x)]. \qquad (8.132)$$

The results of numerical integration of (8.132) are shown in Fig. 8.1. This contribution to the stress decreases rapidly from the massless value (8.127) to zero as $\mu a \to \infty$. For $m \neq 0$ we use the uniform asymptotic expansion for the modified Bessel functions. Doing so with the general expression (8.124) requires only the leading terms for three of the factors there,

$$I_m(x)K_m(x) + \frac{x^2(x^2 - \mu^2 a^2)}{m^2\mu^2 a^2} I'_m(x)K'_m(x) \sim -\frac{mz^2}{2\mu^2 a^2 t}, \qquad (8.133a)$$

$$I_m^2(x) + \frac{x^2(x^2 - \mu^2 a^2)}{m^2 \mu^2 a^2} I_m'^2(x) \sim \frac{m^2 z^2}{\mu^2 a^2 t} \frac{e^{2m\eta}}{2\pi m}, \tag{8.133b}$$

$$K_m^2(x) + \frac{x^2(x^2 - \mu^2 a^2)}{m^2 \mu^2 a^2} K_m'^2(x) \sim \frac{m^2 z^2}{\mu^2 a^2 t} \frac{\pi e^{-2m\eta}}{2m}. \tag{8.133c}$$

(The value of η is, evidently, irrelevant here.) The fourth factor requires that we go out to next-to-leading order:

$$I_m(x) K_m'(x) + K_m(x) I_m'(x) \sim -\frac{z}{2m^2} t^3. \tag{8.134}$$

Here, as in (4.30)

$$t = (1 + z^2)^{-1/2}, \quad z = \frac{x}{m}. \tag{8.135}$$

We substitute these asymptotic expressions into (8.124) and carry out the sum on m using (8.130), again omitting contact terms. The result is

$$\mathcal{S}_{\mathrm{LT}} \doteq -\frac{1}{2a^2} \frac{1}{(2\pi)^2} \int_{2\pi\mu a}^{\infty} dy \frac{y^2}{\sqrt{y^2 - (2\pi\mu a)^2}} \frac{1}{e^y - 1}$$

$$\times \left\{ 2 - \frac{(2\pi\mu a)^2}{y} \left[\frac{1}{y} + 1 + \frac{1}{e^y - 1} \right] \right\}. \tag{8.136}$$

This result, of course, generalizes the formula in (8.131). Numerical integration of (8.136) yields the contribution to the force also shown in Fig. 8.1. This partial result is extremely interesting, because of the sign change, from attractive to repulsive at about $\mu a = 0.27$. However, the $m = 0$ term given in (8.132) and Fig. 8.1 is much larger, so that these terms together always give an attractive force. The sum of these two terms is also plotted in Fig. 8.1.

We have, of course, worked out the next-to-leading contributions to the force. As noted above, these are finite when $\mu \neq 0$, but are larger than the leading term given by (8.136) and Fig. 8.1 even for large μa. This is a signal that the Casimir effect for a circle is not finite. It now appears from the analysis presented in the following Chapter that the two-dimensional Casimir effect is inherently divergent. We will discuss possible resolutions of this difficulty there.

8.3.3 *High-Temperature Limit*

It is easy, in principle, to see how to extract the finite-temperature Casimir effect. We take the general zero-temperature result (8.124) and replace the continuous frequency variable ζ by the discrete variable $2\pi n/\beta$, where $\beta = 1/kT$ and n is an integer. That is

$$\int_{\mu a}^{\infty} dx\, x f(x) \rightarrow \frac{4\pi^2 a^2}{\beta^2} \sum_{n=0}^{\infty}{}' n f(x_n), \qquad (8.137)$$

$$x_n = \left[\left(\frac{2\pi an}{\beta}\right)^2 + (\mu a)^2\right]^{1/2}, \qquad (8.138)$$

where the prime on the summation sign means that $n = 0$ is counted with half weight.

Given the complicated form of the integrand in (8.124), it is hard to work out the general temperature dependence in this case. The high-temperature limit, however, would seem to be tractable. That is because we anticipate that only the $n = 0$ term contributes when $\beta \rightarrow 0$. Indeed, if $n \neq 0$, $x_n \rightarrow 2\pi an/\beta \rightarrow \infty$, and the corresponding contribution to S coming from (8.124) is

$$S_{n\neq0}^{T\rightarrow\infty} = -\frac{1}{a\beta} \sum_{m=-\infty}^{\infty} \sum_{n=1}^{\infty} x_n \left[\frac{I_m(x_n)}{I'_m(x_n)} + \frac{K_m(x_n)}{K'_m(x_n)}\right]$$

$$= -\frac{kT}{a} \sum_{m=-\infty}^{\infty} \sum_{n=1}^{\infty} 1 , \qquad (8.139)$$

where we have used the asymptotic behavior of the Bessel functions. This divergent contribution, in fact, should be subtracted off, for the constant summand may again be identified with a contact term.

The high-temperature limit thus arises from the $n = 0$ term. We will use the uniform asymptotic approximation employed in Sec. 8.3.2, and hence we will make the replacement (8.137) in (8.132) and (8.136). For the former, $m = 0$, term we have

$$S_0^{T\rightarrow\infty} = -\frac{1}{2a\beta} x \frac{d}{dx} \ln[2x I_1(x) K_1(x)], \quad x = \mu a, \qquad (8.140)$$

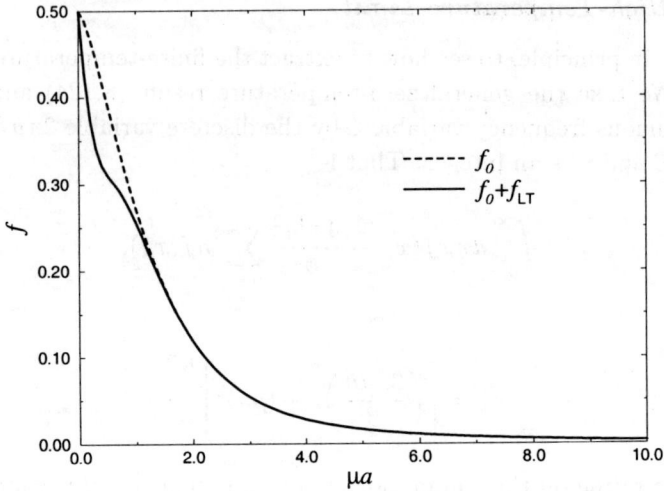

Fig. 8.2   The contribution of $m = 0$ to the Casimir force at high temperature, $\mathcal{S}_0^{T \to \infty}$, given in (8.140). Also shown is the total Casimir stress at high temperature, including the leading uniform asymptotic approximation for $m \neq 0$, (8.141). Plotted is $f^{T \to \infty} = -a\beta \mathcal{S}^{T \to \infty}$ as a function of μa.

which is plotted in Fig. 8.2. This attractive contribution equals $-1/2a\beta$ at $\mu a = 0$, and vanishes as $\mu a \to \infty$. For $m \neq 0$ we have, from (8.136),

$$\mathcal{S}_{LT}^{T \to \infty} \sim -\frac{\pi x}{a\beta} \frac{1}{e^{2\pi x} - 1} \left\{ \frac{1}{2} - \pi x - \frac{\pi x}{e^{2\pi x} - 1} \right\}, \quad x = \mu a. \quad (8.141)$$

This repulsive term vanishes both at $\mu a = 0$ and as $\mu a \to \infty$, and, like (8.136) is rather small compared to $\mathcal{S}_0^{T \to \infty}$. The combined high-temperature Casimir force $\mathcal{S}_0^{T \to \infty} + \mathcal{S}_{LT}^{T \to \infty}$ is also plotted in Fig. 8.2.

8.3.4   *Discussion*

The process of quantization automatically leads to unavoidable vacuum fluctuations. Usually, the vacuum energy of a medium is irrelevant. But the physics changes drastically (i) when a phase transition between two states of the medium can occur; (ii) when the medium is the whole universe and one couples gravity to the vacuum energy (leading to the vexing cosmological constant problem which we will discuss in Chapter 10; or (iii) when geometric boundary effects are taken into account. In this Chapter, we have studied the effect of vacuum fluctuation in the last case, i.e., the Casimir effect. The example of the Casimir effect that we have considered is particularly interesting since it is associated with topology, the non-Abelian generalization of the "photon" mass term being the Chern-Simons secondary characteristic.

In the first section we examined the Casimir effect between parallel lines due to the topologically massive photon in the $(2 + 1)$-dimensional theory of quantum electrodynamics. We found that the Casimir force is attractive and the result is the same as for a massive spin-zero field. The agreement of the respective Casimir forces is not surprising since, like the scalar field, the topologically massive spin-1 field in $(2 + 1)$ dimensions has one (polarization) degree of freedom. We anticipated that this agreement would not persist for other geometries.

In this section we have calculated the Casimir self-stress for a circle. *A priori*, it is hard to guess the sign of the self-stress in this case, since the $(3 + 1)$-dimensional analogue of a circle can be a spherical shell (for which the stress is repulsive [Chapter 4]) or a cylindrical shell (for which the stress is attractive [Chapter 7]). We have found, in fact, the Casimir stress to be divergent, although the leading approximations yield results which, at zero and at high temperatures, are attractive. We have also found that the respective Casimir stresses are not the same for the spin-1 field (8.124) and for the spin-0 field (8.160) discussed in the following section, in accordance with our expectation.

Actually, there is a way to read off the Casimir stress for the *massless* helicity-1 field for the circle from that for the cylindrical shell and the result for the massless scalar field for the circle. We see this by returning to (8.121) and taking the $\mu \to 0$ limit:

$$\Delta\langle t^{11}\rangle = \frac{i\lambda}{4\pi}\left[\frac{J_m''(z)}{J_m'(z)} + \frac{H_m''(z)}{H_m'(z)} + \frac{2}{z}\right], \qquad (8.142)$$

where $z = \lambda a \to \omega a$ and use has been made of the equation of motion and the Wronskian (7.17). We can, in fact, read this off directly from Chapter 7, if, there, we make appropriate $(2 + 1)$-dimensional restrictions. That is, we set the momentum k along the cylinder axis equal to zero, and include only H_z, E_r, and E_θ:

$$T_{rr} = \frac{1}{2}(H_z^2 - E_r^2), \tag{8.143}$$

where the discontinuities are

$$\Delta\langle H_z^2\rangle = \frac{\lambda^2}{2\pi i z}\left(\frac{H_m(z)}{H_m'(z)} + \frac{J_m(z)}{J_m'(z)}\right), \tag{8.144a}$$

$$\Delta\langle E_r^2\rangle = \frac{1}{2\pi i z}\frac{m^2\omega^2}{z^2}\left(\frac{J_m(z)}{J_m'(z)} + \frac{H_m(z)}{H_m'(z)}\right). \tag{8.144b}$$

These follow from (7.14a), (7.14b) and (7.11a), (7.11b). This agrees precisely with (8.142) when we use the Bessel equation (8.122) and recognize that here T_{rr} is a tensor density, with $\sqrt{-g} = r$.

Furthermore, in the next section we calculate the Casimir self-stress for a scalar field vanishing on a circular boundary, (8.158), which when combined with (8.142) yields the form of the $(3+1)$-dimensional result of (7.19):

$$\Delta\langle t_{rr}\rangle|_v + \Delta\langle t_{rr}\rangle|_s = \frac{i\lambda}{4\pi}\left(\frac{J_m''(z)}{J_m'(z)} + \frac{J_m'(z)}{J_m(z)} + \frac{H_m''(z)}{H_m'(z)} + \frac{H_m'(z)}{H_m(z)} + \frac{2}{z}\right). \tag{8.145}$$

Schematically, we write the $k = 0$, $\mu = 0$ correspondence found here as

$$(3+1)_v = (2+1)_v + (2+1)_s. \tag{8.146}$$

We can understand this directly from the $(3+1)_v$ equation of motion

$$\partial_\mu\sqrt{-g}F^{\mu\nu} = 0. \tag{8.147}$$

When there is no z dependence, the $\nu = 0, 1, 2$ components coincide with the $(2+1)_v$ equations of motion, while the $\nu = 3$ component can be written as

$$\left(\frac{1}{r}\frac{\partial}{\partial r}r\frac{\partial}{\partial r} + \omega^2 - \frac{m^2}{r^2}\right)A^3 = 0, \tag{8.148}$$

subject to the boundary condition $A_3 = 0$, so this is the massless scalar problem solved in the next section. And explicitly, the $(2+1)_s$ contribution to the stress tensor is

$$T_{rr} = \frac{1}{2}\langle H_\theta^2\rangle, \quad H_\theta = -\partial_r A_3, \tag{8.149}$$

where

$$\langle A_3(x)A_3(x')\rangle = \frac{1}{i}G(x,x'), \qquad (8.150)$$

in terms of the scalar Green's function, or in terms of the reduced scalar Green's function (8.152),

$$\langle t_{rr}\rangle = \frac{1}{2i}\frac{\partial}{\partial r}\frac{\partial}{\partial r'}g(r,r')\bigg|_{r=r'=a}. \qquad (8.151)$$

This is (apart from a factor of a) just the scalar result (8.157).

No such decomposition occurs when $k \neq 0$ or when $\mu \neq 0$.

8.4   Scalar Casimir Effect on a Circle

Consider a scalar field in $(2 + 1)$ dimensions with a circular boundary of radius a on which the field vanishes. We write the Green's function in Fourier-transformed form as

$$G(x,x') = \int \frac{d\omega}{2\pi}e^{-i\omega(t-t')}\sum_{m=-\infty}^{\infty}e^{im(\theta-\theta')}g(r,r'), \qquad (8.152)$$

where we have suppressed the dependence of the reduced Green's function g on m and ω. The reduced Green's function satisfies the differential equation

$$\left(\frac{\partial^2}{\partial r^2} + \frac{1}{r}\frac{\partial}{\partial r} + \omega^2 - \mu^2 - \frac{m^2}{r^2}\right)g(r,r') = -\frac{1}{2\pi r}\delta(r-r'). \qquad (8.153)$$

We solve this equation subject to the boundary condition

$$g(a,r') = 0. \qquad (8.154)$$

The solution is

$$r,r' < a: \quad g(r,r') = \frac{1}{4i}\left[\frac{H_m(\lambda a)}{J_m(\lambda a)}J_m(\lambda r)J_m(\lambda r') - J_m(\lambda r_<)H_m(\lambda r_>)\right],$$
$$(8.155a)$$

$$r,r' > a: \quad g(r,r') = \frac{1}{4i}\left[\frac{J_m(\lambda a)}{H_m(\lambda a)}H_m(\lambda r)H_m(\lambda r') - J_m(\lambda r_<)H_m(\lambda r_>)\right],$$
$$(8.155b)$$

where $\lambda^2 = \omega^2 - \mu^2$. Then we calculate T_{rr} from

$$T^{\mu\nu} = \sqrt{-g}\left[\partial^\mu\phi\partial^\nu\phi - g^{\mu\nu}\frac{1}{2}(\partial^\lambda\phi\partial_\lambda\phi + \mu^2\phi^2)\right]. \tag{8.156}$$

The vacuum expectation value of the product of fields is taken according to (2.23). Employing the boundary condition (8.154) we find for the Fourier transform for the stress tensor on the circle

$$\langle t_{rr}\rangle = \frac{a}{2i}\frac{\partial}{\partial r}\frac{\partial}{\partial r'}g(r,r')|_{r=r'=a}. \tag{8.157}$$

Using the solution (8.155a), (8.155b), and the Wronskian (7.17) we find

$$\Delta\langle t_{rr}\rangle = -\frac{\lambda}{4\pi i}\left[\frac{J'_m(\lambda a)}{J_m(\lambda a)} + \frac{H'_m(\lambda a)}{H_m(\lambda a)}\right], \tag{8.158}$$

so the stress on the circle is

$$\mathcal{S} = -\frac{1}{4\pi a^2}\frac{1}{i}\int_{-\infty}^{\infty}dz\,\frac{z^2}{\sqrt{z^2 + \mu^2 a^2}}\sum_{m=-\infty}^{\infty}\frac{d}{dz}\ln J_m(z)H_m(z). \tag{8.159}$$

To evaluate this, we perform an imaginary frequency rotation and introduce the modified Bessel functions:[§]

$$\mathcal{S} = -\frac{1}{2\pi a^2}\int_{\mu a}^{\infty}dx\,\frac{x^2}{\sqrt{x^2 - \mu^2 a^2}}\sum_{m=-\infty}^{\infty}\frac{d}{dx}\ln I_m(x)K_m(x). \tag{8.160}$$

For $m = 0$ we can easily evaluate the integral numerically, after we insert the appropriate contact term. For example, for $\mu = 0$ we find upon integrating by parts that

$$-\int_0^{\infty}dx\,\ln 2x I_0(x)K_0(x) = 0.0880137, \tag{8.161}$$

corresponding to a very small attractive stress

$$\mathcal{S}_0 = -\frac{0.0140078}{a^2}. \tag{8.162}$$

For $m \neq 0$ we content ourselves with the leading uniform asymptotic expansion:

$$\frac{d}{dx}\ln I_m(x)K_m(x) \sim -\frac{x}{m^2 + x^2}. \tag{8.163}$$

[§]This formula, for $\mu = 0$, was first given by Sen [299, 300].

We carry out the sum using (8.130) and find

$$\mathcal{S}_{\mathrm{LT}} \sim \frac{1}{a^2} \frac{1}{(2\pi)^2} \int_{2\pi\mu a}^{\infty} \frac{dy\, y^2}{\sqrt{y^2 - (2\pi\mu a)^2}} \frac{1}{e^y - 1}, \qquad (8.164)$$

which is a repulsive stress, precisely the negative of the first term in (8.136). In particular, at $\mu = 0$, we have the negative of (8.131),

$$\mathcal{S}_{\mathrm{LT}} = \frac{1}{24a^2}, \qquad (8.165)$$

which overwhelms $\mathcal{S}_0$ above.

This numerical equivalence at $\mu = 0$ is no coincidence. It is a consequence of the theorem (8.146), because the $(3+1)_v$ result described in Chapter 7 has only higher-order contributions in the uniform asymptotic expansion. However, going beyond the leading approximation, we find that the scalar stress on a circle also diverges, an effect which we shall attempt to understand in the next Chapter—see Sec. 9.3. This divergence was first noted by Sen [299, 300] twenty years ago. See also Nesterenko and Pirozhenko [244], who show the divergence structure, in the zeta-function context, for the Dirichlet (D) and Neumann (N) contributions:

$$s \to 0: \quad E^D \sim -\frac{1}{128a} \frac{1}{s}, \quad E^N \sim -\frac{5}{128a} \frac{1}{s}. \qquad (8.166)$$

Clearly this is one area where experimental input would be extremely valuable. Do quantum fluctuations destabilize a two-dimensional circular boundary? As repeatedly demonstrated in physics, observations will drive the theory to make sense of the phenomena.

Chapter 9

Casimir Effect on a D-dimensional Sphere

Because of the rather mysterious dependence of the sign and magnitude of the Casimir stress on the topology and dimensionality of the bounding geometry, we have carried out calculations of TE and TM modes bounded by a spherical shell in D spatial dimensions [32, 33]. We first consider massless scalar modes satisfying Dirichlet boundary conditions on the surface, which are equivalent to electromagnetic TE modes. Then, in Sec. 9.2, we treat the corresponding TM problem, for modes which satisfy mixed boundary conditions. In three dimensions, the sum of these two contibutions (less the $l = 0$ term) yields the electromagnetic Casimir energy found in Chapter 4. Of course, in D dimensions, the number of photon polarization states is $D - 1$, so electromagnetism is not defined off integer dimensions. The modes considered in this chapter are smooth functions of D, however.

Again we calculate the vacuum expectation value of the stress on the surface, or the energy density, from the Green's function.

9.1   Scalar or TE Modes

The Green's function $G(\mathbf{x}, t; \mathbf{x}', t')$ satisfies the inhomogeneous Klein-Gordon equation (2.13), or

$$\left(\frac{\partial^2}{\partial t^2} - \nabla^2 \right) G(\mathbf{x}, t; \mathbf{x}', t') = \delta^{(D)}(\mathbf{x} - \mathbf{x}')\delta(t - t'), \qquad (9.1)$$

where ∇^2 is the Laplacian in D dimensions. We will solve the above Green's function equation by dividing space into two regions, the interior of a sphere of radius a and the exterior of the sphere. On the sphere we will impose

Dirichlet boundary conditions

$$G(\mathbf{x}, t; \mathbf{x}', t')\big|_{|\mathbf{x}|=a} = 0. \tag{9.2}$$

In addition, in the interior region we will require that G be finite at the origin, $\mathbf{x} = 0$, and in the exterior region we will require that G satisfy outgoing-wave boundary conditions at $|\mathbf{x}| = \infty$.

The radial Casimir force per unit area $\mathcal{F}$ on the sphere is obtained from the radial-radial component of the vacuum expectation value of the stress-energy tensor:

$$\mathcal{F} = \langle 0| T_{\text{in}}^{rr} - T_{\text{out}}^{rr} |0\rangle \big|_{r=a}. \tag{9.3}$$

To calculate $\mathcal{F}$ we exploit the connection between the vacuum expectation value of the stress-energy tensor $T^{\mu\nu}(\mathbf{x}, t)$ and the Green's function at equal times $G(\mathbf{x}, t; \mathbf{x}', t)$, which follows from (2.23):

$$\mathcal{F} = \frac{1}{2i} \left[\frac{\partial}{\partial r} \frac{\partial}{\partial r'} G(\mathbf{x}, t; \mathbf{x}', t)_{\text{in}} - \frac{\partial}{\partial r} \frac{\partial}{\partial r'} G(\mathbf{x}, t; \mathbf{x}', t)_{\text{out}} \right] \Bigg|_{\mathbf{x}=\mathbf{x}', |\mathbf{x}|=a}. \tag{9.4}$$

To evaluate the expression in (9.4) it is necessary to solve the Green's function equation (9.1). We begin by taking the time Fourier transform of G:

$$\mathcal{G}_\omega(\mathbf{x}; \mathbf{x}') = \int_{-\infty}^{\infty} dt\, e^{i\omega(t-t')} G(\mathbf{x}, t; \mathbf{x}', t'). \tag{9.5}$$

The differential equation satisfied by $\mathcal{G}_\omega$ is

$$-\left(\omega^2 + \nabla^2\right) \mathcal{G}_\omega(\mathbf{x}; \mathbf{x}') = \delta^{(D)}(\mathbf{x} - \mathbf{x}'). \tag{9.6}$$

To solve this equation we introduce polar coordinates and seek a solution that has cylindrical symmetry; i.e., we seek a solution that is a function only of the two variables $r = |\mathbf{x}|$ and θ, the angle between $\mathbf{x}$ and $\mathbf{x}'$ so that $\mathbf{x} \cdot \mathbf{x}' = rr' \cos\theta$. In terms of these polar variables (9.6) becomes

$$\left(\omega^2 + \frac{\partial^2}{\partial r^2} + \frac{D-1}{r} \frac{\partial}{\partial r} + \frac{\sin^{2-D}\theta}{r^2} \frac{\partial}{\partial \theta} \sin^{D-2}\theta \frac{\partial}{\partial \theta} \right) \mathcal{G}_\omega(r, r', \theta)$$

$$= -\frac{\Gamma\left(\frac{D-1}{2}\right)}{2\pi^{(D-1)/2} r^{D-1} \sin^{D-2}\theta} \delta(r - r')\delta(\theta). \tag{9.7}$$

Note that the D-dimensional delta function on the right side of (9.6) has been replaced by a cylindrically-symmetric delta function having the property that its volume integral in D-dimensional space is unity. The D-dimensional volume integral of a cylindrically-symmetric function $f(r, \theta)$ is

$$\frac{2\pi^{(D-1)/2}}{\Gamma\left(\frac{D-1}{2}\right)} \int_0^\infty dr \, r^{D-1} \int_0^\pi d\theta \, \sin^{D-2}\theta f(r,\theta). \tag{9.8}$$

We solve (9.7) using the method of separation of variables. Let

$$\mathcal{G}_\omega(r, r', \theta) = A(r)B(z), \tag{9.9}$$

where $z = \cos\theta$. The equation satisfied by $B(z)$ is then

$$\left[(1 - z^2)\frac{d^2}{dz^2} - z(D-1)\frac{d}{dz} + n(n+D-2)\right]B(z) = 0, \tag{9.10}$$

where we have anticipated a convenient form for the separation constant. The equation satisfied by $A(r)$ is

$$\left[\frac{d^2}{dr^2} + \frac{D-1}{r}\frac{d}{dr} - \frac{n(n+D-2)}{r^2} + \omega^2\right]A(r) = 0 \quad (r \neq r'). \tag{9.11}$$

The solution to (9.10) that is regular at $|z| = 1$ is the ultraspherical (Gegenbauer) polynomial [111]

$$B(z) = C_n^{(-1+D/2)}(z) \quad (n = 0, 1, 2, 3, \ldots). \tag{9.12}$$

The solution in the interior region to (9.11) that is regular at $r = 0$ involves the Bessel function [111] ($k = |\omega|$)

$$A(r) = r^{1-D/2}J_{n-1+\frac{D}{2}}(kr). \tag{9.13}$$

In Eq. (9.13) we assume that $D \geq 2$ in order to eliminate the linearly independent solution $A(r) = r^{1-D/2}Y_{n-1+\frac{D}{2}}(kr)$, which is singular at $r = 0$ for all n. The solution in the exterior region to Eq. (9.11) that corresponds to an outgoing wave* at $r = \infty$ involves a Hankel function of the first kind [111]

$$A(r) = r^{1-D/2}H^{(1)}_{n-1+\frac{D}{2}}(kr). \tag{9.14}$$

---

*Recall the discussion in Sec. 4.1. The wave we are considering propagates as $e^{ikr-i\omega t}$, corresponding to the usual causal or Feynman propagator.

Using a few properties of the ultraspherical polynomials, namely, orthonormality [111]

$$\int_{-1}^{1} dz\, (1-z^2)^{\alpha-1/2} C_n^{(\alpha)}(z) C_m^{(\alpha)}(z) = \frac{2^{1-2\alpha}\pi\Gamma(n+2\alpha)}{n!\,(n+\alpha)\Gamma^2(\alpha)}\delta_{nm} \quad (\alpha \neq 0),$$

(9.15)

and the value of the ultraspherical polynomial at $z = 1$,

$$C_n^{(\alpha)}(1) = \frac{\Gamma(n+2\alpha)}{n!\,\Gamma(2\alpha)} \quad (\alpha \neq 0),$$

(9.16)

as well as the duplication formula (2.33), we solve for the Green's function in the two regions. Adding the interior and the exterior contributions,

$$\mathcal{F}_{\text{in}} = i \sum_{n=0}^{\infty} \frac{w(n,D)}{2^{D+1}\pi^{\frac{D+1}{2}} a^D \Gamma\left(\frac{D-1}{2}\right)}$$
$$\times \int_{-\infty}^{\infty} d\omega \left[\frac{ka J_\nu'(ka)}{J_\nu(ka)} + 1 - \frac{D}{2}\right],$$

(9.17a)

$$\mathcal{F}_{\text{out}} = i \sum_{n=0}^{\infty} \frac{w(n,D)}{2^{D+1}\pi^{\frac{D+1}{2}} a^D \Gamma\left(\frac{D-1}{2}\right)}$$
$$\times \int_{-\infty}^{\infty} d\omega \left[\frac{ka H_\nu^{(1)'}(ka)}{H_\nu^{(1)}(ka)} + 1 - \frac{D}{2}\right],$$

(9.17b)

and performing the usual imaginary frequency rotation, we obtain the final expression for the stress [32]:

$$\mathcal{F} = -\sum_{n=0}^{\infty} \frac{w(n,D)}{2^D \pi^{\frac{D+1}{2}} a^{D+1} \Gamma\left(\frac{D-1}{2}\right)}$$
$$\times \int_0^{\infty} dx \left[x\frac{d}{dx}\ln\left(I_\nu(x)K_\nu(x)x^{2-D}\right)\right].$$

(9.18)

Here we have used the abbreviations

$$w(n,D) = \frac{(2n+D-2)\Gamma(n+D-2)}{n!},$$

(9.19)

and

$$\nu = n - 1 + \frac{D}{2}.$$

(9.20)

It is easy to check that this reduces to the known case at $D = 1$, for there the series truncates—only $n = 0$ and 1 contribute, and we easily find

$$\mathcal{F} = \frac{\mathcal{S}}{2} = -\frac{\pi}{96a^2}, \tag{9.21}$$

which agrees with (1.35) and with (2.35) for $d = 0$ and $a \to 2a$, and corresponds to the Lüscher potential of QCD[†] [83].

In general, however, although each x integral can be made finite, the sum over n still diverges. We can make the integrals finite by replacing the x^{2-D} factor in the logarithm by simply x, which we are permitted to do if $D < 1$ because (N represents a positive integer)

$$\sum_{n=0}^{\infty} \frac{\Gamma(n+\alpha)}{n!} \equiv 0 \quad (\alpha < 0, \alpha \neq -N). \tag{9.22}$$

Then the total stress on the sphere[‡] is obtained by multiplying by the area of a hypersphere,

$$A_D = \frac{2\pi^{D/2}}{\Gamma(D/2)} a^{D-1}, \tag{9.23}$$

$$\mathcal{S} = -\frac{1}{2\pi a^2} \sum_{n=0}^{\infty} \frac{w(n, D)}{\Gamma(D-1)} Q_n, \tag{9.24}$$

where the integral is

$$Q_n = -\int_0^{\infty} dx \, \ln \left(2x I_\nu(x) K_\nu(x)\right). \tag{9.25}$$

We proceed as follows:

- Analytically continue to $D < 0$, where the sum converges, although the integrals become complex.
- Add and subtract the leading asymptotic behavior of the integrals.
- Continue back to $D > 0$, where everything is now finite.

[†] See also Nesterenko and Pirozhenko [264] for a detailed discussion of the Casimir energy of a string, including its temperature dependence.

[‡] Note that this result at $D = 2$ agrees with the $\mu = 0$ result in (8.160). The $n = 0$ term appears there with half weight because we must set $n = 0$ before taking the $D \to 2$ limit.

We used two methods to carry out the numerical evaluations, which gave the same results. In the first method we use the uniform asymptotic expansions to yield

$$Q_n \sim \frac{\nu\pi}{2} + \frac{\pi}{128\nu} - \frac{35\pi}{32768\nu^3} + \frac{565\pi}{1048576\nu^5} + \dots \qquad (9.26)$$

Then using the identities

$$\sum_{n=0}^{\infty} \frac{\Gamma(n+D-2)}{n!} = 0 \quad \text{for } D < 2, \qquad (9.27\text{a})$$

$$\sum_{n=0}^{\infty} \frac{\Gamma(n+D-2)}{n!} \nu^2 = 0 \quad \text{for } D < 0, \qquad (9.27\text{b})$$

we obtain the following expression convergent for $D < 4$ (further subtractions can be made for higher dimensions):

$$S \approx -\frac{1}{2a^2\pi} \frac{1}{\Gamma(D-1)} \left\{ \sum_{n=0}^{[N-D/2+1]} w(n,D)\tilde{Q}_n \right.$$

$$\left. + \sum_{n=[N-D/2+2]}^{\infty} w(n,D) \left[-\frac{35\pi}{32768\nu^3} + \frac{565\pi}{1048576\nu^5} \right] \right\}, \qquad (9.28)$$

where $\tilde{Q}_\nu = Q_\nu - \nu\pi/2 - \pi/128\nu$, and the square brackets in the summation limits denote the largest integer less or equal to its argument. The infinite sums are easily evaluated in terms of gamma functions, according to

$$\sum_{n=0}^{\infty} \frac{\Gamma(n+\alpha)}{n!\,(n+\alpha/2)^2} = \frac{\pi^2}{2} \frac{\Gamma(\alpha/2)}{\Gamma(1-\alpha/2)} \frac{1}{\sin^2 \pi\alpha/2}. \qquad (9.29)$$

In the second method we carry out an asymptotic expansion of the summand in (9.24) in n, $n \to \infty$

$$w(n,D)Q_n \sim n^{D-1} + \frac{(D-1)(D-2)}{2} n^{D-2}$$

$$+ \frac{24D^4 - 176D^3 + 504D^2 - 688D + 387}{192} n^{D-3} + \dots \qquad (9.30)$$

The sums on n in the terms in the asymptotic expansion are carried out according to

$$\sum_{n=1}^{\infty} n^{D-k} = \zeta(k-D), \quad D-k < -1. \tag{9.31}$$

In this way we obtain the following formula suitable for numerical evaluation:

$$
\begin{aligned}
S = -\frac{1}{2a^2} \Bigg\{ &\frac{1}{\pi} Q_0 \\
&+ \frac{1}{\Gamma(D-1)} \sum_{n=1}^{\infty} \left[w(n,D)\frac{1}{\pi} Q_n - \sum_{k=1}^{K} n^{D-k} c_k(D) \right] \\
&+ \frac{1}{\Gamma(D-1)} \sum_{k=1}^{K} \zeta(k-D) c_k(D) \Bigg\},
\end{aligned}
\tag{9.32}
$$

where the $c_k(D)$ are polynomials in D, where the optimal number of terms K chosen depends on D.

We can use the $K = 1$ version of this expression to find the stress on a zero-dimensional sphere:

$$S\big|_{D=0} = -\frac{1}{2a^2}, \tag{9.33}$$

where we have used the fact that $\zeta(z) \sim \frac{1}{z-1}$ as $z \to 1$.

Both methods give the same results [32], which are shown in Fig. 9.1. Note the following salient features:

- Poles occur at $D = 2n$, $n = 1, 2, 3, \ldots$.
- As we will see in the next plot for negative D, branch points occur at $D = -2n$, $n = 0, 1, 2, 3, \ldots$, and the stress is complex for $D < 0$.
- The stress vanishes at negative even integers, $S(-2n) = 0$, $n = 1, 2, 3, \ldots$, but is nonzero at $D = 0$: $S(0) = -1/2a^2$.
- The case of greatest physical interest, $D = 3$, has a finite stress, but one which is much smaller than the corresponding electrodynamic one:

$$S(3) = +0.0028168/a^2. \tag{9.34}$$

(This result was confirmed in Ref. [187, 188, 246].)

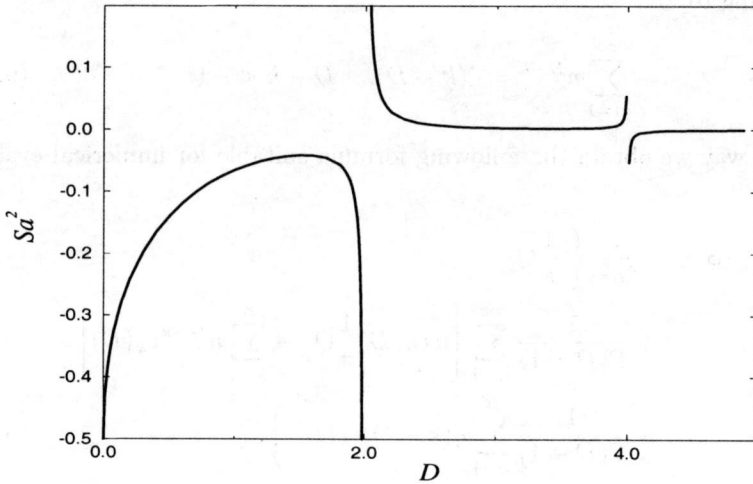

Fig. 9.1   Scalar Casimir stress S for $0 < D < 5$ on a spherical shell.

9.2   TM Modes

The same kind of calculation can be carried out for the TM modes [33]. The TM modes are modes which satisfy mixed boundary conditions on the surface [148, 301],

$$\frac{\partial}{\partial r} r^{D-2} G(\mathbf{x}, t; \mathbf{x}', t') \bigg|_{|\mathbf{x}|=r=a} = 0. \qquad (9.35)$$

In this case when we solve the inhomogeneous Klein-Gordon equation (9.1) we easily find the Green's function to be, in the interior of the sphere,

$$\mathcal{G}_\omega(r, r', \theta) = \sum_{n=0}^{\infty} \frac{2\nu \Gamma\left(\frac{D}{2} - 1\right)}{8(\pi r r')^{D/2-1} \sin \pi \nu} C_n^{(D/2-1)}(\cos \theta)$$

$$\times \left[J_\nu(k r_<) J_{-\nu}(k r_>) - \beta J_\nu(k r) J_\nu(k r') \right], \quad (9.36a)$$

where

$$\beta = \frac{\left(\frac{D}{2} - 1\right) J_{-\nu}(ka) + ka J'_{-\nu}(ka)}{\left(\frac{D}{2} - 1\right) J_\nu(ka) + ka J'_\nu(ka)}, \qquad (9.36b)$$

and, outside the sphere,

$$\mathcal{G}_\omega(r,r',\theta) = -i \sum_{n=0}^{\infty} \frac{2\nu\Gamma\left(\frac{D}{2}-1\right)}{16(\pi r r')^{D/2-1}} C_n^{(D/2-1)}(\cos\theta)$$

$$\times \left[H_\nu^{(1)}(kr_<)H_\nu^{(2)}(kr_>) - \gamma H_\nu^{(1)}(kr)H_\nu^{(1)}(kr')\right],$$

(9.37a)

where

$$\gamma = \frac{\left(\frac{D}{2}-1\right)H_\nu^{(2)}(ka) + ka H_\nu^{(2)\prime}(ka)}{\left(\frac{D}{2}-1\right)H_\nu^{(1)}(ka) + ka H_\nu^{(1)\prime}(ka)},$$

(9.37b)

and ν is given by (9.20).

It is a bit more subtle to calculate the force per area for the TM modes than it was for the TE modes. For a given frequency, we write for the vacuum expectation value of the radial-radial component of the stress tensor

$$\langle t_{rr} \rangle = \frac{i}{2} \left[\nabla_r \nabla_{r'} + \omega^2 - \boldsymbol{\nabla}_\perp \cdot \boldsymbol{\nabla}_{\perp'} \right] \mathcal{G}_\omega,$$

(9.38)

where, if we average over all directions, we can integrate by parts on the transverse derivatives,

$$-\boldsymbol{\nabla}_\perp \cdot \boldsymbol{\nabla}_{\perp'} \rightarrow \nabla_\perp^2 \rightarrow -\frac{n(n+D-2)}{r^2},$$

(9.39)

where the last replacement, involving the eigenvalue of the Gegenbauer polynomial, is appropriate for a given mode n [see (9.10)]. As for the radial derivatives, they are[§]

$$\nabla_r = r^{2-D}\partial_r r^{D-2}, \quad \nabla_{r'} = r'^{2-D}\partial_{r'} r'^{D-2},$$

(9.40)

which, by virtue of (9.35), implies that the $\nabla_r \nabla_{r'}$ term does not contribute to the stress on the sphere. In this way, we easily find the following formula for the contribution to the force per unit area for interior modes,

$$\mathcal{F}_{\text{in}}^{\text{TM}} = -\frac{i}{\pi^{(D+1)/2} 2^D a^{D+1} \Gamma(\frac{D-1}{2})} \int_0^\infty \frac{dx}{x} \sum_{n=0}^{\infty} w(n,D)$$

$$\times (x^2 - n(n+D-2)) \frac{\tilde{s}_n(x)}{\tilde{s}_n'(x)},$$

(9.41a)

[§]In the TM mode, the radial derivatives correspond to tangential components of **E**, which must vanish on the surface. See [301].

and for exterior modes,

$$\mathcal{F}_{\text{out}}^{\text{TM}} = -\frac{i}{\pi^{(D+1)/2}2^D a^{D+1}\Gamma(\frac{D-1}{2})} \int_0^\infty \frac{dx}{x} \sum_{n=0}^\infty w(n,D)$$
$$\times (x^2 - n(n+D-2))\frac{\tilde{e}_n(x)}{\tilde{e}'_n(x)}, \tag{9.41b}$$

where $w(n,D)$ is given by (9.19), $x = ka$, and the generalized Ricatti-Bessel functions are

$$\tilde{s}_n(x) = x^{D/2-1}J_\nu(x), \quad \tilde{e}_n(x) = x^{D/2-1}H_\nu^{(1)}(x). \tag{9.42}$$

It is a small check to observe that for $D = 2$ we recover the known result (8.142)

$$\mathcal{F}_{D=2}^{\text{TM}} = -\frac{i}{8\pi^2 a^3} \int_{-\infty}^\infty dx\, x \sum_{m=-\infty}^\infty \left(1 - \frac{m^2}{x^2}\right)\left(\frac{J_m(x)}{J'_m(x)} + \frac{H_m^{(1)}(x)}{H_m^{(1)'}(x)}\right),$$
$$\tag{9.43}$$

where the half-weight at $n = 0$ is a result of the *limit $D \to 2$*. In two dimensions, the vector Casimir effect consists of only the TM mode contribution.

In general, we can combine the TE mode contribution, given in (9.17a) and (9.17b), and the TM mode contribution, found here, into the following simple formula¶

$$\mathcal{F}^{\text{TM+TE}} = \frac{i}{\pi^{(D+1)/2}2^D a^{D+1}} \sum_{n=0}^\infty \frac{w(n,D)}{\Gamma(\frac{D-1}{2})}$$
$$\times \int_0^\infty dx\, x \left\{\frac{\tilde{s}'_n(x)}{\tilde{s}_n(x)} + \frac{\tilde{e}'_n(x)}{\tilde{e}_n(x)} + \frac{\tilde{s}''_n(x)}{\tilde{s}'_n(x)} + \frac{\tilde{e}''_n(x)}{\tilde{e}'_n(x)}\right\}. \tag{9.44}$$

It will be noted that, for $D = 3$, this result agrees with that found for the usual electrodynamic Casimir force/area, when the $n = 0$ mode is properly excluded. [See (4.23) with the cutoff $\delta = 0$.] Of course, this only coincides with electrodynamics in three dimensions. The number of electrodynamic modes changes discontinuously with dimension, there being only one in $D = 2$, the TM mode, and none in $D = 1$, in general there being $D - 1$ modes. Equation (9.44) is of interest in a mathematical sense, because significant cancellations do occur between TE and TM modes in general.

¶We will not concern ourselves with a constant term in the integrand, which we will deal later.

The integrals in (9.41a) and (9.41b) are oscillatory and therefore very difficult to evaluate numerically. Thus, as usual, it is advantageous to perform a rotation of 90 degrees in the complex-ω plane. The resulting expression for $\mathcal{F}^{\text{TM}}$ is

$$
\mathcal{F}^{\text{TM}} = -\sum_{n=0}^{\infty} \frac{w(n,D)}{2^D \pi^{\frac{D+1}{2}} a^{D+1} \Gamma\left(\frac{D-1}{2}\right)} \int_0^{\infty} dx\, x
$$
$$
\times \frac{d}{dx} \ln\left[x^{2(3-D)} \left(x^{D/2-1} K_\nu(x) \right)' \left(x^{D/2-1} I_\nu(x) \right)' \right]. \quad (9.45)
$$

9.2.1 *Energy Derivation*

As a check of internal consistency, it would be reassuring to derive the same result by integrating the energy density due to the field fluctuations. The latter is computable from the vacuum expectation value of the stress tensor, which in turn is directly related to the Green's function, $\mathcal{G}_\omega$:

$$
\langle T_{00} \rangle = \frac{i}{2} \int_{-\infty}^{\infty} \frac{d\omega}{2\pi} (\omega^2 + \boldsymbol{\nabla} \cdot \boldsymbol{\nabla}') \mathcal{G}_\omega \Big|_{\mathbf{r}=\mathbf{r}'}. \quad (9.46)
$$

Again, because we are going to integrate this over all space, we can integrate by parts, replacing, in effect,

$$
\boldsymbol{\nabla} \cdot \boldsymbol{\nabla}' \to -\nabla^2 \to \omega^2, \quad (9.47)
$$

which uses the Green's function equation (9.6). [Point splitting is always implicitly assumed, so that delta functions may be omitted.] Then, using the area of a unit sphere in D dimensions (9.23), we find the Casimir energy to be given by

$$
E = \frac{i 2\pi^{D/2}}{\Gamma(D/2)} \int_{-\infty}^{\infty} \frac{d\omega}{2\pi} \omega^2 \int_0^{\infty} r^{D-1} dr\, \mathcal{G}_\omega(r,r). \quad (9.48)
$$

So, from the form for the Green's function given in (9.36a) and (9.37a), we see that we need to evaluate integrals such as

$$
\int_0^a r\, dr\, J_\nu(kr) J_{-\nu}(kr), \quad (9.49)
$$

which are given in terms of the indefinite integral

$$
\int dx\, x\, Z_\nu(x) \mathcal{Z}_\nu(x) = \frac{x^2}{2} \left[\left(1 - \frac{\nu^2}{x^2}\right) Z_\nu(x) \mathcal{Z}_\nu(x) + Z_\nu'(x) \mathcal{Z}_\nu'(x) \right], \quad (9.50)
$$

valid for any two Bessel functions Z_ν, $\mathcal{Z}_\nu$ of order ν. Thus we find for the Casimir energy of the TM modes the formula

$$E^{\mathrm{TM}} = -\frac{i}{2\pi\Gamma(D-1)a} \sum_{n=0}^{\infty} w(n,D)$$

$$\times \int_0^\infty \frac{dx}{x} \left[(x^2 - n(n+D-2)) \left(\frac{\tilde{s}_n(x)}{\tilde{s}'_n(x)} + \frac{\tilde{e}_n(x)}{\tilde{e}'_n(x)} \right) + (2-D)x \right].$$

$$(9.51)$$

We obtain the stress on the spherical shell by differentiating this expression with respect to a (which agrees with (9.41a) and (9.41b), apart from the constant in the integrand), followed again doing the complex frequency rotation, which yields

$$S_{\mathrm{TM}} = \frac{1}{2\pi a^2 \Gamma(D-1)} \sum_{n=0}^{\infty} w(n,D) Q_n, \qquad (9.52)$$

where the integrals are

$$Q_n = -\int_0^\infty dx\, x \frac{d}{dx} \ln q(x), \qquad (9.53)$$

where

$$q(x) = \left[\left(\frac{D}{2} - 1 \right) I_\nu(x) + \frac{x}{2} (I_{\nu+1}(x) + I_{\nu-1}(x)) \right]$$

$$\times \left[\left(\frac{D}{2} - 1 \right) K_\nu(x) - \frac{x}{2} (K_{\nu+1}(x) + K_{\nu-1}(x)) \right]. \qquad (9.54)$$

This agrees with the form found directly from the force density, (9.45), again, apart from an additive constant in the x integrand.

9.2.2 *Numerical Evaluation of the Stress*

We now need to evaluate the formal expression (9.52) for arbitrary dimension D. We implicitly assumed in its derivation that $D > 2$ and that D was not an even integer, but we will argue that (9.52) can be continued to all D.

9.2.2.1 *Convergent Reformulation of (9.52)*

First of all, it is apparent that as it sits, the integral Q_n in (9.53) does not exist. [The form in (9.45) does exist for the special case of $D = 3$.] As in the scalar case discussed in Sec. 9.1, we argue that since

$$\sum_{n=0}^{\infty} w(n, D) = 0 \quad \text{for} \quad D < 1, \tag{9.55}$$

we can add an arbitrary term, independent of n, to Q_n in (9.52) without effect as long as $D < 1$. In effect then, we can multiply the quantity in the logarithm in (9.53) by an arbitrary power of x without changing the value for the force for $D < 1$. We choose that multiplicative factor to be $-2/x$ because then a simple asymptotic analysis shows that the integrals converge. Then, we analytically continue the resulting expression·to all D. The constant -2 is, of course, without effect in (9.53), but allows us to integrate by parts, ignoring the boundary terms. The result of this process is that the expression for the Casimir force is still given by (9.52), but with Q_n replaced by

$$Q_n = \int_0^\infty dx \ln\left[-\frac{2}{x}q(x)\right], \tag{9.56}$$

$q(x)$ being given by (9.54).

Now the individual integrals in (9.52) converge, but the sum still does not. We can see this by making the uniform asymptotic approximations for the Bessel functions in (9.54) [111], which leads to ($n \to \infty$)

$$Q_n \sim \frac{\pi\nu}{2}\left(1 + \frac{-101 + 80D - 16D^2}{64\nu^2}\right.$$
$$\left. + \frac{-5861 + 11152D - 7680D^2 + 2304D^3 - 256D^4}{16384\nu^4} + \ldots\right). \tag{9.57}$$

[Note that the coefficients in this expansion depend on the dimension D, unlike the scalar case, given in (9.26).] Because of this behavior, it is apparent that the series diverges for all positive D, except for $D = 1$, where the series truncates.

Recall that two procedures were used to turn the corresponding sum in the scalar case into a convergent series, and to extract numerical results. In

the second procedure described in Sec. 9.1 we subtract from the summand
the leading terms in the $1/n$ expansion, derived from (9.57), identifying
those summed subtractions with Riemann zeta functions:

$$
\mathcal{S}^{\mathrm{TM}} \approx \frac{1}{2\pi a^2} \left\{ Q_0 + \frac{1}{\Gamma(D-1)} \sum_{n=1}^{N} \left[w(n,D)Q_n - \pi n^{D-1} \left(1 + \sum_{k=1}^{K} \frac{b_k}{n^k} \right) \right] \right.
$$

$$
\left. + \frac{\pi}{\Gamma(D-1)} \left[\zeta(1-D) + \sum_{k=1}^{K+1} b_k \zeta(k+1-D) - b_{K+1} \sum_{n=1}^{N} n^{D-K-2} \right] \right\}.
$$

$$(9.58)$$

Here b_k are the coefficients in the asymptotic expansion of the summand in
(9.52), of which the first two are

$$
b_1 = \frac{(D-2)(D-1)}{2}, \tag{9.59a}
$$

$$
b_2 = \frac{81 - 448D + 456D^2 - 176D^3 + 24D^4}{192}. \tag{9.59b}
$$

In (9.58) we keep K terms in the asymptotic expansion, and, after N terms
in the sum, we approximate the subtracted integrand by the next term in
the large n expansion. The series converges for $D < K + 1$, so more and
more terms in the asymptotic expansion are required as D increases.

The method described first in Sec. 9.1 gives identical results, and is, in
fact, more convergent. The results given there were, in fact, first computed
by this procedure, which is based on analytic continuation in dimension.
Here, we simply subtract from Q_ν the first two terms in the asymptotic
expansion (9.57), and then argue, as a generalization of (9.55), that the
identities (9.27a), (9.27b) hold. Therefore, by continuing from negative
dimension, we argue that we can make the subtraction without introducing
any additional terms. Thus, if we define

$$
\hat{Q}_n = Q_n - \frac{\pi \nu}{2} \left(1 + \frac{-101 + 80D - 16D^2}{64\nu^2} \right), \tag{9.60}
$$

we have

$$
\mathcal{S}^{\mathrm{TM}} = \frac{1}{2\pi a^2 \Gamma(D-1)} \sum_{n=0}^{\infty} w(n,D)\hat{Q}_n \approx \frac{1}{2\pi a^2 \Gamma(D-1)}
$$

$$
\times \left(\sum_{n=0}^{N} w(n,D)\hat{Q}_n + \frac{\pi g(D)}{2} \sum_{n=N+1}^{\infty} \frac{w(n,D)}{\nu^3} \right), \tag{9.61}
$$

where $g(D)$ is the coefficient of ν^{-4} in (9.57). The last sum in (9.61) can be evaluated according to (9.29). The approximation given in (9.61) converges for $D < 4$.

9.2.3 *Casimir Stress for Integer $D \leq 1$*

The case of integers ≤ 1 is of special note, because, for those cases, the series truncates. For example, for $D = 0$ only the $n = 0, 2$ terms appear, where the integrals cancel by virtue of the symmetry of Bessel functions,

$$K_\nu(x) = K_{-\nu}(x), \quad I_n(x) = I_{-n}(x), \tag{9.62}$$

for n an integer. However, using the first procedure (9.58), we have a residual zeta function contribution:

$$\mathcal{S}_{D=0}^{\mathrm{TM}} = \frac{1}{2\pi a^2}\left(Q_0 - Q_2 + \pi\right) = \frac{1}{2a^2}, \tag{9.63}$$

because both $\zeta(1 - D)$ and $\Gamma(D - 1)$ have simple poles, with residue -1, at $D = 0$. This result for $D = 0$ is the negative of the result found in the scalar case, (9.33), which is a direct consequence of the fact that the n^{D-2} term in the asymptotic expansion cancels when the TE and TM modes are combined [compare (9.59a) with the corresponding term in (9.30).] The continuation in D method gives the same result, because then

$$\mathcal{S}_{D=0}^{\mathrm{TM}} = \frac{1}{2\pi a^2}(\hat{Q}_0 + D\hat{Q}_1 - \hat{Q}_2) = \frac{1}{2a^2}\left(1 - \frac{101}{64} + \frac{101}{64}\right) = \frac{1}{2a^2}, \tag{9.64}$$

where a limiting procedure, $D \to 0$, is employed to deal with the singularity which occurs for $n = 1$, where $\nu \to 0$.

For the negative even integers we achieve a similar cancellation between pairs of integers, with no zeta function residual because the ζ functions no longer have poles there. For example, for $D = -2$ we have

$$\mathcal{S}_{D=-2}^{\mathrm{TM}} = \frac{1}{2\pi a^2}(Q_0 - 2Q_1 + 2Q_3 - Q_4) = 0 \tag{9.65}$$

because $Q_0 = Q_4$ and $Q_1 = Q_3$. Again, the other method of regularization gives the same result when a careful limit is taken.

For odd integers ≤ 1, trucation occurs without cancellation, because $I_\nu \neq I_{-\nu}$. For example, for $D = 1$,

$$a^2 \mathcal{S}_{D=1}^{\mathrm{TM}} = \frac{1}{2\pi}(Q_0 + Q_1) = -0.2621 + 0.6032i. \tag{9.66}$$

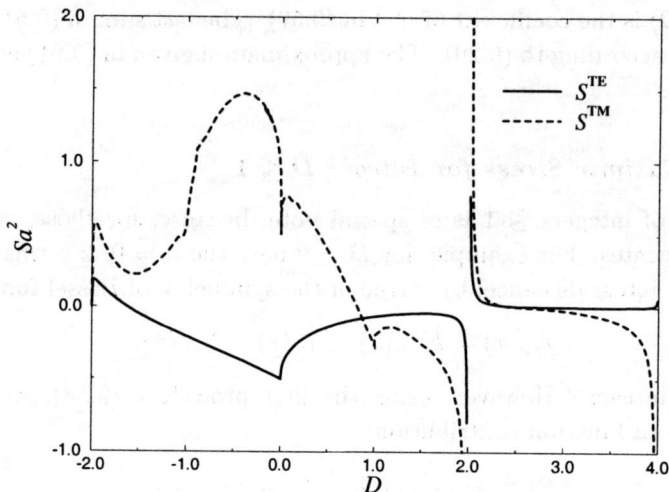

Fig. 9.2   A plot of the TM and TE Casimir stress for $-2 < D < 4$ on a spherical shell. For $D < 2$ ($D < 0$) the stress $\mathcal{S}^{\mathrm{TM}}$ ($\mathcal{S}^{\mathrm{TE}}$) is complex and we have plotted the real part.

9.2.4   *Numerical results*

We have used both methods described above to extract numerical results for the stress on a sphere due to TM fluctuations in the interior and exterior. Results are plotted in Fig. 9.2. Salient features are the following:

- As in the scalar case, poles occur for positive even dimension.
- The integrals become complex for $D < 2$ because the function $q(x)$, (9.54), occurring in the logarithm develops zeros. (This phenomenon started at $D = 0$ for the scalar case.) Correspondingly, there are logarithmic singularities, and cusps, occurring at 2, 1, 0, -1, -2, ..., rather than just at the nonpositive even integers.
- The sign of the Casimir force changes dramatically with dimension. Here this is even more striking than in the scalar case, where the sign was constant between the poles for $D > 0$. For the TM modes, the Casimir force vanishes for $D = 2.60$, being repulsive for $2 < D < 2.60$ and attractive for $2.60 < D < 4$.
- Also in Fig. 9.2, the results found here are compared with those found in Sec. 9.1 for the the scalar or TE case. The correspondence

is quite remarkable. In particular, for $D < 2$ the qualitative structure of the curves are very similar when the scale of the dimensions
in the TE case is reduced by a factor of 2; that is, the interval
$0 < D < 2$ in the TE case corresponds to the interval $1 < D < 2$
in the TM, $-2 < D < 0$ for TE corresponds to $0 < D < 1$ for TM,
etc.

- Physically, the most interesting result is at $D = 3$. The TM mode
calculated here has the value $S_{D=3}^{TM} = -0.02204/a^2$. However, if
we wish to compare this to the electrodynamic result described in
Chapter 4, we must subtract off the $n = 0$ mode, which is given
in terms of the integral $Q_0 = 0.411233 = \pi^2/24$, which displays
the accuracy of our numerical integration. Similarly removing the
$n = 0$ mode $(= -\pi/24)$ from the result quoted in (9.34), $S_{D=3}^{TE} =$
$0.0028168/a^2$, gives agreement with the familiar result (4.40) [13,
76, 14, 15, 187, 188, 189, 190, 191]:

$$S_{n>0}^{TM+TE}\bigg|_{D=3} = \frac{0.0462}{a^2}, \qquad (9.67)$$

9.3   Toward a Finite $D = 2$ Casimir Effect

The truly disturbing aspect of the results given in this Chapter are the poles
in even dimensions. In particular many very interesting condensed matter
systems are well-approximated by being two dimensional, as discussed in
Chapter 8. Are we to conclude that the Casimir effect does not exist in two
dimensions?

One trivial way to extract a finite answer from our expressions, which
have simple poles at $D = 2$ (setting aside the logarithmic singularity there
in the TM mode, because that only occurs in one integral, Q_0) is to average
over the singularity. If we do so for the scalar result in (9.32), we obtain

$$S_{D=2}^{TE} = -\frac{0.01304}{a^2}, \qquad (9.68)$$

while for the TM result in (9.58), we find

$$S_{D=2}^{TM} = -\frac{0.340}{a^2}, \qquad (9.69)$$

which numbers, incidentally, are remarkably close to the leading Q_0 term, as stated in (8.162), (8.127), which are -0.0140 and -0.254. But, there seems to be no reason to have any belief in these numbers.

However, something remarkable does happen in the scalar case. If we use the first procedure, (9.58), we note that the poles can arise both from the integrals and from the explicit zeta functions. For the latter, let the dependence on D be given by $r(D)/(D-2)$ which has a pole at $D=2$. When we average over the pole, we obtain

$$\lim_{\epsilon \to 0} \frac{1}{2} \left(\frac{r(2+\epsilon)}{\epsilon} - \frac{r(2-\epsilon)}{\epsilon} \right) = r'(2), \qquad (9.70)$$

where the prime denotes differentiation. For the scalar modes it is easy to verify that $r'(2) = 0$. Thus, there is no contribution from those subtracted terms. In other words, they might just as well be omitted, which is what we would do if we inserted a cutoff and simply dropped the divergent terms. (This procedure does give the correct $D=3$ results.) This provides some evidence for the validity of the procedure which yields (9.68).

Unfortunately, the same effect does not occur for the TM modes, $r'(2) \neq 0$. Nor does it occur for higher dimensions, $D=4, 6, \ldots$, even for scalars. And, even for scalars, it is not clear how the divergences of a massive $(2+1)$ theory can be removed. So we are no closer to solving the divergence problem in even dimensions.$^\parallel$ It is clear there is much more work to do on Casimir phenomena.

$^\parallel$ For a discussion of the inadequacies of the dubious procedure of attempting to extract a finite result in Ref. [31], and described in Chapter 8, see Ref. [188, 225, 302]. Recall that the divergences which occur in 2 dimensions where first discovered by Sen [299, 300]. See also Ref. [244].

Chapter 10

Cosmological Implications

In Chapter 6 we considered some hadronic implications of quantum vacuum energy. Here we turn to the opposite extreme, the cosmological scale. Significant issues arise when we consider gravitation, because the absolute scale of energy presumably is now meaningful as the source of gravity. In particular, one might think that the cosmological constant would have its origin in quantum fluctuations of the gravitational and other fields, yet naive estimates give far too large a value.

In this Chapter we will give a modest introduction to this subject by considering quantum fluctuations in a gravitational regime where they are surely relevant.* The original higher-dimensional theory was that of Kaluza and Klein [303, 304, 305, 306, 307]. There an extra fifth dimension allowed the unification of electromagnetism and gravity. Higher dimensional generalizations allow the inclusion of Yang-Mills fields [308, 309]. Of course, superstring theory is of necessity formulated in at least ten dimensions [310, 311, 312]. Where are these extra dimensions? Barring recent speculations [313, 314, 315, 316, 317, 318] that the extra dimensions may be large, $\sim 1\,\text{mm}$, the presumption has been that they are curled up on a scale of the Planck length. If so, the resulting field confinement should give rise to an observable Casimir effect, which might stabilize (or destabilize) the compact geometry. We will consider this phenomenon, for the case of N compact dimensions on a hypersphere, in the following section. We will conclude this Chapter with some speculations concerning the cosmological constant, which now appears to be nonzero but small [319, 320, 321, 322,

---

*For a classic reference to quantum field theory in curved space, see Ref. [101].

323, 324, 325, 326, 327, 328, 329].

10.1 Scalar Casimir Energies in $M^4 \times S^N$

The idea of Casimir compactification, that zero-point fluctuations stablize
the geometry of the extra dimensions, has been explored by several authors,
starting with Appelquist and Chodos [28, 330, 331, 29, 332, 333, 334, 335,
336, 337, 338, 339, 340, 341, 342, 343, 344, 345]. Candelas and Weinberg
[29] showed that stability could result for large number of scalar and Fermi
fields in $4 + N$ dimensions, $N = 3(\text{mod}4)$, while Chodos and Myers [339,
340] explored graviton fluctuations and found unstable tachyonic behavior. The early work on this subject was restricted to odd N, because when
the number of compact dimensions are even, ultraviolet logarithmic divergences remain after all legitimate subtractions. This is associated with the
fact that one-loop counterterms can only be constructed for the Einstein-
Hilbert action in even numbers of dimensions [346]. Myers [341] numerically computed this logarithmic term for gravity fluctuations using a zeta
function technique. The general N case for scalars was first considered by
Kantowski and Milton [25], which we follow below.

We will follow that latter development here. Following the formalism of
Sec. 2.3, we can write the Casimir energy density of a massless scalar field
in a $M^4 \times S^N$ manifold, the N-sphere having radius a and volume V_N, as

$$u(a) = V_N \langle T^{00} \rangle = V_N \lim_{(x,y)\to(x',y')} \partial^0 \partial'^0 \text{Im}\, G(x,y;x',y'), \qquad (10.1)$$

where the x are the coordinates in the Minkowski space M^4, while the S^N
coordinates are denoted by y. For definiteness, we understand the point-
splitting limit in (10.1) to be taken with a spacelike separation. Because
of translational invariance in x, we can express G as a four-dimensional
Fourier transform,

$$G(x,y;x',y') = \int \frac{d^4k}{(2\pi)^4} e^{-ik_\mu(x-x')^\mu} g(y,y'; k^\mu k_\mu), \qquad (10.2)$$

in terms of which the vacuum energy can be simply expressed as

$$u(a) = -\frac{iV_N}{2(2\pi)^4} \int d^3k \int_c d\omega\, \omega^2 g(y,y; k^2 - \omega^2), \qquad (10.3)$$

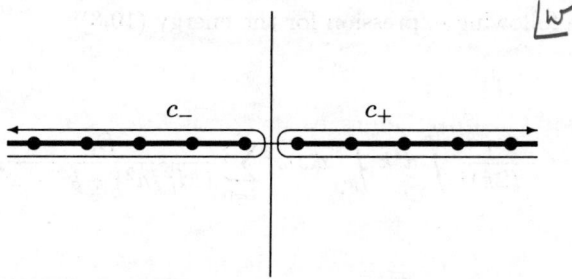

Fig. 10.1　The ω plane for odd N showing the Green's function contours in the complex ω plane. Shown schematically are the poles in the Green's function, and the branch cuts starting at $\beta = 0$, where β is given by (10.15). The corresponding branch points occur at $\omega = \pm\sqrt{k^2 - (N-1)^2/4a^2}$. Note that if $k < (N-1)/2a$ the branch point and pole there lie on the imaginary axis. In that case, c_+ encloses the branch point on the positive imaginary axis, while c_- encloses the branch point on the negative imaginary axis.

where the contour c of the ω integration consists of c_- and c_+, c_+ encircling the poles on the positive real axis in a clockwise sense, and c_- encircling those on the negative real axis in a counterclockwise sense. See Fig. 10.1. The reduced Green's function satisfies

$$(\nabla_N^2 + k^2 - \omega^2)g(y, y'; k^2 - \omega^2) = -\delta(y - y'), \qquad (10.4)$$

where ∇_N^2 is the Laplacian on S^N and $\delta(y-y')$ is the appropriate δ function.

In general we find g for arbitrary N by expanding in N-dimensional spherical harmonics:

$$\nabla_N^2 Y_l^m(y) = -\frac{M_l^2}{a^2} Y_l^m(y), \qquad (10.5)$$

where the eigenvalues and degeneracies are

$$M_l^2 = l(l + N - 1), \qquad (10.6a)$$

$$D_l = \frac{(2l + N - 1)(l + N - 2)!}{(N-1)!\, l!}. \qquad (10.6b)$$

(The dependence of the hyperspherical harmonic on the polar angle is given by the Gegenbauer polynomials discussed in Chapter 9.) Use of the generalized addition theorem for the hyperspherical harmonics,

$$\sum_m Y_l^m(y) Y_l^{m*}(y) = \frac{D_l}{V_N}, \qquad (10.7)$$

leads to the following expression for the energy (10.3):

$$u(a) = -\frac{i}{(2\pi)^4} \int d^3k \int_{c_+} d\omega \, \omega^2 \sum_{l=0}^{\infty} \frac{D_l}{(M_l^2/a^2) + k^2 - \omega^2}, \qquad (10.8)$$

where the integrand's dependence on ω^2 has been used to combine the two parts of the c contour to one, the right-hand contour c_+. [Compare the form of (10.8) with the generic form (1.17).]

As is obvious from (10.8), the vacuum energy of the massless scalar in $M^4 \times S^N$ is a linear sum of vacuum energies of massive scalars in 4 dimensions. The mode sum on l diverges for $N > 1$ and the momentum integrals diverge for all N. To obtain finite Casimir energies we can subtract off divergences identifiable as contact or cosmological terms from the outset. Because the l sum is finite for the $N = 1$ case, we consider that situation first.

10.1.1 $N = 1$

In that case, the masses are $M_l^2 = l^2$, and the degeneracies are $D_0 = 1$, $D_{l \geq 1} = 2$, so that the mode sum gives, from (8.130)

$$\sum_{l=0}^{\infty} \frac{D_l}{M_l^2/a^2 + k^2 - \omega^2} = \frac{a\pi}{(k^2 - \omega^2)^{1/2}} \coth\left[a\pi(k^2 - \omega^2)^{1/2}\right]$$

$$= \frac{a\pi}{(k^2 - \omega^2)^{1/2}} \left(1 + \frac{2}{e^{2\pi a(k^2 - \omega^2)^{1/2}} - 1}\right). \qquad (10.9)$$

This sum has been written as an asymptotic part plus a remainder. The asymptotic part produces an infinite "cosmological term" in the energy, and is either subtracted off completely, or regulated by inserting a cutoff $\omega_{\max} \sim b^{-1}$, b presumably at the Planck scale, resulting in a cosmological energy density,[†]

$$u_{\text{cosmo}}(a) = \frac{1}{(2\pi)^4} \int d^3k \, 2 \int_k^\infty d\omega \, \omega^2 \frac{a\pi}{(\omega^2 - k^2)^{1/2}}$$

$$= \frac{V_1}{80\pi^2 b^5}, \tag{10.10}$$

where $V_1 = 2\pi a$ is the "volume" of a circle. Here we have taken an abrupt cutoff in ω; however, any other technique also yields $u_{\text{cosmo}} \sim V_1/b^5$. It is important to notice that the sum in (10.9) has only simple poles; however, the part that we identify as a cosmological term and subtract off has branch points at $\omega = \pm k$. For odd N, including the $N = 1$ case, the branch cuts are drawn away from each other on the real ω axis out to $\pm\infty$ (see Fig. 10.1). The remainder of (10.9) produces the unique Casimir energy and is easily evaluated by (i) integrating over the 4π solid angle in the momentum element d^3k, (ii) distorting the contour c_+ to one lying along the imaginary ω axis, $\omega = i\zeta$ $(-\infty < \zeta < \infty)$, and (iii) replacing ζ and k by plane polar coordinates, $k = \kappa \sin\theta$, $\zeta = \kappa \cos\theta$, and integrating first over θ and then over κ, just as we did in Sec. 2.3. The result is

$$u_{\text{Casimir}} = -\frac{1}{64\pi^2 a^4} \int_0^\infty (\kappa a)^4 d(\kappa a)^2 \frac{2\pi}{\kappa a} \frac{1}{e^{2\pi\kappa a} - 1}$$

$$= -\frac{3\zeta(5)}{64\pi^6 a^4} = -\frac{5.0558077 \times 10^{-5}}{a^4}, \tag{10.11}$$

which uses (2.34). This is exactly the result first obtained by Appelquist and Chodos [28, 330].

10.1.2 *The General Odd-N Case*

The Casimir energy for arbitrary odd N can be extracted similarly. We first define a new mode index m,

$$m \equiv l + \frac{N-1}{2}, \tag{10.12}$$

[†]If b is the Planck scale, $b^{-4} \sim 10^{76}$ GeV^4 $\sim 10^{108}$ GeV/cm^3, so even if $a/b \sim 1$ this is over a hundred orders of magnitude larger than the observed mass density of the universe, or of the current inferred value of the cosmological constant. [The critical mass density of the universe is $1.05 \times 10^{-5} h_0^2$ GeV/cm^3, where in terms of the Hubble constant, $h_0 = H_0/100 \, \text{km s}^{-1} \text{Mpc}^{-1}$.] This is the cosmological constant problem, which we shall discuss further in Sec. 10.3. [For a review of the cosmological constant problem, see Weinberg, [347, 348].]

in terms of which M_l^2 and D_l of (10.6a) and (10.6b) can be written as functions of m^2:

$$M_m^2 = m^2 - \left(\frac{N-1}{2}\right)^2, \tag{10.13a}$$

$$D_m' = \frac{2}{(N-1)!}\left[m^2 - \left(\frac{N-3}{2}\right)^2\right]\left[m^2 - \left(\frac{N-5}{2}\right)^2\right]\cdots$$
$$\cdots(m^2 - 1)m^2. \tag{10.13b}$$

The sum in (10.8) becomes

$$\sum_{l=0}^{\infty}\frac{D_l}{M_l^2/a^2 + k^2 - \omega^2} = \sum_{m=0}^{\infty}\frac{D_m'}{m^2 - \beta^2}$$
$$= \sum_{m=0}^{\infty}\left(\frac{D_\beta'}{m^2 - \beta^2} + \text{polynomial in } m^2 \text{ and } \beta^2\right), \tag{10.14}$$

where

$$\beta^2 = [(N-1)/2]^2 + a^2\omega^2 - a^2k^2. \tag{10.15}$$

The polynomial terms make no contribution to (10.8) and can be discarded. Notice from (10.12) that the m sum should start at $(N-1)/2$; however, since $D_m' = 0$ for $0 \le m \le (N-3)/2$ we can start at $m = 0$. Once again, the sum of the pole terms in (10.14) can be evaluated using (2.45),

$$\sum_{m=0}^{\infty}\frac{1}{m^2 - \beta^2} = -\frac{\pi}{2\beta}\cot(\pi\beta) - \frac{1}{2\beta^2}. \tag{10.16}$$

Since D_β'/β^2 is a polynomial in β^2, the $-1/2\beta^2$ term in (10.16) makes no net contribution to (10.8) and can be discarded. The remaining $\cot\pi\beta$ term contains both the divergent cosmological energy and the finite Casimir energy. In Fig. 10.1 the branch cuts for β are shown as well as the c_+ contour which avoids the $\beta = 0$ branch points. As in (10.9) we write the $\cot\pi\beta$ term as an asymptotic part plus a remainder; the former yields the cosmological energy

$$u_{\text{cosmo}} \propto \frac{V_N}{b^{N+4}}. \tag{10.17}$$

The remainder yields the finite Casimir energy. It is best evaluated by following the three steps described after (10.10):

$$u(a) = -\frac{1}{64\pi^2 a^4} \text{Re} \int_0^\infty (a\kappa)^4 d(a\kappa)^2 D'_\beta \frac{\pi i}{\beta(e^{-2\pi i \beta} - 1)}, \qquad (10.18)$$

where now

$$\beta^2 = \left(\frac{N-1}{2}\right)^2 - a^2 \kappa^2. \qquad (10.19)$$

The integral in (10.18) is most easily evaluated by changing from κ to β as the integration variable. The contour for β which comes from $0 < \kappa < \infty$ is not suited for extracting the Casimir energy. This energy is most easily computed by integrating β first vertically,

$$\beta = \frac{N-1}{2} + iy, \quad 0 \le y < \infty, \qquad (10.20a)$$

and then horizontally,

$$\beta = x + i\infty, \quad \frac{N-1}{2} \ge x \ge 0. \qquad (10.20b)$$

The nondivergent part of the integral on the horizontal part of the new contour is imaginary and hence does not contribute to the Casimir energy (10.18). By making the substitution (10.20a), so $d(a\kappa)^2 = -2i\beta dy$, we obtain

$$u_{\text{Casimir}}(a) = -\frac{1}{64\pi^2 a^4} \text{Re} \int_0^\infty dy [y^2 - i(N-1)y]^2 D_{iy} \frac{2\pi}{e^{2\pi y} - 1}. \qquad (10.21)$$

For $N \ge 1$ this is nothing more than a sum of Riemann zeta functions at odd integer values $4 + N$, $4 + N - 2$, $\ldots$, 3. For example, the $N = 3$ and $N = 5$ expressions are

$$u_3(a) = \frac{1}{32\pi a^4} \left[\frac{\Gamma(7)\zeta(7)}{(2\pi)^7} - \frac{13\Gamma(5)\zeta(5)}{(2\pi)^5} + \frac{4\Gamma(3)\zeta(3)}{(2\pi)^3} \right]$$

$$= \frac{1}{a^4}(7.5687046\ldots 10^{-5}), \qquad (10.22a)$$

$$u_5(a) = \frac{1}{384\pi a^4} \left[\frac{\Gamma(9)\zeta(9)}{(2\pi)^9} - \frac{103\Gamma(7)\zeta(7)}{(2\pi)^7} + \frac{604\Gamma(5)\zeta(5)}{(2\pi)^5} - \frac{192\Gamma(3)\zeta(3)}{(2\pi)^3} \right]$$

$$= \frac{1}{a^4}(4.2830381\ldots 10^{-4}), \qquad (10.22b)$$

N	$a^4 u_N$
1	$-5.0558077 \times 10^{-5}$
3	7.5687046×10^{-5}
5	4.2830382×10^{-4}
7	8.1588536×10^{-4}
9	1.1338947×10^{-3}
11	1.3293159×10^{-3}
13	1.3740262×10^{-3}
15	1.2524870×10^{-3}
17	9.5591579×10^{-4}
19	4.7935196×10^{-4}
21	$-1.7990889 \times 10^{-4}$
23	$-1.0231947 \times 10^{-3}$
25	$-2.0509729 \times 10^{-3}$
27	$-3.2631628 \times 10^{-3}$
29	$-4.6593317 \times 10^{-3}$
31	$-6.2388216 \times 10^{-3}$
33	$-8.0008299 \times 10^{-3}$
35	$-9.9444650 \times 10^{-3}$
37	$-1.2068783 \times 10^{-2}$
39	$-1.4372813 \times 10^{-2}$

Table 10.1   Casimir energy density $u_N(a)$ for a massless scalar in $M^4 \times S^N$, with N odd. The radius of the sphere is a.

respectively. Results for larger N are tabulated in Table 10.1 and graphed in Fig. 10.2; they agree with the findings of Candelas and Weinberg [29]. Note that the Casimir energy is attractive for $N = 1$, repulsive for odd N, $3 \leq N \leq 19$, and attractive for odd N thereafter.

10.1.3   *The Even-N Case*

When the internal space is a sphere S^N with N even, we find a cutoff-dependent divergent cosmological energy exactly as in the odd-N case [see (10.17)]; however, now the Casimir energy also diverges. The divergence is logarithmic in a/b and, fortunately, the coefficient of the logarithm is independent of the actual cutoff technique used. We evaluate $u(a)$ of (10.8) by shifting the index mode as was done in the previous subsection, with m

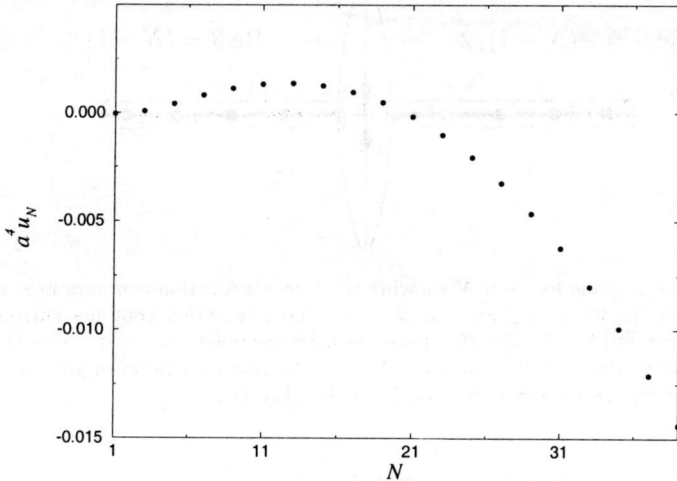

Fig. 10.2   Plot of $a^4 u_N$ for odd N.

and M_m^2 defined as in (10.12) and (10.13a), and

$$D'_m = \frac{2}{(N-1)!} \left[m^2 - \left(\frac{N-3}{2} \right)^2 \right] \left[m^2 - \left(\frac{N-5}{2} \right)^2 \right] \cdots$$

$$\cdots \left[m^2 - \left(\frac{1}{2} \right) \right] m. \qquad (10.23)$$

The mode index m is now half-integral, and, because of the vanishing of D'_m, the sum on m can start from $1/2$. If $D'_m/m(m^2 - \beta^2)$ is again factored into a pole term plus a polynomial as in (10.14), the polynomial contributions vanish and the remaining energy can be written as

$$u(a) = -\frac{ia^2}{(2\pi)^4} \int d^3k \, I(c_+, \Sigma), \qquad (10.24)$$

where the functional I is the contour integral in the ω plane

$$I(c, F) \equiv \int_c d\omega \, \omega^2 \frac{D'_\beta}{\beta} F(\omega). \qquad (10.25)$$

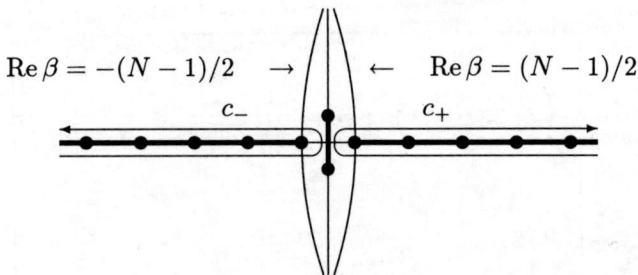

Fig. 10.3 The ω plane for even N showing the Green's function contours used and the branch cut on the imaginary axis for $\beta = 0$. The integration contours surround the branch cuts for $\ln[(N-1)/2 \pm \beta]$. Here we have assumed that $k < (N-1)/2a$. If $k > (N-1)/2a$ the $\beta = 0$ branch cut lies on the real axis between the c_+ and c_- contours. The curves $\mathrm{Re}\,\beta = \pm(N-1)/2$ are also sketched.

The function Σ in (10.24) is the infinite sum

$$\Sigma(\omega) \equiv \sum_{m=1/2}^{\infty} \frac{m}{m^2 - \beta^2}, \tag{10.26}$$

β^2 is again given by (10.15), and c_+ is the right-hand contour shown in Fig. 10.3. The m sum in (10.26) necessarily diverges unless we regulate it in some manner. Here we simply subtract a constant, $1/(m+1/2)$, from each mode, i.e., we write

$$\Sigma(\omega) = \frac{1}{2} \sum_{m'=1}^{\infty} \left[\frac{1}{m' + (\beta - \frac{1}{2})} - \frac{1}{m'} + \frac{1}{m' - (\beta + \frac{1}{2})} - \frac{1}{m'} + \frac{2}{m'} \right]$$
$$= \frac{1}{2}\left[-\psi\left(\frac{1}{2} + \beta\right) - \psi\left(\frac{1}{2} - \beta\right) - 2\gamma + 2\zeta(1) \right], \tag{10.27}$$

where $m' = m+1/2$ and $\zeta(1)$ is an infinite constant. The digamma function is defined by

$$\psi(z) = \frac{d}{dz} \ln \Gamma(z), \tag{10.28}$$

and $\gamma = -\psi(1) \approx 0.57721$ is Euler's constant. The digamma function $\psi(z)$ is analytic over the entire complex plane except at $z = 0, -1, -2, \ldots$, where it has simple poles with residues all equal to -1. To evaluate (10.24) we use the identities

$$\psi(1/2 \pm \beta) = \psi\left(\frac{N-1}{2} \pm \beta\right) - \frac{1}{(N-3)/2 \pm \beta} - \frac{1}{(N-5)/2 \pm \beta} - \cdots$$
$$- \frac{1}{1/2 \pm \beta},\tag{10.29a}$$
$$\psi(1/2 - \beta) = \psi(1/2 + \beta) - \pi \tan \beta\pi,\tag{10.29b}$$

and the representation

$$\psi\left(\frac{N-1}{2} \pm \beta\right) = \ln\left(\frac{N-1}{2} \pm \beta\right) - \frac{1}{N-1 \pm 2\beta}$$
$$- 2 \int_0^\infty \frac{dt\,t}{e^{2\pi t} - 1}\left(\frac{1}{t^2 + [(N-1)/2 \pm \beta]^2}\right),$$
$$\tag{10.29c}$$

which is valid for $\operatorname{Re}\left[(N-1)/2 \pm \beta\right] > 0$. We connect the branch points $\beta = 0$ by the branch cut as shown in Fig. 10.3. We choose the branch cut in $\ln[(N-1)/2 + \beta]$ to run along the negative real ω axis starting at $\beta = -(N-1)/2$ (i.e., for $\omega = -k$) as also shown in Fig. 10.3. For the $+$ sign this choice makes (10.29c) valid in the ω plane to the right of the line $\operatorname{Re}\beta = -(N-1)/2$. The branch cut for $\ln[(N-1)/2 - \beta]$ is drawn to the right from $\beta = (N-1)/2$ (i.e., for $\omega = k$) on the positive real ω axis making (10.29c) valid for the $-$ case to the left of $\operatorname{Re}(N-1)/2$.

Using (10.29a), (10.29b), and (10.29c) we can rewrite (10.27) as

$$\Sigma(\omega) = \frac{1}{(N-3)/2 + \beta} + \frac{1}{(N-5)/2 + \beta} + \cdots + \frac{1}{1/2 + \beta}$$
$$+ \frac{1/2}{(N-1)/2 + \beta} - \ln\left(\frac{N-1}{2} + \beta\right) + \frac{\pi}{2} \tan \pi\beta$$
$$+ 2 \int_0^\infty \frac{dt\,t}{e^{2\pi t} - 1}\left(\frac{1}{t^2 + [(N-1)/2 + \beta]^2}\right) - \gamma + \zeta(1),$$
$$\tag{10.30}$$

valid for $\operatorname{Re}\beta > -(N-1)/2$ (see Fig. 10.3). The infinite constant $-\gamma + \zeta(1)$ makes no contribution to (10.24) and will be discarded. We rewrite the logarithm term in (10.30) as

$$-\ln\left(\frac{N-1}{2}+\beta\right) = -\frac{1}{2}\ln\left[\left(\frac{N-1}{2}\right)^2 - \beta^2\right] + \frac{1}{2}\ln\left[\frac{(N-1)/2-\beta}{(N-1)/2+\beta}\right],$$

(10.31)

and identify the first term as producing most of the cosmological energy density: (the discontinuity of the first logarithm across the cut on the positive real axis is $-2\pi i$)

$$
\begin{aligned}
u_{\log} &= -\frac{ia^2}{(2\pi)^4}\int d^3k\, I\left(c_+, -\frac{1}{2}\ln(a^2k^2 - a^2\omega^2)\right) \\
&= \frac{a^2}{(2\pi)^2}\int_0^\infty k^2 dk \int_k^\infty d\omega\, \omega^2 \frac{D'_\beta}{\beta} \\
&= \frac{a^2}{(2\pi)^2}\int_0^{1/b} d\omega\, \omega^2 \int_0^\omega dk\, k^2 \frac{D'_\beta}{\beta} \\
&= \frac{1}{2\pi^2(N+4)(N+1)[(N-1)!!]^2}\frac{a^N}{b^{N+4}} \\
&\quad + \frac{1}{24\pi^2}\frac{N(N-1)}{(N+2)[(N-1)!!]^2}\frac{a^{N-2}}{b^{N+2}} \\
&\quad + \cdots + \frac{1}{72\pi^2(N-1)/2}\frac{a^2}{b^6}.
\end{aligned}
$$

(10.32)

The constants in (10.32) depend on the cutoff technique used. We include them because they are easy to compute and because they illustrate what kind of subtractions must be made in the final dynamical equation for a. The remaining terms of (10.30)

$$\Sigma'(\omega) = \Sigma(\omega) + \gamma - \zeta(1) + \frac{1}{2}\ln\left[\left(\frac{N-1}{2}\right)^2 - \beta^2\right]$$

(10.33)

contribute two more terms to the cosmological energy, proportional to $1/b^4$ and $1/a^2b^2$, as well as produce the Casimir energy, proportional to $1/a^4$. To evaluate these we integrate $\Sigma'(\omega)$ along the c_v contour and subtract its integral along the two-part $c_{+\infty}$ contour [see Fig. 10.4(a)]:

$$I(c_+, \Sigma') = I(c_v, \Sigma') - I(c_{+\infty}, \Sigma').$$

(10.34)

The expression given for $\Sigma'(\omega)$ by (10.33) and (10.30) is valid along $c_{+\infty}$.

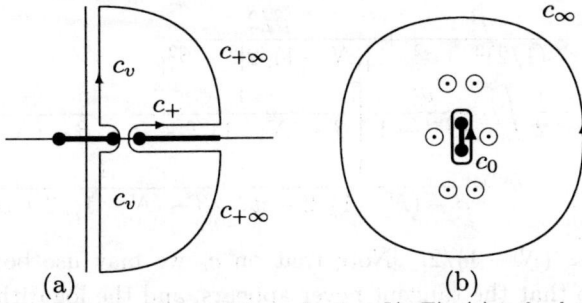

Fig. 10.4 Contours in the ω plane used in evaluating $I(c, \Sigma)$. (a) is for $k > (N-1)/2a$ and shows the contours c_v and $c_{+\infty}$. (b) is for $k < (N-1)/2a$ and shows the six poles in Σ_v as given in (10.41a) and (10.41b).

There, the remaining ln and tan terms combine as

$$\frac{1}{2} \ln \left[\frac{(N-1)/2 - \beta}{(N-1)/2 + \beta} \right] + \frac{\pi}{2} \tan \pi\beta = \frac{\mp i\pi}{e^{\mp 2\pi i\beta} + 1}, \qquad (10.35)$$

with the $-$ ($+$) sign being valid in the first (fourth) quadrant. [The imaginary part of $\log((N-1)/2 - \beta)$ is $-i\pi$ above the cut on the positive real axis, and $+i\pi$ below.] The combined terms vanish exponentially fast away from the real ω axis and contribute at most a polynomial in k^2 to (10.34). As usual, such terms correspond to derivatives of δ functions, e.g., $\nabla^2\delta(\mathbf{x} - \mathbf{x}')$, and vanish before the point-split limit is taken in (10.1). The remaining terms of $\Sigma'(\omega)$ are

$$\Sigma_{+\infty} = \frac{1}{(N-3)/2 + \beta} + \frac{1}{(N-5)/2 + \beta} + \cdots + \frac{1}{1/2 + \beta}$$
$$+ \frac{1/2}{(N-1)/2 + \beta} + 2 \int_0^\infty \frac{dt\, t}{e^{2\pi t} - 1} \left(\frac{1}{t^2 + [(N-1)/2 + \beta]^2} \right). \qquad (10.36)$$

These terms are analytic to the right of c_v and hence can simply be subtracted from $\Sigma'(\omega)$ in the c_v integral of (10.34),

$$I(c_v, \Sigma_v) = I(c_v, \Sigma') - I(c_{+\infty}, \Sigma_{+\infty}), \qquad (10.37)$$

where, according to (10.27), (10.29a), (10.29c), and (10.36),

$$\Sigma_v \equiv \Sigma' - \Sigma_{+\infty} = \frac{\beta}{[(N-3)/2]^2 - \beta^2} + \frac{\beta}{[(N-5)/2]^2 - \beta^2} + \cdots$$

$$+ \frac{\beta}{(1/2)^2 - \beta^2} + \frac{\beta/2}{[(N-1)/2]^2 - \beta^2}$$

$$-\frac{i}{2} \int_0^\infty \frac{dt}{e^{2\pi t} - 1} \left[\frac{1}{\beta + (N-1)/2 + it} - \frac{1}{\beta + (N-1)/2 - it} \right.$$

$$\left. + \frac{1}{\beta - (N-1)/2 - it} - \frac{1}{\beta - (N-1)/2 + it} \right] \qquad (10.38)$$

for $|\mathrm{Re}\,\beta| < (N-1)/2$. [Note that on c_v we may use both versions of (10.29c), so that the tangent never appears, and the logarithm in (10.29c) is cancelled by that in (10.33). As we will see, Σ_v now possesses singularities in the first and fourth quadrants, unlike $\Sigma_{+\infty}$.]

The part of $u(a)$ in (10.24) not given by (10.32) is

$$u(a) - u_{\log}(a) = -\frac{ia^2}{(2\pi)^4} \int d^3k\, I(c_v, \Sigma_v). \qquad (10.39)$$

Because Σ_v is odd in β the only contribution to $I(c_v, \Sigma_v)$ comes when c_v skirts the branch cut in $\beta(\omega)$ [see Fig 10.4(a)]. If we replace c_v by a contour completely encircling the cut we get twice (10.39), or

$$u(a) - u_{\log}(a) = -\frac{ia^2}{2(2\pi)^4} \int d^3k\, I(c_0, \Sigma_v), \qquad (10.40)$$

where c_0 is shown in Fig. 10.4(b). This integral is evaluated by distorting c_0 to c_∞, a complete circle at $\omega = \infty$, and picking up the residues of the poles of $(D_\beta'/\beta)\Sigma_v(\omega)$. The integral around c_∞ results in polynomials in k^2, which again are contact terms which are to be disregarded. There are only six poles in Σ_v which are not canceled by zeroes in D_β'/β; they are (each $\pm$ sign in independent)

$$\beta = \pm\frac{N-1}{2}, \qquad (10.41a)$$

$$\beta = \pm\frac{N-1}{2} \pm it. \qquad (10.41b)$$

The subtracted energy is

$$u(a) - u_{\log}(a) = -\frac{1}{(2\pi)^2} \int_0^\infty dk\, k^2 \left[-\frac{1}{2} D_0 k \right.$$

$$\left. + 2\,\mathrm{Im} \int_0^\infty \frac{dt}{e^{2\pi t} - 1} D_{it} \left\{ k^2 + [(N-1)it - t^2]/a^2 \right\}^{1/2} \right]$$

$$= \frac{1}{32\pi^2 b^4} - \frac{1}{8\pi^2} \int_0^\infty \frac{dt}{e^{2\pi t}-1} \mathrm{Im} \left\{ \frac{D_{it}}{b^4} + \frac{D_{it}}{a^2 b^2}[(N-1)it - t^2] \right.$$
$$\left. - \frac{1}{2a^4} \left[\ln\frac{a}{b} - \frac{1}{2}\ln\left(\frac{(N-1)it - t^2}{4}\right) - \frac{1}{4} \right] D_{it}[(N-1)it - t^2]^2 \right\}.$$
$$(10.42)$$

The first three terms on the right side of this equality are additional contributions to the cosmological energy density, while the last term, proportional to a^{-4} is the Casimir energy,

$$u_{\mathrm{Casimir}} = \frac{1}{a^4}[\alpha_N \ln(a/b) + \gamma_N], \qquad (10.43)$$

where

$$\alpha_N = \frac{1}{16\pi^2} \int_0^\infty \frac{dt}{e^{2\pi t}-1} \mathrm{Im}\, D_{it}[(N-1)it - t^2]^2, \qquad (10.44a)$$

$$\gamma_N = -\frac{1}{32\pi^2} \int_0^\infty \frac{dt}{e^{2\pi t}-1} \left(\frac{1}{2}\mathrm{Im}\,\{D_{it}[(N-1)it - t^2]^2\} \right.$$
$$\times \left[\ln\frac{(N-1)^2 t^2 + t^4}{16} + 1 \right]$$
$$\left. + \mathrm{Re}\,\{D_{it}[(N-1)it - t^2]^2\} \arctan\frac{N-1}{t} \right). \qquad (10.44b)$$

Here we have given meaning to the divergent logarithms by cutting off the frequency integration at $1/b$, where b is presumably the Planck scale. Although the "constant" term γ_N depends on the regularization scheme, the coefficient of the logarithm α_N does not. For example the same result is obtained by a ζ-function technique, as we will see in the next subsection. The coefficient of the log is easily expressed in terms of Bernoulli numbers since

$$\int_0^\infty \frac{dt\, t^{2k-1}}{e^{2\pi t}-1} = \frac{|B_{2k}|}{4k}. \qquad (10.45)$$

The first few values are

$$\alpha_2 = \frac{1}{8\pi^2}\left(\frac{1}{12}|B_6| - \frac{1}{4}|B_4| \right)$$
$$= -\frac{1}{1260\pi^2} = -8.0413637 \times 10^{-5}, \qquad (10.46a)$$

N	α_N
2	$-8.0413637 \times 10^{-5}$
4	$-4.9923466 \times 10^{-4}$
6	$-1.3144888 \times 10^{-3}$
8	$-2.5052903 \times 10^{-3}$
10	$-4.0355535 \times 10^{-3}$
12	$-5.8734202 \times 10^{-3}$
14	$-7.9931201 \times 10^{-3}$
16	$-1.0373967 \times 10^{-2}$
18	$-1.2999180 \times 10^{-2}$
20	$-1.5844933 \times 10^{-2}$

Table 10.2   Coefficient α_N for the divergent logarithm for the Casimir energy (10.43) for a massless scalar in $M^4 \times S^N$.

$$
\begin{aligned}
\alpha_4 &= -\frac{1}{3!\,8\pi^2}\left(\frac{1}{16}|B_8| - \frac{85}{24}|B_6| + \frac{153}{16}|B_4|\right) \\
&= -\frac{149}{30240\pi^2} = -4.9923466 \times 10^{-4}, \qquad\qquad\text{(10.46b)} \\
\alpha_6 &= -\frac{1}{5!\,8\pi^2}\left(\frac{1}{20}|B_{10}| - \frac{105}{8}|B_8| + 252|B_6| - \frac{4325}{8}|B_4|\right) \\
&= -\frac{137}{10560\pi^2} = -1.3144888 \times 10^{-3}. \qquad\qquad\text{(10.46c)}
\end{aligned}
$$

Values for larger N are tabulated in Table 10.2 and graphed in Fig. 10.5. Even though the γ_N coefficient is not uniquely calculable, the α_N term may be sufficient for practical purposes, because we might expect, if the extra dimensions are large, $a/b \sim 10^{16}$ (see below) so the logarithmic term should dominate.

10.1.4 *A Simple ζ-Function Technique*

The results found in the previous subsection by the rigorous and physically transparent Green's function technique can be quickly and easily reproduced by a simple ζ-function method, which, as usual with such methods, sweeps divergence difficulties under the rug and does not reveal their interpretation as contact terms or cosmological-type terms. The scheme described in this section, however, is extremely simple to implement insofar as the Casimir energy is concerned. It is far simpler, in fact, than the method

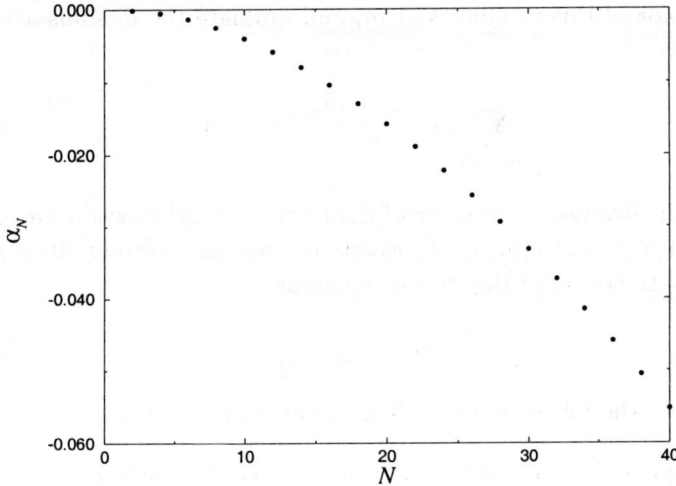

Fig. 10.5 The coefficient of the divergent logarithm, α_N, for the Casimir energy (10.43) for $M^4 \times S^N$, N even.

given in Refs. [29, 339, 340, 341]. The starting point is the expression (10.8) for the energy

$$u = -\frac{ia^2}{(2\pi)^4} \int d^3k \int_{c_+} d\omega\, \omega^2 \sum_{m=1/2}^{\infty} \frac{D'_m}{m^2 - \beta^2}. \qquad (10.47)$$

We regulate the integrals here by replacing the denominator by $(m^2 - \beta^2)^{1+s}$, where ultimately s will be taken to approach 0. If s is large enough, we can exchange summation and integration, distort the c_+ contour to the imaginary ω axis, and introduce polar coordinates, $\omega = i\kappa\cos\theta$, $k = \kappa\sin\theta$. By first integrating over θ, then over κ we find

$$u(a) = -\frac{a^2}{64\pi^2} \sum_{m=1/2}^{\infty} D'_m \int_0^\infty \frac{d\kappa^2\, \kappa^4}{[m^2 + \kappa^2 a^2 - (N-1)^2/4]^{1+s}}$$

$$= -\frac{1}{64\pi^2 a^4} \sum_{m=1/2}^{\infty} D'_m \left(\frac{1}{s} - \frac{2}{s-1} + \frac{1}{s-2}\right) [m^2 - (N-1)^2/4]^{2-s}.$$

$$(10.48)$$

We next expand this in powers of m, and evaluate the m sums according to (4.39), or

$$\sum_{m=1/2}^{\infty} m^z = (2^{-z} - 1)\zeta(-z). \qquad (10.49)$$

As $s \to 0$ the divergent terms are of the form $\alpha_N/(2a^4 s)$, where we identify $1/s$ with $\ln(a^2/b^2)$ in (10.43). To isolate α_N we can multiply (10.48) by $2s$ and set $s = 0$, not forgetting terms involving

$$s\zeta(1+2s) \to \frac{1}{2}. \qquad (10.50)$$

This leads to the following easily implemented algorithm for α_N:

(1) Expand $D'_m(m^4 - 2m^2 x + x^2)$ in powers of m, where $x = (N-1)^2/4$.
(2) Make the replacement (10.49), that is, replace m^n by $(1/2^n - 1)\zeta(-n)$.
(3) In the expansion of D'_m replace m^n by (n is necessarily odd)

$$m^n \to x^{(n+5)/2} \frac{[(n-1)/2]!}{[(n+5)/2]!}. \qquad (10.51)$$

(4) Add the replacements in 2 and 3. [Steps 1 and 2 overlook terms of the form (10.50), step 3 includes them.]

The results of this simply implemented algebraic scheme coincide with those found earlier, and given in Table 10.2. Let us illustrate it for the simple case of $N = 2$. We simply expand $D'_m[m^2 - (N-1)^2/4]^{2-s}$:

$$2m\left[m^2 - \frac{1}{4}\right]^{2-s} = 2m\left[m^{4-2s} - \frac{2-s}{4}m^{2-2s} + \frac{(2-s)(1-s)}{4^2 2!}m^{-2s}\right.$$
$$\left. - \frac{(2-1)(1-s)(-s)}{4^3 3!}m^{-2-2s} + \dots\right]. \qquad (10.52)$$

The first three terms here correspond to step 2 above, while the last term corresponds to step 3. We thus have for the Casimir energy

$$u_2 = -\frac{1}{64\pi^2 a^4}\frac{1}{s}\left[2\left(\frac{1}{2^5} - 1\right)\zeta(-5) - \left(\frac{1}{2^3} - 1\right)\zeta(-3)\right.$$
$$\left. + \frac{1}{8}\left(\frac{1}{2} - 1\right)\zeta(-1) + \frac{1}{3!}\frac{1}{32}\right] = -\frac{1}{2520\pi^2 a^4}\frac{1}{s}, \qquad (10.53)$$

which coincides with (10.46a) when it is remembered that $1/s$ corresponds to $\ln a^2/b^2$.

10.2 Discussion

It is easy to extend this analysis to massive particles, and to higher spins. All that is necessary is to generalize the denominator in (10.8) to

$$(M_l^2 + c)/a^2 + k^2 - \omega^2, \qquad (10.54)$$

where $M_l^2 = l(l + N - 1)$ for scalars, vectors, and tensors, as above, but $M_l^2 = l(l + N)$ for spinors. Mass effects are included through the constant c. For a minimally coupled scalar of mass μ, $c = \mu^2 a^2$, while for a fermion, $c = \mu^2 a^2 + (N/2)^2$. The calculation of the Casimir effect for such fields on $M^4 \times S^N$ is worked out in detail in Ref. [26], using both the Green's function and the zeta-function methods described here. The generalization the internal space being the product of spheres, that is $S^{N_1} \times S^{N_2} \times \cdots$, for both scalars and spinors, was investigated in Ref. [27], in the hope of finding a stable configuration, but no physically acceptable solutions were found. The fundamental difficulty which has stymied progress now for over a decade is the inability to include graviton fluctuations. It is well-understood that the so-called Vilkovisky-DeWitt correction must be incorporated in order to achieve gauge and parameterization independence [349, 90, 350, 91]. But because this correction does not correspond to a quadratic term in the Lagrangian, its implementation in the Casimir context remains incomplete. As Cho and Kantowski recently noted [351], because the determination of the V-D effective action for gravity involved evaluating determinants of complicated nonlocal operators, progress in compactification "slowed to a snail's pace." After years of work they were able to evaluate the divergent part of the effective action for even-dimensional geometries of the form $M^4 \times S^N$, $N = 2$, 4, 6, but no stable configurations were found.

10.2.1 *Other Work*

Kuo and Ford [352] discussed how large quantum fluctuations in the stress tensor

$$(\Delta T_{\mu\nu})^2 = \langle T_{\mu\nu}^2 \rangle - \langle T_{\mu\nu} \rangle^2 \qquad (10.55)$$

could cause large metric fluctuations even far away from the Planck scale. This could occur for a scalar field periodic in one dimension where

$$\langle T_{00} \rangle = -\frac{\pi^2}{90a^4}. \tag{10.56}$$

[See (11.8b) with $a \to a/2$.] However, Phillips and Hu [353] have recently argued persuasively that a large ratio of $(\Delta T_{00})^2/\langle T_{00}^2 \rangle$ is not necessarily a good measure of the invalidity of semiclassical gravity.

We conclude this section by remarking that there has been recent interest in treating the Casimir effect dynamically (in an adiabatic, i.e., slowly-varying approximation) in cosmological models. The effects could be significant, in both 4-dimensional and brane-world [317, 318] scenarios. For examples of such calculations see Refs. [354, 355]. A survey ot the rapidly exploding literature on this subject would be inappropriate, given the rapid state of flux in the field. We merely cite a few representative papers to give the reader some flavor of the field [356, 357, 358, 359, 360, 361].

10.3 The Cosmological Constant

In the above, terms were identified as "cosmological," which were then discarded. That was because they are much too large; they are more than 100 orders of magnitude bigger than any observed cosmological constant, and if such terms were present, the universe would have rapidly expanded to zero density today. Here we wish to propose that perhaps the Casimir energy, proportional to $1/a^4$, where a is the size of the compact dimension, could be the observed cosmological constant, if a is of a suitable value.

We first note that the "Casimir energy" calculated here has the correct structure to be a cosmological constant. In (10.3) we gave an expression for the energy proportional to

$$u \propto \int d^3k \int_c d\omega \, \omega^2 g(y, y; k^2 - \omega^2). \tag{10.57}$$

The ω^2 came from the two time derivatives in T^{00}. If we were to calculate T^{11} we would obtain

$$\langle T^{11} \rangle \propto \int d^3k \int_c d\omega \, \frac{1}{3} k^2 g(y, y; k^2 - \omega^2), \tag{10.58}$$

since all three spatial directions are on the same footing. Recall that we may evaluate the integrals here by first making a Euclidean rotation, $\omega \to i\zeta$, and then adopt polar coordinates,

$$\zeta = \kappa \cos\theta, \quad k = \kappa \sin\theta, \tag{10.59}$$

so

$$T^{00} \propto - \int_0^{2\pi} d\theta \sin^2\theta \cos^2\theta = -\frac{\pi}{4}, \tag{10.60a}$$

$$T^{11} \propto \frac{1}{3} \int_0^{2\pi} d\theta \sin^4\theta = \frac{\pi}{4}. \tag{10.60b}$$

Thus the vacuum expectation value of the energy-momentum tensor has the required form (of course, nothing else is possible, from relativistic covariance):

$$\langle T^{\mu\nu}\rangle = -u(a)g^{\mu\nu} = -\frac{\Lambda}{8\pi G}g^{\mu\nu}. \tag{10.61}$$

[That this argument is not merely formal, but holds for the finite regulated terms as well, follows from the approach given in Sec. 10.1.4, for example, for even N.]

What now seems to result from the analysis of distant type Ia supernovæ [319, 320, 321], consistent with balloon measurements of the anisotropy of the cosmic microwave background [323, 324, 325, 326, 327, 328, 329], is a positive cosmological constant of the same order as the critical density,

$$u(a) \sim \rho_c \sim 10^{-5}\,\mathrm{GeV/cm^3}. \tag{10.62}$$

If we take $u(a)$ to be of the order of magnitude as given for the maximum in in Fig. 10.2, $u \sim 10^{-3}/a^4$, we find, restoring units ($\hbar c = 2 \times 10^{-14}\,\mathrm{GeV\,cm}$)

$$a^4 \sim 10^2 \frac{\mathrm{cm^3}}{\mathrm{GeV}}\hbar c \sim 10^{-12}\,\mathrm{cm^4}, \tag{10.63}$$

or

$$a \sim 10^{-3}\,\mathrm{cm} = 10\,\mu\mathrm{m}. \tag{10.64}$$

Thus, "large" extra dimensions are necessary to understand the cosmological constant. Indeed, were they much smaller, an unacceptably large cosmological constant would result, while if they were bigger than a fraction of a millimeter, they would be phenomenologically irrelevant in this

context. However, very recent measurements now show no deviations from
Newton's law down to the 100 μm level [362], so this interesting regime is
on the point of being ruled out experimentally. In particular, the graviton
fluctuation energy of Cho and Kantowski [351] for S^6 is too large to be
consistent with Newton's law, while that for S^4 is of the wrong sign to be
consistent with a positive cosmological constant. For further background
on the experimental limits, see Refs. [363, 364, 365].

Chapter 11

Local Effects

11.1 Parallel Plates

Heretofore, we have considered the global Casimir effect: the total energy of a field configuration or the force per unit area on a bounding surface.* But one can also consider the local energy density, or, more generally $\langle T^{\mu\nu} \rangle$, which will reveal new information about the divergence structure of the theory. Such a quantity is also relevant to the coupling to gravity, as we saw in the previous Chapter.

We begin by considering a scalar field subject to Dirichlet boundary conditions on parallel plates at $z = 0$ and $z = a$, for which the vacuum expectation value of the energy density is given by (2.38), or

$$\langle t^{00} \rangle = -\frac{1}{2i\lambda \sin \lambda a}[\omega^2 \cos \lambda a - k^2 \cos \lambda(2z - a)], \qquad (11.1)$$

where $\lambda^2 = \omega^2 - k^2$. The energy per unit volume is obtained from this by integrating over frequency and wavevectors,

$$\langle T^{00} \rangle(z) = \int \frac{d\omega \, d^2 k}{(2\pi)^3} \langle t^{00} \rangle. \qquad (11.2)$$

We evaluate this by making a Euclidean rotation,

$$\omega \to i\zeta, \quad \lambda \to i\kappa, \qquad (11.3)$$

and, as in Sec. 2.3, introducing polar coordinates in the ζ, k plane,

$$\zeta = \kappa \cos \theta, \quad k = \kappa \sin \theta, \qquad (11.4)$$

*Although local, the energy densities computed in the last Chapter were constant.

223

so

$$
\begin{aligned}
\langle T^{00}\rangle(z) &= -\frac{1}{4\pi^2}\int_0^\infty \kappa\,d\kappa \int_0^{\pi/2} d\theta\,\kappa^2 \frac{\sin\theta}{\sinh\kappa a}[\cos^2\theta\cosh\kappa a \\
&\qquad\qquad + \sin^2\theta\cosh\kappa(2z-a)] \\
&= -\frac{1}{12\pi^2}\int_0^\infty d\kappa\,\kappa^3 \frac{1}{\sinh\kappa a}[\cosh\kappa a + 2\cosh\kappa(2z-a)] \\
&= -\frac{1}{6\pi^2}\int_0^\infty d\kappa\,\kappa^3 \left(\frac{1}{e^{2\kappa a}-1} + \frac{1}{2} + \frac{e^{2\kappa z}+e^{2\kappa(a-z)}}{e^{2\kappa a}-1}\right).
\end{aligned}
$$

$$(11.5)$$

Notice that the second term in the last integrand here corresponds to a constant energy density, independent of a, so it may be discarded as irrelevant. If we integrate the third term over z,

$$
\int_0^a dz\left[e^{2\kappa z}+e^{2\kappa(a-z)}\right] = \frac{1}{\kappa}\left[e^{2\kappa a}-1\right], \qquad (11.6)
$$

we obtain another (divergent) constant term, so the only part of the vacuum energy corresponding to an observable force is that coming from the first term:

$$
\int_0^a dz\,\langle T^{00}\rangle(z) = -\frac{a}{6\pi^2}\int_0^\infty d\kappa\frac{\kappa^3}{e^{2\kappa a}-1} = -\frac{\pi^2}{1440a^3}, \qquad (11.7)
$$

just as found in Chapter 2—see (2.9).

In general, we have

$$
\langle T^{00}\rangle(z) = u + g(z), \qquad (11.8\text{a})
$$

$$
u = -\frac{\pi^2}{1440a^4} \qquad (11.8\text{b})
$$

where

$$
g(z) = -\frac{1}{6\pi^2}\frac{1}{16a^4}\int_0^\infty dy\,y^3\frac{e^{yz/a}+e^{y(1-z/a)}}{e^y-1}. \qquad (11.9)
$$

If we expand the denominator in a geometric series,

$$
\frac{1}{e^y-1} = \frac{e^{-y}}{1-e^{-y}} = \sum_{n=1}^\infty e^{-ny}, \qquad (11.10)
$$

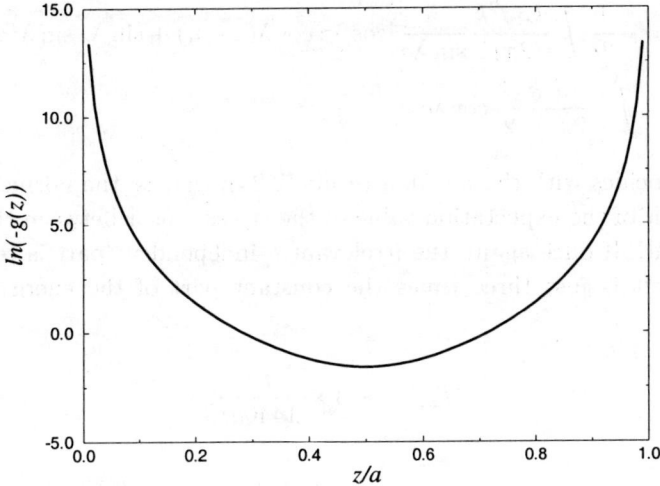

Fig. 11.1 The singular part of the local energy density between parallel plates at $z = 0$ and $z = a$.

we can express g in terms of the generalized or Hurwitz zeta function,

$$\zeta(s, a) \equiv \sum_{n=0}^{\infty} \frac{1}{(n+a)^s}, \quad a \neq \text{ a nonnegative integer}, \qquad (11.11)$$

as follows:

$$g(z) = -\frac{1}{16\pi^2 a^4}[\zeta(4, z/a) + \zeta(4, 1 - z/a)]. \qquad (11.12)$$

This function is plotted in Fig. 11.1, where it will be observed that it diverges quartically as $z \to 0, a$. (Its z integral over the region between the plates diverges cubically.) As we have seen, this badly behaved function does not contribute to the force on the plates.

Next, we turn to $\langle T_{zz} \rangle$. According to the stress tensor (2.21) and the Green's function (2.20), that is given by

$$\langle T_{zz} \rangle = \frac{1}{2i}(\partial_z \partial_{z'} - \partial_x \partial_{x'} - \partial_y \partial_{y'} + \partial_0 \partial_{0'})G(x, x')$$

$$= \frac{1}{2i} \int \frac{d\omega\, d^2k}{(2\pi)^3}(\partial_z \partial_{z'} + \lambda^2)\left[-\frac{1}{\lambda \sin \lambda a}\sin \lambda z_< \sin \lambda(z_> - a)\right]$$

$$= -\frac{1}{2i} \int \frac{d\omega \, d^2k}{(2\pi)^3} \frac{\lambda}{\sin \lambda a} [\cos \lambda z \cos \lambda(z-a) + \sin \lambda z \sin \lambda(z-a)]$$

$$= \int \frac{d\omega \, d^2k}{(2\pi)^3} \frac{i\lambda}{2} \cot \lambda a, \qquad (11.13)$$

which coincides with the $z = 0$, a result (2.24); that is, the normal-normal component of the expectation value of the stress tensor between the plates is constant! If once again, the irrelevant a-independent part is removed,[†] what is left is just three times the constant part of the energy density (11.8b),

$$\langle T_{zz} \rangle = -3 \times \frac{\pi^2}{1440a^4}. \qquad (11.15)$$

The remaining nonzero components of the stress tensor are

$$\langle T_{xx} \rangle = \langle T_{yy} \rangle = \frac{1}{2i}[\partial_x \partial_{x'} - \partial_y \partial_{y'} - \partial_z \partial_{z'} + \partial_0 \partial_{0'}]G(x, x')$$

$$= -\frac{1}{2i} \int \frac{d\omega \, d^2k}{(2\pi)^3} \frac{1}{\lambda \sin \lambda a} [\omega^2 \sin \lambda z \sin \lambda(z-a)$$

$$- \lambda^2 \cos \lambda z \cos \lambda(z-a)]$$

$$= -u - g(z), \qquad (11.16)$$

where, once again, we have introduced polar coordinates in the frequency-wavenumber plane, and dropped the infinite (a-independent) constant in u. Thus the tensor structure of stress-tensor is

$$\langle T^{\mu\nu} \rangle(z) = u \begin{pmatrix} 1 & 0 & 0 & 0 \\ 0 & -1 & 0 & 0 \\ 0 & 0 & -1 & 0 \\ 0 & 0 & 0 & 3 \end{pmatrix} + g(z) \begin{pmatrix} 1 & 0 & 0 & 0 \\ 0 & -1 & 0 & 0 \\ 0 & 0 & -1 & 0 \\ 0 & 0 & 0 & 0 \end{pmatrix}, \qquad (11.17)$$

where u is given by (11.8b) and g by (11.12). Because u is constant, this vacuum expectation value is divergenceless, because $g(z)$ does not contribute

[†]The infinite parts of $\langle T_{00} \rangle$ and $\langle T_{zz} \rangle$ are related by the same factor of three:

$$\langle T_{zz} \rangle^{\text{inf}} = -\int \frac{d\kappa \, \kappa^3}{4\pi^2}, \qquad \langle T_{00} \rangle^{\text{inf}} = -\int \frac{d\kappa \, \kappa^3}{12\pi^2}. \qquad (11.14)$$

to $\langle T^{zz} \rangle$:

$$\partial_\mu \langle T^{\mu\nu} \rangle = \partial_z \langle T^{zz} \rangle = 0. \tag{11.18}$$

The second term in (11.17) diverges at the boundaries, $z = 0, a$, and has a integral over the volume which diverges; yet as we have seen, it is physically irrelevant because its integral is constant, and it has no normal component. Is there a natural way in which it simply does not appear in the local formulation?

The affirmative answer hinges on the ambiguity in defining the stress tensor.[‡] In Chapter 2 we noted that this ambiguity was without effect as far as the total stress or the total energy was concerned. Now, however, we see the virtue of the conformal stress tensor [120]:

$$\tilde{T}^{\mu\nu} = \partial^\mu \phi \partial^\nu \phi - \frac{1}{2} g^{\mu\nu} \partial_\lambda \phi \partial^\lambda \phi - \frac{1}{6} (\partial^\mu \partial^\nu - g^{\mu\nu} \partial^2) \phi^2, \tag{11.19}$$

which, by virtue of the equation of motion $\partial^2 \phi = 0$ has a vanishing trace,

$$\tilde{T}^\mu_\mu = 0. \tag{11.20}$$

If we use this stress tensor rather than the canonical one, we merely need supplement the above computations by that of the vacuum expectation value of the extra term. Thus to obtain $\langle \tilde{T}^{xx} \rangle$ we add to (11.16)

$$\begin{aligned}
\frac{1}{6i} (\partial_y^2 &+ \partial_z^2 - \partial_0^2) G(x, x) \\
&= \frac{1}{6i} \int \frac{d\omega\, d^2k}{(2\pi)^3} \partial_z^2 \left[-\frac{1}{\lambda \sin \lambda a} \sin \lambda z \sin \lambda(z - a) \right] \\
&= -\frac{1}{6i} \int \frac{d\omega\, d^2k}{(2\pi)^3} \frac{2\lambda}{\sin \lambda a} \cos \lambda(2z - a) \\
&= g(z),
\end{aligned} \tag{11.21}$$

which just cancels the extra term in (11.16). Again, because $G(x, x)$ only depends on z, there is no extra contribution to $\langle T_{zz} \rangle$:

$$-\frac{1}{6} (\partial_z^2 - g_{zz} \partial^2) \langle \phi^2 \rangle = \frac{1}{6i} (\partial_x^2 + \partial_y^2 - \partial_0^2) G(x, x) = 0. \tag{11.22}$$

[‡]For a rather complete discussion of this see Ref. [119], Secs. 3-7, 3-17.

The extra term for $\langle T_{00}\rangle$ is just the negative of that in (11.21),

$$-\frac{1}{6i}\partial_z^2 G(x,x) = -g(z),\qquad(11.23)$$

which cancels the second term in (11.8a). Thus, the conformal stress tensor has the following vacuum expectation value for the region between the parallel plates:

$$\langle \tilde{T}^{\mu\nu}\rangle = u\begin{pmatrix} 1 & 0 & 0 & 0 \\ 0 & -1 & 0 & 0 \\ 0 & 0 & -1 & 0 \\ 0 & 0 & 0 & 3 \end{pmatrix},\qquad(11.24)$$

which is traceless, thereby respecting the conformal invariance of the massless theory. This is just the result found by Brown and Maclay by general considerations [100], who argued that

$$\langle \tilde{T}^{\mu\nu}\rangle = u[4\hat{z}^\mu\hat{z}^\nu - g^{\mu\nu}],\qquad(11.25)$$

where $\hat{z}^\mu$ is the unit vector in the z direction.

11.2   Local Casimir Effect for Wedge Geometry

In Sec. 7.1.3 we briefly discussed the Casimir effect in wedge-shaped geometries, which are relevant for cosmic strings. Using the formalism developed in Sec. 7.1, as first stated in Ref. [24], Brevik and Lygren [254] (see also Ref. [260]) considered the region between two conducting planes making a dihedral angle of α. The vacuum expectation value of the electromagnetic stress tensor is in cylindrical coordinates (t, r, θ, z)

$$\langle T_{\mu\nu}(r)\rangle = -\frac{1}{720\pi^2 r^4}\left(\frac{\pi^2}{\alpha^2}+11\right)\left(\frac{\pi^2}{\alpha^2}-1\right)\begin{pmatrix} 1 & 0 & 0 & 0 \\ 0 & -1 & 0 & 0 \\ 0 & 0 & 3 & 0 \\ 0 & 0 & 0 & -1 \end{pmatrix}.\qquad(11.26)$$

Note that this reduces to twice the scalar result for parallel plates, (11.24), in the limit $\alpha \to 0$, as expected. For a cosmic string, the line element is

$$ds^2 = -dt^2 + dr^2 + \beta^{-2}r^2 d\theta^2 + dz^2,\quad \beta = (1-4\mu G)^{-1},\qquad(11.27)$$

where μ is the mass per unit length of the string. The stress tensor in the presence of the cosmic string is obtained from (11.26) by the replacement $\pi/\alpha \to \beta$.

11.3 Other Work

Local effects on a scalar quantum field induced by ideal point and line boundaries were discussed by Actor [366]. Actor also considered the local effect on a scalar field confined within a rectangular cavity in Ref. [250]. He makes a conjecture concerning the extraction of the finite effective potential, or global Casimir energy, which seems rather dubious, however, given the neglect of external modes.

11.4 Quark and Gluon Condensates in the Bag Model

The considerations in Sec. 11.1 were quite clear cut, as is usual in the essentially one-dimensional geometry of parallel plates. Three-dimensional configurations are of much more interest, yet they provide extra difficulties. In fact, as we have already seen in this monograph, the general divergence structure of the Casimir effect is still not well understood.

Here, however, we will hazard a prolegomenon of a calculation, given nearly two decades ago [21] which is of great importance in hadronic physics. This has to do with the quark and gluon condensates in the QCD vacuum. The phenomenological input is the value of the gluon condensate derived from an analysis of QCD sum rules [234, 235, 367]

$$\langle \frac{\alpha_s}{\pi} G^2 \rangle = 0.012 \,\text{GeV}^4, \tag{11.28}$$

which is said to signify a chromomagnetic vacuum because

$$\frac{1}{4} G^2 = \frac{1}{2} (\mathbf{B}^a \cdot \mathbf{B}^a - \mathbf{E}^a \cdot \mathbf{E}^a), \tag{11.29}$$

a being the color index. Theoretically, this result is supported by the observation [368, 369] that the one-loop effective potential [368, 369, 370, 371] (in the absence of quarks)

$$V\left(\langle \frac{1}{4} G^2 \rangle\right) = \langle \frac{1}{4} G^2 \rangle - \frac{1}{2} b_0 g^2 \langle \frac{1}{4} G^2 \rangle \left(\ln \langle \frac{1}{4} G^2 \rangle + \text{const.} \right) \tag{11.30}$$

possesses a minimum with a nonzero value of $\langle G^2 \rangle$, the condensate,

$$\langle \frac{1}{4}G^2 \rangle = \left(\frac{\Lambda^4}{g^2} \right) \exp(2/b_0 g^2 - 1), \tag{11.31}$$

where the scale parameter Λ incorporates the constant in (11.30). Numerically ($b_0 = -11/16\pi^2$, if we ignore quarks, and we take $\alpha_s = g^2/4\pi \approx 0.2$) this gives

$$\langle G^2 \rangle \approx \Lambda^4 \times 6 \times 10^{-6}, \tag{11.32}$$

which, combined with the phenomenological result (11.28) yields for Λ

$$\Lambda \approx 13 \, \text{GeV}, \tag{11.33}$$

much larger that the scale parameter of QCD,

$$\mu \approx 300 \, \text{MeV}. \tag{11.34}$$

Part of this discrepancy is a renormalization group effect.[§] For if we regard Λ as the scale at which α_s has the value 0.2, (11.31) would read (e is the base of the natural logarithms)

$$\langle \frac{\alpha_s}{\pi} G^2 \rangle = \frac{\mu^4}{\pi^2 e}, \tag{11.36}$$

some 30 times too small. In this sense the gluon condensate is large.

Similarly, the vacuum is characterized by a quark condensate,

$$-\langle \bar{q}q \rangle \approx 0.4\mu^3 \approx 0.01 \, \text{GeV}^3 \tag{11.37}$$

for each light quark, as we know from chiral symmetry breaking [372]. This second phenomenon has not been derived from an effective potential, renormalization-group type of argument, because of the difficulty of including quarks in that analysis.

Two obvious questions are raised by the above remarks: (i) Is it possible to shed further light on the condensation mechanism of quarks and gluons? In particular, the occurrence of the correct sign for both condensates is nontrivial. (ii) Beyond this it would be of great interest if the numerical

[§]The renormalization group equation for the running strong coupling is, at the one-loop level,

$$\mu \frac{dg}{d\mu} = b_0 g^3. \tag{11.35}$$

This is the basis of (11.30).

values of $\langle G^2 \rangle$ and of $\langle \bar{q}q \rangle$ could be understood, particularly their great difference in scale. Of course, there remains a question of a rather higher order: What does the condensation phenomenon tell us about the dynamics of QCD, especially confinement?

Olaussen and Ravndal [373, 374] were the first to attempt to calculate the gluon condensate in terms of vacuum fluctuations in Johnson's bag model of the vacuum [17], described in Chapter 6. Here we will use the machinery described in that chapter. We will first calculate the free ($g = 0$) quantum fluctuation value of $\langle G^2(r) \rangle$ inside a spherical cavity of radius a, the surface of which is a perfect magnetic conductor. Because of the dual connection between the chromomagnetic field and the Green's dyadic $\boldsymbol{\Gamma}$, we have from (4.8)

$$\langle B^2(r) \rangle = \frac{1}{i} \int \frac{d\omega}{2\pi} e^{-i\omega\tau} \sum_{lm} \{\omega^2 F_l(r,r) |\mathbf{X}_{lm}(\Omega)|^2$$

$$- \text{tr}|\boldsymbol{\nabla} \times [G_l(r,r') \mathbf{X}_{lm}(\Omega) \mathbf{X}_{lm}^*(\Omega')] \times \overleftarrow{\boldsymbol{\nabla}}'|_{\mathbf{r}=\mathbf{r}'}\}, \quad (11.38)$$

where in the interior of the bag, $r, r' < a$,

$$F_l(r,r'), G_l(r,r') = ikj_l(kr_<)[h_l(kr_>) - A_{F,G}j_l(kr_>)]. \quad (11.39)$$

Here, in order the to satisfy the boundary conditions (6.1), the constants A_F, A_G are given by (6.5a), (6.5b). To eliminate the m sum over the vector spherical harmonics we use (4.6) and (when f_l, g_l are spherical Bessel functions of order l)

$$\sum_{m=-l}^{l} [\boldsymbol{\nabla} \times f_l(r) \mathbf{X}_{lm}(\Omega)] \cdot [\boldsymbol{\nabla} \times g_l(r) \mathbf{X}_{lm}(\Omega)]^*$$

$$= \frac{2l+1}{4\pi} \left(\frac{1}{r^2} \frac{\partial}{\partial r}(rf_l) \frac{\partial}{\partial r}(rg_l) + l(l+1) \frac{f_l g_l}{r^2} \right)$$

$$= \frac{2l+1}{4\pi} k^2 \left[f_l g_l + \frac{1}{k^2 r^2} \frac{d}{dr} \left(rf_l \frac{d}{dr}(rg_l) \right) \right]. \quad (11.40)$$

Thus the vacuum expectation value of the square of the magnetic field within the bag is

$$\langle B^2(r) \rangle = \frac{1}{i} \int \frac{d\omega}{2\pi} e^{-i\omega\tau} \sum_{l=1}^{\infty} \frac{2l+1}{4\pi} \left[\omega^2 F_l(r,r) + \omega^2 G_l(r,r) \right.$$

$$+ \frac{1}{r^2} \frac{d}{dr} r \left(\frac{\partial}{\partial r'} r' G_l(r,r') \right)_{r'=r} \right]. \quad (11.41)$$

By interchanging $F_l \leftrightarrow G_l$ we obtain the expectation value of E^2:

$$\langle E^2(r) \rangle = \frac{1}{i} \int \frac{d\omega}{2\pi} e^{-i\omega\tau} \sum_{l=1}^{\infty} \frac{2l+1}{4\pi} \left[\omega^2 G_l(r,r) + \omega^2 F_l(r,r) \right.$$

$$\left. + \frac{1}{r^2} \frac{d}{dr} r \left(\frac{\partial}{\partial r'} r' F_l(r,r') \right)_{r'=r} \right]. \quad (11.42)$$

Evidently $4\pi \int_0^a r^2 \, dr \, \langle \frac{1}{2}[E^2(r) + B^2(r)] \rangle$ is the form of the gluon zero-point energy given, for example, in (5.27b).

It is interesting to note that in the QED case of a perfectly conducting spherical shell, considered in Sec. 4.1, where both interior and exterior modes are present, the surface term in the total energy vanishes, so that from (11.41) and (11.42) we see that the electric and magnetic fields make equal net contributions.[¶] This is not the case for the QCD bag since then the surface term makes a crucial contribution. For the quantity of interest here, $G^2 = G_{\mu\nu}G^{\mu\nu} = 2(B^2 - E^2)$, the surface term is everything:

$$\langle G^2 \rangle = \frac{1}{4\pi^2 i} \int d\omega \, e^{-i\omega\tau} \sum_{l=1}^{\infty} (2l+1)$$

$$\times \frac{1}{r^2} \frac{d}{dr} r \left(\frac{\partial}{\partial r'} r' [G_l(r,r') - F_l(r,r')] \right)_{r'=r}. \quad (11.43)$$

Substituting (6.5a) and (6.5b), introducing the modified Bessel function (4.25), and using the Wronskian (4.26), we easily find the following simple expression for the vacuum expectation value of G^2:

$$\langle G^2(r) \rangle = \frac{1}{4\pi^2 a} \frac{1}{r^2} \frac{d}{dr} \sum_{l=1}^{\infty} (2l+1)2 \int_0^{\infty} dx \, e^{ix\delta} \frac{s_l(xr/a)s_l'(xr/a)}{s_l(x)s_l'(x)},$$

$$(11.44)$$

where the limit $\delta \to 0$ is understood. If we do not integrate over r, we may set $\delta = 0$.

Before proceeding with the numerical evaluation of this, let us derive the corresponding expression for $\langle \bar{q}q \rangle$ in the bag. For a single color and

---

[¶]The surface term contributions from $\langle E^2 \rangle$ and $\langle B^2 \rangle$ separately are not zero, but are equal and opposite and hence cancel.

flavor we have

$$\langle \bar{q}q(r) \rangle = -\frac{1}{i} \text{tr}\, G(x,x), \qquad (11.45)$$

where the fermion Green's function was determined in Sec. 4.2. Using the completeness relation (4.57) for the angular momentum eigenstates given there, we find immediately

$$\langle \bar{q}q(r) \rangle = -\frac{1}{4\pi^2 a r^2} \sum_{l=0}^{\infty} 2(l+1) \int_{-\infty}^{\infty} dx\, e^{ix\delta} \frac{s_l^2(xr/a) + s_{l+1}^2(xr/a)}{s_l^2(x) + s_{l+1}^2(x)}. \quad (11.46)$$

The numerical evaluation of formulæ (11.44) and (11.46) is straight-forward using the uniform asymptotic expansion for large l, (4.29a). The leading term, which is probably quite accurate[||] gives

$$l \to \infty: \quad \frac{s_l(xr/a)s_l'(xr/a)}{s_l(x)s_l'(x)} \sim \exp\{-2\nu[\eta(z) - \eta(zr/a)]\}, \qquad (11.47)$$

where z, ν and η are given by (4.30). The l sum can then be carried out exactly, and the z integral is easily evaluated numerically. The result is shown in Fig. 11.2, and coincides, for all r, to better than 1% with an exact numerical evaluation of (11.44). Especially of interest are the values at the center of the bag,[**]

$$r = 0: \quad \langle G^2 \rangle = \frac{0.80}{a^4}, \qquad (11.48a)$$

and near the surface,

$$r \to a: \quad \langle G^2 \rangle = \frac{3}{4\pi^2} \frac{1}{(a-r)^4} = \frac{0.076}{(a-r)^4}. \qquad (11.48b)$$

These agree with the results of Olaussen and Ravndal [373].

[||] As shown in Sec. 6.2.1.3, the leading nonvanishing term in the asymptotic expansion for the energy is already very good for the lowest modes, $l = 1$ and $l = 0$, for gluons and quarks, respectively.

[**] To within about 1%, this may be obtained directly from the lowest mode contribution, $l = 1$, which is the only mode contributing at $r = 0$. Exact numerical integration gives $\langle G^2(0) \rangle = 0.797/a^4$. This shows that the asymptotic expansion is valid all the way down to $l = 1$. The same remark applies to the $l = 0$ quark contribution.

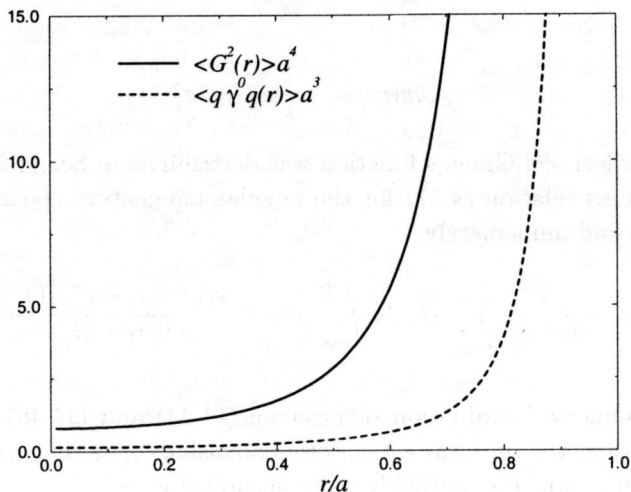

Fig. 11.2 Magnitude of the quark and gluon condensates as functions of the distance from the center of the bag. The values shown are for a single color, and, for the quarks, for a single flavor and helicity state.

The quark condensate (11.46) is only slightly harder to evaluate. It is helpful to use the recursion relation

$$s_{l+1}(x) = s'_l(x) - \frac{l+1}{x} s_l(x), \qquad (11.49)$$

and then we see as $l \to \infty$:

$$\frac{s_l^2(xr/a) + s_{l+1}^2(xr/a)}{s_l^2(x) + s_{l+1}^2(x)} \sim \frac{a}{r} \frac{(1 + z^2 r^2/a^2)^{1/2} - 1}{(1 + z^2)^{1/2} - 1}$$
$$\times \exp\{-2\nu[\eta(z) - \eta(zr/a)]\}. \quad (11.50)$$

The resulting numerical integral is also given in Fig. 11.2. Here the limits are

$$r = 0: \quad \langle \bar{q}q \rangle = -\frac{0.15}{a^3}, \qquad (11.51a)$$

$$r \to a: \quad \langle \bar{q}q \rangle = -\frac{0.025}{(a-r)^3}. \qquad (11.51b)$$

It is now clear where the problem lies in applying these results to nature: Both condensates give divergent results when integrated over the bag, because of the singularity near the surface. (This reminds us of the similar behavior seen in Fig. 11.1 for the canonical scalar energy density near the wall in the parallel plate geometry.) In fact, when $a - r$ is sufficiently small, the cutoff δ plays a crucial role. It is especially interesting to look at the total gluon condensate, since it is entirely a surface term,

$$4\pi \int_0^a r^2 \, dr \langle G^2 \rangle = \frac{1}{\pi a} \sum_{l=1}^{\infty} (2l + 1) \int_{-\infty}^{\infty} dx \, e^{ix\delta}; \qquad (11.52)$$

it is very tempting to identify the last integral with $2\pi\delta(\delta)$, and then interpret this as a physically uninteresting contact term. Does this mean that the gluon condensate, in the bag model at least, vanishes? Most probably this is the wrong interpretation.

We prefer instead to regard the surface divergence in (11.48b) and (11.51b) as a defect of the bag model, as a reflection of the over-idealized nature of the bag boundary conditions. One might suppose that if the boundary conditions were suitably softened, the singularity would disappear. To get some notion of the expected meaningful results, suppose we take the values at the center of the bag as typical, presumably rather insensitive to the boundary conditions. Then, with $a = 2.6 \, \mathrm{GeV}^{-1}$ appropriate for an empty bag [17], we obtain from (11.48a) for the gluon condensate summed over the eight gluons ($\alpha_s = 0.2$)

$$\langle \frac{\alpha_s}{\pi} G^a G^a \rangle = 0.01 \, \mathrm{GeV}^4, \qquad (11.53)$$

which is really remarkably close to the sum rule value (11.28). And multiplying (11.51a) by 3 for the color trace give

$$\langle \bar{q}^a q^a \rangle = -0.03 \, \mathrm{GeV}^3, \qquad (11.54)$$

quite consistent with (11.37) given the crudeness of our model.

To conclude, we see that quark and gluon condensation of the right magnitude emerge naturally from elementary considerations of quantum fluctuations. The bag model is an artifice for incorporating our notions of asymptotic freedom and infrared slavery and evidently needs to be improved to eliminate spurious surface divergences. All of this suggests the intimate connection between these condensation phenomena and color confinement.

11.5 Surface Divergences

We have seen in several examples that in general the Casimir energy density diverges in the neighborhood of a surface. For flat surfaces and conformal theories (such as the conformal scalar theory considered in Sec. 11.1, or electromagnetism) those divergences are not present.[tt] However, as Deutsch and Candelas [102] showed many years ago, for conformally invariant theories, $\langle T_{\mu\nu} \rangle$ diverges as ϵ^{-3}, where ϵ is the distance from the surface, with a coefficient proportional to the sum of the principal curvatures of the surface. In particular they obtain the result, in the vicinity of the surface,

$$\langle T_{\mu\nu} \rangle \sim \epsilon^{-3} T_{\mu\nu}^{(3)} + \epsilon^{-2} T_{\mu\nu}^{(2)} + \epsilon^{-1} T_{\mu\nu}^{(1)}, \tag{11.55}$$

and obtain explicit expressions for the coefficient tensors $T_{\mu\nu}^{(3)}$ and $T_{\mu\nu}^{(2)}$ in terms of the extrinsic curvature of the boundary.

For example, for the case of a sphere, the leading surface divergence has the form, for conformal fields, for $r = a + \epsilon$, $\epsilon \to 0$

$$\langle T_{\mu\nu} \rangle = \frac{A}{\epsilon^3} \begin{pmatrix} 2/a & 0 & 0 & 0 \\ 0 & 0 & 0 & 0 \\ 0 & 0 & 1 & 0 \\ 0 & 0 & 0 & \sin\theta \end{pmatrix}, \tag{11.56}$$

in spherical polar coordinates, where the constant is

$$A = \frac{1}{1440\pi^2} \quad \text{for a scalar, or} \quad A = \frac{1}{120\pi^2} \quad \text{for the electromagnetic field.} \tag{11.57}$$

Note that (11.56) is properly traceless. The cubic divergence in the energy density near the surface translates into the quadratic divergence in the energy found, for example, in Chapter 5, cf. (5.93). The corresponding quadratic divergence in the stress corresponds to the absence of the cubic divergence in $\langle T_{rr} \rangle$.

This is all completely sensible. However, in their paper Deutsch and Candelas [102] express a certain skepticism about the validity of the result of Ref. [15] for the spherical shell case [described in Chapter 4] where the

---

[tt]In general, this need not be the case. For example, Romeo and Saharian [375] show that with mixed boundary conditions (9.35) the surface divergences need not vanish for parallel plates. For additional work on local effects with mixed boundary conditions, applied to spheres and cylinders, and corresponding global effects, see Refs. [376, 377].

divergences cancel. That skepticism is reinforced in a later paper by Candelas [213], who criticizes the authors of Ref. [15] for omitting δ function terms, and constants in the energy. These objections seem utterly without merit. A certain degree of shrillness in present there, and in a later critical paper by the same author [201], in which it is asserted that errors were made, instead of a conscious removal of unphysical divergences.

Of course, surface curvature divergences are present. As Candelas notes [213, 201], they have the form

$$E = E^S \int dS + E^C \int dS \left(\kappa_1 + \kappa_2 \right)$$

$$+ E_I^C \int dS \left(\kappa_1 - \kappa_2 \right)^2 + E_{II}^C \int dS \kappa_1 \kappa_2 + \ldots, \qquad (11.58)$$

where κ_1 and κ_2 are the principal curvatures of the surface. The question is to what extent are they observable. After all, as we saw in the first section of this Chapter, we can drastically change the local structure of the vacuum expectation value of the energy-momentum tensor by merely exploiting the ambiguity in the definition of that tensor, yet each yields the same finite, observable (and observed!) energy of interaction between the plates. For curved boundaries, much the same is true. *A priori*, we do not know which energy-momentum tensor to employ, and the local vacuum-fluctuation energy density is to a large extent meaningless. It is the global energy, or the force between distinct bodies, that has an unambiguous value. It is the belief of the author that divergences in the energy which go like a power of the cutoff are probably unobservable, being subsumed in the properties of matter. Moreover, the coefficients of the divergent terms depend on the regularization scheme. Logarithmic divergences, of course, are of another class [41].

Chapter 12

Sonoluminescence and the Dynamical Casimir Effect

12.1 Introduction

Single-bubble sonoluminescence* [380, 381, 382, 383, 384, 385, 386, 387, 34] remains a curiously poorly understood subject. Recall, from the brief discussion in the Introduction, that if a small air bubble, of radius $\sim 10^{-3}$ cm, is injected into water and subjected to a strong acoustic field (over-pressure ~ 1 atm, frequency $\sim 2 \times 10^4$ Hz) at minimum radius $\sim 10^{-4}$ cm the bubble emits an intense flash of light, of total energy ~ 10 MeV in the optical. The flash duration has recently been determined to be on the order of 100 ps [388, 389, 390]. Shock wave emission has also been observed [391]. The process is sufficiently non-catastrophic that a single bubble may continue to undergo collapse and emission for periods of times up to months. Many still unexplained properties have been observed, including sensitivity to small impurities, strong temperature dependence, and the necessity of small amounts of noble gases being present.[†]

Although there have been many interesting theoretical explanations starting from Ref. [396], none of them seem, in the words of Putterman et al., "falsifiable" [34]. Among the ideas proposed are Bremsstrahlung from ionized regions [397, 398, 399, 400, 401, 389], fluid dynamical models [402, 403], blackbody radiation [404], molecular interactions [405, 406], inelastic collisions between atoms [407], a coherent lasing process [408, 409], and a coherent QED interaction with the water vapor-liquid phase transition [410].

---

*Multiple-bubble sonoluminescence was discovered in the 1930s [378, 379].

[†]In fact, the bubbles may be entirely composed of noble gas [392, 393, 394, 395].

Perhaps the most intriguing idea, initially proposed by Julian Schwinger, was the notion that this phenomenon had its origin in zero-point fluctuations of the electromagnetic field, or the Casimir effect. In a series of papers in the last three years of his life, Schwinger proposed [35, 36] that the "dynamical Casimir effect" could provide the energy that drives the copious production of photons in this puzzling phenomenon. In fact, however, he guessed an approximate (static) formula for the Casimir energy of a spherical bubble in water, based on a general, but incomplete, analysis [103, 104], leading, for a spherical bubble of radius a in a medium with index of refraction n, to a Casimir energy proportional to the volume of the bubble:

$$E_{\text{bulk}} = \frac{4\pi a^3}{3} \int \frac{(d\mathbf{k})}{(2\pi)^3} \frac{1}{2} k \left(1 - \frac{1}{n}\right). \qquad (12.1)$$

which is just the difference between the zero-point energy $\sum \frac{1}{2}\hbar\omega$ of the medium from that of the vacuum. Of course, this expression is quartically divergent. If he put in a suitable ultraviolet cutoff, Schwinger could indeed obtain the needed 10 MeV per flash. On the other hand, one might have serious reservations about the physical meaning of such a divergent result. Schwinger apparently was unaware that I had, in the late 1970s, completed the analysis of the Casimir force for a dielectric ball [16]. The generalization to a bubble of permittivity ϵ' and permeability μ' in a medium of permittivity ϵ and permeability μ is rather immediate, as was first described in Ref. [411, 37] and described in Chapter 5. (The divergence structure of the calculation had also been investigated by Candelas [213].)

Of course, the calculation given there is not directly relevant to sonoluminescence, which is anything but static, since the frequency of the bubble collapse and re-expansion is ~ 20 kHz. It is offered as only a preliminary step, but it should give an idea of the orders of magnitude of the energies involved. It is a significant improvement over the crude estimation used in Ref. [35]. Attempts at dynamical calculations exist [412, 413, 414, 415, 416, 417, 418, 419, 420, 421]; but they are subject to possibly serious methodological objections, some of which will be discussed below, and probably cannot be trusted beyond the adiabatic approximation in any event. In fact, we anticipate that because the relevant scale of the electromagnetic Casimir effect is in the optical region, with characteristic time scale $t \sim 10^{-15}$s, and the scale of the bubble collapse is of order $\tau \sim 10^{-6}$s (undoubtedly, more relevant is the duration of each flash, which has been measured to be on the order of 10^{-10}s [388, 389,

390]), the adiabatic approximation of treating the bubble as static for calculating the Casimir energy should be very accurate.

We recall from Chapter 5 that the Casimir energy so constructed, even with physically required subtractions, and including both interior and exterior contributions, is divergent, but that if one supplies a plausible contact term, a finite result (at least for a dilute medium) follows. This finite result agrees with that found using ζ-function regularization. (Physically, we expect that the divergence is regulated by including dispersion in both frequency and wavevector, i.e., angular momentum, and the divergences are absorbed in a renormalization of physical parameters.) This finite result is the same as that obtained from the sum of van der Waals interactions, see Sec. 5.9 and the following section. Numerical estimates of both the divergent and finite terms are given in the next section, and comparison is made with the calculations of Schwinger and others.

But although these arguments against the relevance of the Casimir energy to sonoluminescence may seem persuasive, a number of authors have continued to insist of the viability of the scheme. In particular, Carlson et al. [422, 423, 424] insist on the relevance of the bulk energy (12.1). So this issue is examined in detail in Sec. 12.4. Subsequently, Liberati et al. [417, 418, 419, 420, 421] argued, rather more persuasively, that, following Schwinger, it is the dynamical Casimir effect that is relevant.[‡] The problem here is that Casimir energies in the presence of dynamically changing boundaries are effectively unknown, so these authors are forced to use the other extreme simplification, that of the sudden approximation, and estimate plausible photon production rates from the overlap of two static configurations. The difficulty with this approach is discussed in the last section of this Chapter. There, a simple estimate is given which suggests that any macroscopic electromagnetic phenomenon such as the Casimir effect cannot possibly supply the energy required for sonoluminescence.

Eberlein [413] had earlier proposed a version of the dynamical Casimir mechanism (perhaps more properly called the Moore-Unruh [426, 427, 428, 429] mechanism) as an explanation of the observed radiation.[§] We have noted [37] technical difficulties associated with her work, especially those

---

[‡] Jensen and Brevik [425] further elaborate the ideas of Liberati et al. by supposing that a thin gas of electrons somehow forms within a sonoluminescent bubble so that transition radiation also occurs. A similar effect could occur with excited atomic electrons.

[§] A still earlier, one-dimensional precursor, was that of Sassaroli et al. [412].

related to the use of superluminal velocities. See also Refs. [430, 431]. In Sec. 12.3 we discuss the form of the force on the surface due to the fluctuating electric and magnetic fields, and make a comparison with the results of Ref. [413]. If, in fact, reasonable numbers are used in her result, the energies involved are too small by 18 orders of magnitude, and even if her superluminal velocities are employed, only 10^{-3} MeV is available. So, qualitatively, her results are not inconsistent with ours.[¶]

12.2   The Adiabatic Approximation

As noted, it is plausible that the adiabatic approximation should be valid, that is, that the time scale of the flash is long compared to the optical time scale, so that it should be adequate to use the static result of Chapter 5. There, a finite result was obtained for the zero-point energy of a vacuum bubble in a medium of dielectric constant ϵ, $|\epsilon - 1| \ll 1$, as given in (5.83):

$$E = \frac{23}{1536\pi a}(\epsilon - 1)^2. \tag{12.2}$$

This value is ten orders of magnitude too small to be relevant to sonoluminescence, as well as being of the wrong sign: that is, with this form of the energy, the collapse process would be endothermic. (See below.)

However, our finite result was obtained by a perhaps not fully justifiable method; that is, we deleted a cubically divergent term, or, equivalently, employed a zeta-function scheme that suppresses such divergences. Alternatively, one could argue that dispersion should be included [212, 432, 213, 201, 186], crudely modeled by

$$\epsilon(\omega) - 1 = \frac{\epsilon_0 - 1}{1 - \omega^2/\omega_0^2}. \tag{12.3}$$

If this rendered the expression for the stress finite [we consider the stress, not the energy, for it is not necessary to consider the dispersive factor $d(\omega\epsilon(\omega))/d\omega$ there], we could drop the cutoff δ and the sign of the force would be positive. From the first line of (5.74) with $\epsilon' = 1$ we find for the

[¶]A strange variant of the vacuum radiation idea was given by Chodos and Groff [415, 416], who adopt an *ad hoc* QED Lagrangian that violates parity, which apparently can yield exponential growth of radiation.

force per unit area

$$\mathcal{F} \sim + \frac{(\epsilon_0 - 1)^2}{128\pi^2 a^4} \sum_{l=1}^{\infty} \nu^2 \int_{-\infty}^{\infty} dz \, \frac{1}{(1+z^2)^2} \frac{1}{(1+z^2/z_0^2)^2}, \qquad (12.4)$$

where $z_0 = \omega_0 a/\nu$. As $\nu \to \infty$, $z_0 \to 0$, and the integral here approaches $\pi z_0/2$, and so

$$\mathcal{F} \sim \frac{(\epsilon_0 - 1)^2}{256\pi a^3} \omega_0 \sum_{l=1}^{\nu_c} \nu \sim \frac{(\epsilon_0 - 1)^2}{512\pi a} \omega_0^3, \qquad (12.5)$$

if we take as the cutoff[||] of the angular momentum sum $\nu_c \sim \omega_0 a$. The corresponding energy is obtained by integrating $-4\pi a^2 \mathcal{F}$,

$$E \sim -\frac{(\epsilon_0 - 1)^2}{256} \omega_0^3 a^2, \qquad (12.6)$$

which is of the form of (5.108) with $1/\delta \to \omega_0 a/4$.

So what does this say about sonoluminescence? To calibrate our remarks, let us recall (a simplified version of) the argument of Schwinger [35]. On the basis of a provocative but incomplete analysis he argued that a bubble ($\epsilon' = 1$) in water ($\epsilon \approx (4/3)^2$) possessed a positive Casimir energy[**] given by (12.1) or

$$E_{\text{bulk}} \sim \frac{a^3 K^4}{12\pi} \left(1 - \frac{1}{\sqrt{\epsilon}} \right), \qquad (12.7)$$

where K is a wavenumber cutoff. Putting in his estimate, $a \sim 4 \times 10^{-3}$cm, $K \sim 2 \times 10^5$cm^{-1} (in the UV), we find a large Casimir energy, $E_c \sim 13$ MeV, and something like 3 million photons would be liberated if the bubble collapsed.

What does our full (albeit static) calculation say? If we believe the subtracted result, the form in (12.2), and say that the bubble collapses from an initial radius $a_i = 4 \times 10^{-3}$cm to a final radius $a_f = 4 \times 10^{-4}$cm, as suggested by experiments [34], we find that the change in the Casimir

---

[||]Inconsistently, for then $z_0 \sim 1$. If $z_0 = 1$ in (12.4), however, the same angular momentum cutoff gives 5/12 of the value in (12.5).

[**]Note, for small $\epsilon - 1$, Schwinger's result goes like $(\epsilon - 1)$, rather than $(\epsilon - 1)^2$, indicating that he had not removed the "vacuum" contribution corresponding to (5.23). This is the essential physical reason for the discrepancy between his results and ours. We discuss this further in Sec. 12.4.

energy is $\Delta E \sim -10^{-4}$eV. This is far to small to account for the observed emission. The sign further indicates that energy is absorbed, not released, in the process.

On the other hand, perhaps we should retain the divergent result (12.6), and put in reasonable cutoffs. If we do so, we have

$$E = -\frac{(\epsilon - 1)^2}{256} a^2 K^3 \sim -10^4 \text{eV}, \qquad (12.8)$$

100 times too small, and still of the *wrong* sign. (The emission occurs *during* or at the end of the collapse.)

So we are unable to see how the Casimir effect could possibly supply energy relevant to the copious emission of light seen in sonoluminescence. Of course, dynamical effects could change this conclusion, but elementary arguments suggest that this is impossible unless superluminal velocities and incredibly small time scales are achieved. See Sec. 12.5.

Further experimental results have made it even more difficult to accommodate any explanation based on macroscopic considerations. In particular, Hiller and Putterman [386] found a remarkably strong isotope effect when water (H_2O) is replaced by heavy water (D_2O), where the dielectric properties change by no more than 10%. Subsequently, they published an erratum [387] reporting exceedingly strong sample dependence, thus warning that "interpretation in terms of an isotope effect should be regarded as premature." This, together with the already known strong temperature dependence, and strong dependence on gas concentration and the gas mixture, may rule out Casimir effect explanations entirely. Yet the subject of vacuum energy is sufficiently subtle that surprises could be in store.

12.3   Discussion of Form of Force on Surface

There seems, in the literature, to be some confusion about the correct form for the stress on a surface due to electromagnetic fields (here, of course, we are interested in vacuum expectation values of those fields). The definitive discussion seems to be given in Stratton [184]. See also Ref. [99].

In Chapter 5, we computed the force on the surface by considering the discontinuity of the stress tensor,

$$T_{nn} = \frac{1}{2}\epsilon(E_\perp^2 - E_n^2) + \frac{1}{2}\mu(H_\perp^2 - H_n^2), \qquad (12.9)$$

across the surface, where **n** denotes the direction normal to the surface, and $\perp$ directions tangential to the surface. This follows directly from a consideration of the interpretation to T_{nn} as the flow of nth component of momentum in the direction **n**. Because of the boundary conditions that

$$\mathbf{E}_\perp, \quad D_n, \quad \mathbf{H}_\perp, \quad B_n \tag{12.10}$$

be continuous, the stress per unit area on the surface is

$$\mathcal{F} = T_{nn}(a-) - T_{nn}(a+) = \frac{1}{2}\left[(\epsilon' - \epsilon)E_\perp^2 - \left(\frac{1}{\epsilon'} - \frac{1}{\epsilon}\right)D_n^2\right.$$
$$\left. + (\mu' - \mu)H_\perp^2 - \left(\frac{1}{\mu'} - \frac{1}{\mu}\right)B_n^2\right], \tag{12.11}$$

in terms of fields on the surface, and where a prime denotes quantities on the "$-$" side ("inside") of the surface. This is obviously equivalent to the following form for the force density,

$$\mathbf{f} = -\frac{1}{2}\left(E_\perp^2\boldsymbol{\nabla}\epsilon - D_n^2\boldsymbol{\nabla}\frac{1}{\epsilon} + H_\perp^2\boldsymbol{\nabla}\mu - B_n^2\boldsymbol{\nabla}\frac{1}{\mu}\right) = -\frac{1}{2}(E^2\boldsymbol{\nabla}\epsilon + H^2\boldsymbol{\nabla}\mu), \tag{12.12}$$

which is just what is obtained from the stationary principle for the energy [99].

The controversy seems to center around the additional "Abraham" term (5.44)

$$\mathbf{f}' = (\epsilon - 1)\frac{\partial}{\partial t}(\mathbf{E} \times \mathbf{H}). \tag{12.13}$$

(Henceforward we restrict ourselves to nonmagnetic material, $\mu = 1$.) As noted in Sec. 5.4, this makes no contribution to the Casimir effect, because the vacuum expectation value is stationary. Furthermore, the existence of such a term is dependent upon the (essentially arbitrary) split between field and particle momentum. The Minkowski choice for field momentum

$$\mathbf{G}_M = \mathbf{D} \times \mathbf{H} \tag{12.14}$$

would not imply this additional force density. The analysis of experimental data given by Brevik [203], however, seems to favor the Abraham value (5.43).

If we were calculating a net *force* on the surface, (12.13) would indeed give a further contribution to the force beyond that given by (12.11).

Through use of Maxwell's equations, we easily find

$$f'_n = (\epsilon - 1) \left[-\frac{1}{2\epsilon} \nabla_n B^2 - \frac{1}{2} \nabla_n E^2 + \frac{1}{\epsilon} \nabla \cdot (\mathbf{B} B_n) + \nabla \cdot (\mathbf{E} E_n) \right]. \quad (12.15)$$

If **n** were a fixed direction, the volume integral of this force density would turn into a surface integral, and the result given by Eberlein [413] follows,

$$F'_n = \frac{1}{2} \int dS \left[\left(\frac{1}{\epsilon} - \frac{1}{\epsilon'} \right) (B_n^2 - B_\perp^2) + (\epsilon - \epsilon') E_\perp^2 \right.$$
$$\left. - \left(\frac{1}{\epsilon} - \frac{1}{\epsilon'} \right) \left(1 - \frac{1}{\epsilon} - \frac{1}{\epsilon'} \right) D_n^2 \right]. \quad (12.16)$$

But this result cannot be used to compute the stress. Thus, the formula (C5) given in Appendix C of in [413] is wrong, and, accordingly, so is (3.18) there.[tt] (It should be stressed that this extra force must be zero in statics. For example, if $\mathbf{B} = \mathbf{0}$, $f'_n = 0$ in (12.15) because $\nabla \times \mathbf{E} = \mathbf{0}$.)

Finally, we note there is yet another formula for the force on a dielectric given in terms of polarization charges and currents,

$$\mathbf{F} = \int (d\mathbf{r}) \left[\rho_{\text{pol}} \mathbf{E} + \frac{1}{c} \mathbf{j}_{\text{pol}} \times \mathbf{B} \right], \quad (12.17)$$

where

$$\rho_{\text{pol}} = -\nabla \cdot \mathbf{P}, \quad \mathbf{j}_{\text{pol}} = \frac{\partial}{\partial t} \mathbf{P} + c \nabla \times \mathbf{M}, \quad (12.18)$$

with the polarization and magnetization fields given by

$$\mathbf{P} = \mathbf{D} - \mathbf{E} = \left(1 - \frac{1}{\epsilon} \right) \mathbf{D}, \quad \mathbf{M} = \mathbf{B} - \mathbf{H} = (\mu - 1)\mathbf{H}. \quad (12.19)$$

Again, it is easy to show that if one is calculating the force in a fixed direction, so one can freely integrate by parts, for a nonmagnetic medium, we recover the expected force including the Abraham term:

$$\mathbf{F} = \int (d\mathbf{r}) \left[-\frac{E^2}{2} \nabla \epsilon + \frac{1}{c} \frac{\partial}{\partial t} (\epsilon - 1) \mathbf{E} \times \mathbf{B} \right]. \quad (12.20)$$

[tt]The first derivation there is based incorrectly on the formula for the *force* given in the following paragraph, while the second is based on an obviously incorrect extrapolation from the vacuum stress tensor, which of course gives vanishing stress.

But the integrand in (12.17) is not interpretable as a force density from which the stress may be computed. In effect, it is that interpretation that Eberlein uses in Ref. [413].

12.4 Bulk Energy

More recently there has been a proposal that, indeed, the bulk energy result of Schwinger is relevant (of course, it is correct) [422, 423, 424]. These authors make an issue of the subtraction of the uniform medium contribution, implying, it would seem, that we were in error in Refs. [16, 37]. Since this is a serious issue with experimental consequences, and since, admittedly, there are subtle issues of principle involved here, in this section we wish to return to this point and provide further evidence for the result (12.2). We will explain more fully why this subtraction was made, indicate that it has a rather long history in Casimir effect calculations, and was in fact made by Schwinger in [103, 104] before he abandoned the effort to recalculate the Casimir effect for a dielectric ball. This is supported by the connection established in Sec. 5.9 between the Casimir effect and van der Waal forces, because the same finite Casimir energy (12.2) can be obtained from the latter. Indeed, it is self-evident that pairwise van der Waals forces must give rise to a net force proportional to $(\epsilon - 1)^2$, not proportional to $\epsilon - 1$ as in (12.1).

It is completely manifest that (5.19) does not have a well-defined limit as $\delta \to 0$ —it is quartically divergent. Indeed, it is easy to show as Refs. [422, 423] do, that the quartically divergent term here corresponds precisely to the Schwinger result (12.1) when $\epsilon' = \mu' = 1$, $\mu = 1$. However, it is also quite clear that the calculation is not yet done when we have reached this point. As we stated in Ref. [37], and repeated in Sec. 5.2, "We must remove the term which would be present if either medium filled all space (the same was done in the case of parallel dielectrics [11])." When we look at the latter reference, we see immediately the point. Again to quote, this time from [11]: "These terms [to be subtracted] correspond to the electromagnetic energy required to replace medium 1 by medium 2 in the displacement volume. (Since this term in the energy is already phenomenologically described, it must be cancelled by an appropriate contact term.)" (See also the discussion in Chapter 3.) What we were saying there, in the present context, is that the term in the energy corresponding to the boundary-condition-independent Green's function

$$F_l^{(0)} = ikj_l(kr_<)h_l^{(1)}(kr_>), \tag{12.21}$$

must be removed, because it contributes (a formally infinite amount) to the bulk energy of the material, which is already phenomenologically described in terms of its bulk properties. In fact, we are not creating material, e.g., water, we are simply displacing it when we insert the bubble, and force the bubble to expand and contract. The energy per unit element of medium is therefore not changed. (The density of the gas in the bubble of course changes greatly, but the zero-point energy of that relatively dilute medium is certainly insignificant because $n \approx 1$. In any case, the effect of this density change is also included in the phenomenological description.)

Indeed, the spectacular agreement between the the Lifshitz theory of parallel dielectrics [7, 8], rederived in Ref. [11] and described in Chapter 3, and the beautiful experiment of Sabisky and Anderson [66] seems strong vindication of this subtraction procedure.

Further evidence that we are on the right track is provided by Schwinger himself. In the first reference in [104], where he rederives the result for parallel dielectrics, he explicitly removes volume and surface energies:

> one finds contributions to E that, for example, are proportional ... to the volume enclosed between the slabs. The implied constant energy density—independent of the separation of the slabs—violates the normalization of the vacuum energy density to zero. Accordingly, the additive constant has a piece that maintains the vacuum energy normalization. There is also a contribution to E that is proportional to [the area], energy associated with individual slabs. The normalization to zero of the energy for an isolated slab is maintained by another part of the additive constant.

Admittedly, the situation is more clear-cut in the parallel-plate geometry. However, in the following paper (the second reference in [104]) where Schwinger begins to set up the problem for the spherical geometry (but leaves the details to Harold,[‡‡]) a close reading shows a similar subtraction is

---

[‡‡]Harold—the "Hypothetical Alert Reader of Limitless Dedication"—makes his first appearance in Ref. [119]. He is, of course, a tribute to Schwinger's older brother. See Ref. [147].

implicit. Unfortunately, when Schwinger went on to apply Casimir energy to sonoluminescence in [35], he did not make use of the general analysis in [103, 104]. Instead, needing an immediate result to confront the phenomenology, Schwinger simply jumped to the unsubtracted, unregulated result (12.1)—see the second reference in [35].

12.5 Dynamical Casimir Effect

It might well be thought that it is the "dynamical Casimir effect," not the static Casimir effect, that is relevant to sonoluminescence. Unfortunately, the former phenomenon is not at all well-studied. It seems plausible that the dynamical Casimir effect is closely allied with the so-called Unruh effect (probably more correctly attributed to Moore) [426, 427, 428, 429, 433, 101, 434], wherein an accelerated observer, with acceleration a, sees a bath of photons with temperature T,

$$T = \frac{a}{2\pi}. \tag{12.22}$$

Indeed, the observed radiation in sonoluminescence is consistent with the tail of a blackbody spectrum, with temperature ~20,000 K. That is, kT is about 1 eV. Let us, rather naively, apply this to the collapsing bubble, of radius R, where $a = d^2R/dt^2 \sim R/\tau^2$, where τ is some relevant time scale for the flash. We then have

$$kT \sim \frac{R}{(c\tau)^2}\hbar c, \tag{12.23}$$

or

$$1\,\mathrm{eV} \sim \frac{10^{-4}\mathrm{cm}\,2\times10^{-5}\mathrm{eV\text{-}cm}}{\tau^2(3\times10^{10}\mathrm{cm\,s^{-1}})^2} \sim \frac{10^{-30}\mathrm{eV}}{\tau^2(\mathrm{s}^2)}. \tag{12.24}$$

That is, $\tau \sim 10^{-15}$ s, which seems implausibly short; it implies a characteristic velocity $R/\tau \sim 10^{12}$ cm/s $\gg c$. It is far shorter than the observed flash duration, 10^{-10} s. Indeed, if we use the latter in the Unruh formula (12.22) we get a temperature below 1 micro Kelvin! This conclusions seem consistent with those of Eberlein [413], who indeed stressed the connection with the Unruh effect, but whose numbers required superluminal velocities.

Let us consider this problem from the point of view of elementary radiation theory. For example, consider the (nonrelativistic) Larmor formula, appropriate to dipole radiation; it gives the power radiated, in Gaussian units:

$$P = \frac{2}{3} \frac{(\ddot{\mathbf{d}})^2}{c^3}, \qquad (12.25)$$

where $\mathbf{d}$ is the electric dipole moment. If our bubble, with N atoms, *coherently* emits radiation, we expect

$$|\ddot{\mathbf{d}}| \sim \frac{N d_a}{\tau^2}, \qquad (12.26)$$

where d_a is an atomic or molecular dipole moment, and τ a characteristic collapse time for the bubble. Thus the energy emitted during one collapse of a bubble in water is

$$E \sim \alpha \hbar c \left(10^{23} \left(\frac{a}{\text{cm}} \right)^3 \right)^2 \frac{(d_a/e)^2}{(c\tau)^3}. \qquad (12.27)$$

(We are assuming that it is atoms or molecules in an equivalent dense volume that are radiating, not the relatively small number of gas molecules in the interior.) So with $a \sim 10^{-4}$ cm (the minimum radius of the bubble), $\tau \sim 10^{-10}$ s (the flash duration), and $d_a \sim 10^{-8} e$-cm, we get an energy of only $E \sim 10^{-3}$ eV. This is in spite of the optimistic assumption of coherent radiation![*] The corresponding velocity across the bubble of 10^6 cm/s, well in excess of the speed of sound, thus precluding the presumed coherent radiation process. We would need a much shorter time scale than observed, $\tau \sim 10^{-13}$ s, to obtain the necessary energy, and then the velocity of the bubble walls would only be an order of magnitude less than the speed of light!

We therefore believe that in Eberlein's calculation [413, 414] there is an implicit assumption of superluminal velocities.[†] We note that the short wavelength result of Eberlein, (4.7) of Ref. [413] or (10) of Ref. [414], can

---

[*]Commenting on Schwinger's earlier belief in the phenomenon of cold fusion, and his attempts to explain the result by a coherent interaction with the lattice, Norman Ramsey has remarked that it is easy to be fooled: "Nature does not like coherence." (Interview of Norman Ramsey by K. A. Milton, 8 June 1999, quoted in Ref. [147], p. 550.)

[†]Indeed, if one follows Eberlein and uses a time scale $\gamma \sim 1$ fs (though the experimental value seems to be closer to 100 ps) in her model profile, one finds the maximum speed of the bubble surface to exceed the speed of light by almost two orders of magnitude. Actually, even with such a small γ, we find her result yields an energy output of only 10^{-3} MeV, insufficient to explain sonoluminescence.

be cast in the dipole form (12.25) by integrating by parts. Up to factors nearly equal to one,

$$\left(\frac{d}{e}\right)_E \approx a\frac{\dot{a}}{c}, \tag{12.28}$$

where $a(t)$ is the bubble radius. Because $\dot{a}/c < 1$, we find that emission energy of 10 MeV requires a time scale of $\tau_E \sim 10^{-17}$ s. This seems to be an implausibly short scale unless remarkable relativistic phenomena are involved. (The corresponding speed is $a/\tau \sim 10^{13}$ cm/s.) [Incidentally, the magnitude of our cutoff estimate, (12.8), also agrees with (12.25) if $K \sim 1/\tau$. This demonstrates that there is nothing classical about the estimate (12.27)].

The only plausible origin of such short time scales lies in the formation of a shock. In fact, shock wave have now been visualized [391]. In that case, velocities can remain nonrelativistic, while accelerations, or derivatives thereof, become very large. Classical shock models of sonoluminescence have been proposed by Greenspan and Nadim and by Wu and Roberts [399, 397, 398]. In this case, the radiation is supposed to be emitted by Bremsstrahlung after ionization of the air in the bubble, or by collision-induced emission from a basically neutral environment [405]. But this picture has nothing to do with quantum vacuum radiation.

However, we must remain open to the possibility that discontinuities, as in a shock, could allow changes on such short time scales without requiring superluminal speeds. Indeed, Liberati et al. [417, 418, 419, 420, 421], following Schwinger's earlier suggestion [35], assume an extremely short time scale, so that rather than the adiabatic approximation discussed above being valid, a sudden approximation is more appropriate. We therefore turn to an analysis of that situation.

A free interpretation of the picture offered by Liberati *et al.* [417, 418, 419, 420, 421] is that of the abrupt disappearance of the bubble at $t = 0$, as shown in Fig. 12.1. On the face of it, this picture seems preposterous—the bubble simply disappears and water is created out of nothing.[‡] It may be no surprise that a large energy release would occur in such a case. Further, the static Casimir effect calculations employed in these papers are invalid

---

[‡]In a recent comment, Liberati et al. [435] insist it is not the matter which changes, but the dielectric constant; but since the matter here is characterized by its electric properties, this distinction seems *semantic*.

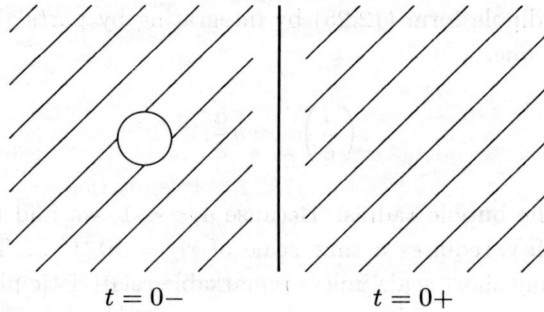

$t = 0-$ $\qquad$ $t = 0+$

Fig. 12.1   The sudden collapse of an otherwise static bubble.

in this instantaneously changing model. Therefore, rather than computing Bogoliubov coefficients from the overlap of states belonging to two static configurations, let us follow the original methodology of Schwinger [35], which appears to be essentially equivalent.

As in Schwinger's papers, let us confine our attention to the magnetic (TE) modes. They are governed by the time-dependent Green's function satisfying [see (5.35)]

$$(\partial_0 \epsilon(x)\partial_0 - \nabla^2)G(x, x') = \delta(x - x').\qquad(12.29)$$

The photon production is given by the effective two-photon source [see (13.10)]

$$\delta(JJ) = i\delta G^{-1} = i\partial_0 \delta\epsilon(x)\partial_0.\qquad(12.30)$$

The effectiveness for producing a photon in the momentum element centered about **k** is (see Ref. [119])

$$J_k = \sqrt{\frac{(d\mathbf{k})}{(2\pi)^3}\frac{1}{2\omega}} \int (dx)e^{-i(\mathbf{k}\cdot\mathbf{r} - \omega t)}J(x), \quad \omega = |\mathbf{k}|.\qquad(12.31)$$

Schwinger considered one complete cycle of appearance and disappearance of the bubble, which we assume disappears for a time τ. From (12.31), the probability of producing two photons is, with V being the bubble volume,

$$|J_k J_{k'}|^2 \propto (\epsilon' - 1)^2 \omega^2 \sin^2 \omega\tau \, \delta(\mathbf{k} + \mathbf{k}')V.\qquad(12.32)$$

The total number of photon pairs emitted is then, if dispersion is ignored,

$$N \approx \left(\frac{4\pi}{3}\right)^2 \left(\frac{R}{\Lambda}\right)^3 \left(\frac{\epsilon - 1}{4}\right)^2, \qquad (12.33)$$

where the cutoff wavelength is given in terms of the cutoff wavenumber in (12.7) by $K = 2\pi/\Lambda$. Such a divergent result should be regarded as suspect. It was Eberlein's laudable goal [413] to put this type of argument on a sounder footing. Nevertheless, if we put in plausible numbers, $\sqrt{\epsilon} = 4/3$, $R = 4 \times 10^{-3}$ cm, and, as in Schwinger's earlier estimate, $\Lambda = 3 \times 10^{-5}$ cm, we obtain the required $N \sim 10^6$ photons per flash.

The problem with this estimate is one of time and length scales—for the instantaneous approximation to be valid, the flash time τ_f must be much less than the period of optical photons, $\tau_o \sim 10^{-15}$ s. This is consistent with the discussion of the Unruh effect above, and is acknowledged by Liberati et al. [417, 418, 419, 420, 421]. On the other hand, the collapse time $\tau_c \sim 10^{-5}$ s is vastly longer than τ_f, and is therefore totally irrelevant to the photon production mechanism. In fact, the observed flash time is of order 10^{-10} s, so as noted above, the adiabatic, not the instantaneous, approximation would seem to be valid. Moreover, the flash occurs near minimum radius, and thus the appropriate value of R in (12.33) would seem to be at least an order of magnitude smaller, $R \sim 10^{-4}$ cm. This would lead to $N < 10^3$ photon pairs, totally insufficient.

We conclude this chapter by candidly admitting that the probability of the relevance of the dynamical Casimir effect to sonoluminescence, although small, remains nonzero. This is because the dynamical Casimir effect, in general, has not in fact been developed at all. As we remarked above, radiation from a moving mirror is well-studied [426, 412, 429, 433], but for cavities, little beyond perturbation theory, valid for "small but arbitrary dynamical changes" [436, 437, 438], hardly relevant to the profound changes seen in sonoluminescence, is known. Photons may be produced resonantly by vibrating mirrors in a cavity [439], but the effects are small. See Ref. [440] for a review of quantum radiation produced by an accelerated dielectric body. Here is an area ripe for development.[§]

[§]Here are some relevant results: Motional Casimir forces were considered by Jaekel and Reynaud [441]; the (very small) mechanical response of the vacuum was studied by Golestaina and Kardar [442]; and Kenneth and Nussinov examine the (very small) Casimir radiation [443] from accelerated objects.

Chapter 13

Radiative Corrections to the Casimir Effect

Everything we have considered so far in this book has been at the one-loop level. This reflects the fact that in general the Casimir effect itself is quite small, so higher loop corrections seem beyond experimental reach. Yet theoretically, such corrections are of great interest, and could potentially be of great importance in hadronic and cosmological applications. Given the 50-year history of the Casimir effect, it is somewhat surprising that rather little has been done in this direction, and that there exists controversy on the appropriate methodology. In this Chapter, we will present a calculation of the simplest two-loop correction to the Casimir effect for the geometry of parallel plates, and discuss the corresponding calculation for a sphere. These calculations were first carried out by Bordag and collaborators [105, 107, 444, 445, 446, 447]; see also Ref. [106] for a self-interacting scalar field inside a cube, with periodic boundary conditions. Methods of calculating loop diagrams in boxes, relevant to bag model, were first discussed by Peterson, Hansson, and Johnson [448].

In the spirit of the preceding Chapters it would be natural to compute the vacuum expectation value of the two-loop correction to the stress-energy tensor. That, however, is unnecessarily complicated. Following Ref. [107], we sketch a formalism in the next section which requires only the computation of the two-loop vacuum graph. The one-loop contribution of this formalism, the familiar trace-log of the propagator, yields the usual 1-loop Casimir energy.

In Sec. 13.2 we then use this formalism to compute the QED radiative correction to the Casimir energy to first order in the fine structure constant $\alpha = e^2/4\pi\hbar c$ for the special case of parallel conducting plates with the property that the photon propagator satisfies perfect conductor boundary conditions on the plates, while the plates are transparent to the electrons.[*] Since for nearly any conceivable experimental situation the plate separation is large compared to the Compton wavelength of the electron, λ_c, the correction is extremely small. The energy per unit area of the plates is through two loops

$$\mathcal{E}_\| = -\frac{\pi^2}{720a^3}\left(1 - \frac{9\alpha}{32\mu a}\right), \tag{13.1}$$

where μ is the mass of the electron.

Since this is so phenomenologically irrelevant, we merely quote the result of Ref. [107] in Sec. 13.3 for the corresponding calculation for a perfectly conducting spherical shell. For that geometry, there is also a logarithmic correction term:

$$E_S = \frac{0.046176}{a}\left(1 - 0.016413\frac{\alpha}{\mu a}\log(\mu a) - 0.14040\frac{\alpha}{\mu a}\right), \tag{13.2}$$

which, like (13.1) is derived in the approximation that $\mu a \gg 1$. Of course, where these corrections might become significant is in the bag model, where the radiative corrections are those due to QCD. Then, not only is the bag radius $a \sim \lambda_c$ for the quarks, but we must take into account the fact that the quark propagators experience the bag boundary, that is satisfy the bag boundary condition (6.26). To my knowledge, such a calculation has yet to be attempted.

A further theoretical issue of some interest is discussed in Sec. 13.4. This is the discrepancy of the above result with that of Kong and Ravndal [449, 450] who attempt to compute the radiative corrections to the Casimir effect with the same boundary conditions but using the Euler-Heisenberg Lagrangian as an effective Lagrangian. For parallel plates they obtain a much smaller correction, $E^{(2)} \sim (\alpha^2/a^3)(\lambda_c/a)^4$. It is clear that this approach does not begin to capture the relevant physics.

---

[*] According to Ref. [445], inclusion of a boundary condition for the spinors would give rise to an exponentially small correction if $\lambda_c \ll a$.

13.1 Formalism for Computing Radiative Corrections

We will develop the formalism in the simple context of a scalar theory; the generalization to electrodynamics is rather immediate. In particular, the radiatively corrected Casimir energy is given for QED by an expression which is merely twice that of the scalar theory for the case of the parallel plate geometry. (Of course, the vacuum polarization operator is different.)

We adopt an approach given, for example, in Schwinger's first Casimir effect paper [2]. Consider the vacuum amplitude, or generating function, given in terms of external sources $K(x)$ in a region characterized by some bounding surfaces:

$$\langle 0_+|0_-\rangle^K = e^{iW[K]}, \tag{13.3}$$

where

$$W[K] = \frac{1}{2}\int (dx)(dx')K(x)G(x,x')K(x'), \tag{13.4}$$

where $G(x,x')$ is the appropriate scalar propagation function for the region in question. We can also introduce an (effective) field according to

$$\phi(x) = \int (dx')G(x,x')K(x'), \tag{13.5}$$

or in terms of the inverse operator G^{-1},

$$\int (dx'')G^{-1}(x,x'')G(x'',x') = \delta(x-x'), \tag{13.6}$$

we can express the source in terms of the field,

$$K(x) = \int (dx')G^{-1}(x,x')\phi(x'). \tag{13.7}$$

Now if the geometry of the region is altered slightly, as through moving one of the bounding surfaces, the vacuum amplitude is altered:

$$\begin{aligned}
\delta W[J] &= \frac{1}{2}\int (dx)(dx')K(x)\delta G(x,x')K(x') \\
&= -\frac{1}{2}\int (dx)(dx')\phi(x)\delta G^{-1}(x,x')\phi(x').
\end{aligned} \tag{13.8}$$

Upon comparison with the two particle term in the expansion of the vacuum amplitude,

$$e^{iW[K]} = e^{i \int (dx) K(x) \phi(x)} = \ldots + \frac{1}{2} \left[i \int (dx) K(x) \phi(x) \right]^2 + \ldots, \quad (13.9)$$

we see that the variation of the inverse Green's function is effectively a two-particle source,

$$iK(x)K(x') \Big|_{\text{eff}} = -\delta G^{-1}(x, x'). \quad (13.10)$$

Therefore, the change in the generating function is given by

$$\begin{aligned} \delta W &= \frac{i}{2} \int (dx)(dx') G(x, x') \delta G^{-1}(x', x) \\ &= -\frac{i}{2} \int (dx)(dx') \delta G(x, x') G^{-1}(x', x). \end{aligned} \quad (13.11)$$

Formally, if we regard space-time coordinates as matrix indices, we can write this in the familiar form of a trace:

$$\delta W = -\frac{i}{2} \operatorname{Tr} \delta \ln G. \quad (13.12)$$

Thus for a static situation we can read off the following expression for the energy of the system, which is to be supplemented by an infinite constant:

$$E = \frac{i}{2T} \operatorname{Tr} \ln G, \quad (13.13)$$

where T is the "infinite" time that the configuration exists. Let us directly evaluate this for the one-loop scalar propagator in the interior of two parallel plates (on which the field vanishes), given by (2.14) and (2.43). That means that the energy per unit area of the plates is given by the expression

$$\mathcal{E} = \frac{E}{A} = a \frac{i}{2} \int \frac{d\omega}{2\pi} \frac{d^2 k}{(2\pi)^2} \frac{1}{a} \sum_{n=1}^{\infty} \ln \frac{1}{(n\pi/a)^2 - \lambda^2}. \quad (13.14)$$

As usual, we perform the Euclidean rotation, and change to polar coordinates as in (2.25), with the result

$$\mathcal{E} = \frac{1}{4\pi^2} \sum_{n=1}^{\infty} \int_0^{\infty} d\kappa\, \kappa^2 \ln[\kappa^2 + (n\pi/a)^2]. \quad (13.15)$$

We now integrate by parts, and carry out the sum over n using (7.26). When terms which are merely nonnegative powers of a are omitted, as being contact terms, we are left with

$$\mathcal{E} = -\frac{a}{6\pi^2} \int_0^\infty d\kappa\, \kappa^3 \frac{1}{e^{2\kappa a} - 1} = -\frac{\pi^2}{1440a^3}, \qquad (13.16)$$

where we encounter just the integral we found earlier in (2.30) for $d = 2$. Of course, this is precisely the scalar Casimir energy (2.9) for the case of parallel plates.

13.2   Radiative Corrections for Parallel Conducting Plates

The "trace-log" formula (13.13) is immediately extended to describe higher-loop corrections. We simply write the Green's function in perturbative form,

$$G = G_0(1 + \Pi G_0), \qquad (13.17)$$

where the polarization operator Π is regarded as small, and then the energy is given by

$$E = \frac{i}{2T}\operatorname{Tr}\ln G_0 + \frac{i}{2T}\operatorname{Tr}\Pi G_0. \qquad (13.18)$$

Here we take G_0 to be the propagator used above, subject to the Dirichlet boundary conditions at $z = 0$, $z = a$ [see (2.20)]:

$$G_0(x, x') = -\int \frac{d^3k}{(2\pi)^3} e^{ik(x-x')} \frac{1}{\lambda} \frac{\sin\lambda z_< \sin\lambda(z_> - a)}{\sin\lambda a}, \qquad (13.19)$$

where the three-dimensional momentum integration is over frequency and the two transverse momenta,

$$\frac{d^3k}{(2\pi)^3} = \frac{d\omega}{2\pi} \frac{(d\mathbf{k}_\perp)}{(2\pi)^2}, \qquad (13.20)$$

the three-dimensional scalar product refers to the same variables,

$$k(x - x') = -\omega(t - t') + \mathbf{k}_\perp \cdot (\mathbf{r} - \mathbf{r}')_\perp, \qquad (13.21)$$

and λ is the three-dimensional (Minkowski) invariant:

$$\lambda^2 = \omega^2 - k_\perp^2. \qquad (13.22)$$

To evaluate the second term in (13.18) we consider the case in which the particles ("electrons") in the vacuum polarization operator Π do not experience the effects of the boundaries, so we can use the usual free-space expression for that operator. Expanding out the meaning of the trace, we have the following expression for the correction to the energy per transverse area:

$$\mathcal{E}^{(2)} = -\frac{i}{2} \int \frac{d^4k}{(2\pi)^4} \Pi(k) \int_0^a dz \int_0^a dz' e^{ik_z(z-z')} \frac{\sin \lambda z_< \sin \lambda (z_> - a)}{\lambda \sin \lambda a}.$$

(13.23)

Carrying out the integration over z and z', we obtain

$$\int_0^a dz \int_0^a dz' \, e^{ik_z(z-z')} \sin \lambda z_< \sin \lambda (z_> - a)$$
$$= -\frac{1}{k^2} \left(\lambda a \sin \lambda a - 2\frac{\lambda^2}{k^2} [\cos \lambda a - \cos k_z a] \right).$$

(13.24)

Here $k^2 = k_z^2 - \lambda^2$ is the four-dimensional Minkowski scalar. When this is inserted into (13.23) we obtain two terms which are nonnegative powers of a, which we omit. We are left with

$$\mathcal{E}^{(2)} = -i \int \frac{d^4k}{(2\pi)^4} \frac{\Pi(k)}{(k^2)^2} \frac{\lambda}{\sin \lambda a} \left[e^{i\lambda a} - \cos k_z a \right].$$

(13.25)

This is, apart from a factor of two, exactly the expression given by Bordag and Lindig [107].

Supplying that factor, we now evaluate this correction in the QED case. The vacuum polarization can be written in the spectral form[†]

$$\Pi(k) = (k^2)^2 \int_{4\mu^2}^\infty dM^2 \frac{a(M^2)}{k^2 + M^2 - i\epsilon},$$

(13.26)

with the spectral function

$$a(M^2) = \frac{\alpha}{3\pi} \frac{1}{M^2} \left(1 - \frac{4\mu^2}{M^2} \right)^{1/2} \left(1 + \frac{2\mu^2}{M^2} \right).$$

(13.27)

(See, for example Ref. [451].) After performing the Euclidean rotation on

---

[†]The $(k^2)^2$ prefactor is required by gauge invariance; in physical terms the structure is such that the photon pole of the propagator at $k^2 = 0$ is not disturbed.

the frequency ω, and introducing the variable κ, we write the correction as

$$\mathcal{E}^{(2)} = \frac{1}{2\pi^3} \int_{4\mu^2}^{\infty} dM^2 a(M^2) \int_0^{\infty} \kappa^2 d\kappa \int_{-\infty}^{\infty} dk_3 \frac{[e^{-\kappa a} - \cos k_3 a]}{k_3^2 + \kappa^2 + M^2} \frac{\kappa}{\sinh \kappa a}.$$

(13.28)

The k_3 integrals are elementary:

$$\int_{-\infty}^{\infty} dk_3 \frac{e^{ik_3 a}}{k_3^2 + b^2} = \frac{\pi}{b} e^{-ab},$$

(13.29)

so we are left with a double integral:

$$\mathcal{E}^{(2)} = \frac{1}{2\pi^2} \int_{4\mu^2}^{\infty} dM^2 a(M^2) \int_0^{\infty} \frac{\kappa^3 d\kappa}{\sinh \kappa a} \frac{1}{\sqrt{\kappa^2 + M^2}} \left(e^{-\kappa a} - e^{-\sqrt{\kappa^2 + M^2} a} \right).$$

(13.30)

If we consider the regime where the plate separation is large compared to the Compton wavelength of the electron, $\mu a \gg 1$, the second exponential factor in the above is negligible, and the κ integral reduces to

$$\int_0^{\infty} \frac{2\kappa^3 d\kappa}{e^{2\kappa a} - 1} \frac{1}{M} = \frac{\pi^4}{120 M a^4}.$$

(13.31)

The remaining spectral mass integral becomes elementary upon making the substitution $v = \sqrt{1 - 4\mu^2/M^2}$:

$$\mathcal{E}^{(2)} = \frac{\alpha\pi}{720\mu a^4} \int_0^1 \frac{dv\, v^2}{\sqrt{1 - v^2}} \left(\frac{3}{2} - \frac{1}{2} v^2 \right)$$

$$= \frac{\alpha\pi^2}{2560\mu a^4},$$

(13.32)

which agrees with (13.1), first found in [105].

In general, it is not difficult to integrate the double integral (13.30) numerically. The result is shown in Fig. 13.1. We see that the asymptotic result (13.32) is only some 20% high at $\mu a = 1$, and becomes extremely accurate for larger μa, but for smaller values of μa instead of diverging as the expression (13.32) indicates, the energy tends to a constant value as $\mu a \to 0$,

$$\mathcal{E}^{(2)}(0) = \frac{\alpha}{a}(0.00557745).$$

(13.33)

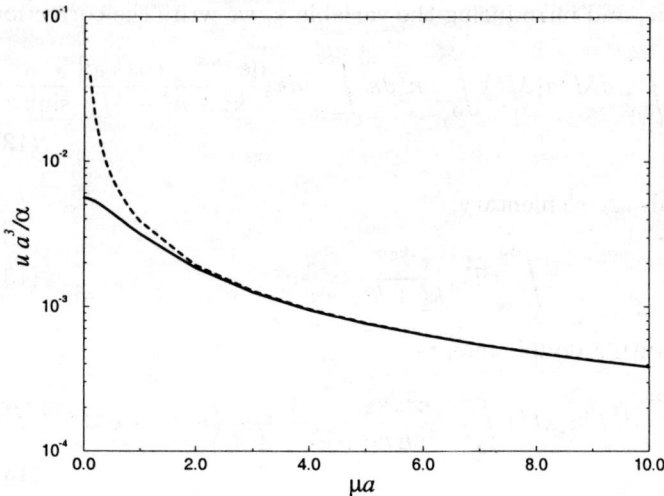

Fig. 13.1 Graph of the QED radiative correction for the Casimir effect (solid line) for parallel plates as a function of the separation of the plates a relative to the Compton wavelength $1/\mu$ of the electron. In this calculation it is assumed that the plates are perfect conductors, but that the boundaries are transparent to the electrons. The asymptotic result for $\mu a \gg 1$ (13.32) is also shown as the dashed line.

13.2.1 *Other Work*

For $\lambda\phi^4$ in $D+1$ dimensions between parallel Dirichlet plates, Albuquerque has shown that the pressure develops a divergence at two loops if D is even [452], presumably a reflection of the divergences seen in general in Casimir calculations in even D.

13.3 Radiative Corrections for a Spherical Boundary

For the case of QED, it appears that the radiative corrections to the Casimir effect are far beyond experimental reach. Therefore, we have merely quoted the result, derived by Bordag and Lindig [107] in (13.2). Where such corrections could be important is in the bag model of hadrons, where the QCD corrections could be large. Unfortunately, corrections in that case have not yet been calculated. The only calculation in a spherical geometry extant

[107] is that for QED in which, as in the previous section, the boundary is transparent to the fermions, which is not the case with quarks. Moreover, the only explicit result given in Ref. [107] is shown in (13.2), which applies only in the case when the radius of the sphere is large compared to the Compton wavelength of the fermion. Again, this condition does not apply to quarks in a hadronic bag.[‡] However, the correction is probably not large. If we evaluate the QED correction (13.2) at $\mu a = 1$ (the limit of applicability of the formula—see Fig. 13.1) we obtain a reduction of the Casimir attraction by 0.1%, while replacing $\alpha = 1/137$ by $\alpha_s \sim 2$ boosts this reduction to 30%. This is comparable to the present uncertainties in the hadronic Casimir correction, so it certainly would be worthwhile computing the correction for QCD.

13.4 Conclusions

Given the still primitive understanding of the Casimir effect, it is perhaps not surprising that few higher-loop calculations exist. Such corrections are certainly not important in electrodynamics. If our understanding of the implications of vacuum fluctuations for hadronic physics improves, however, higher-loop corrections will be important. It would be very much worthwhile to carry out a calculation of the two-loop corrections to the Casimir effect in a spherical bag, where the quarks, as well as the gluons, are subjected to the bag boundary conditions.

There seems to remain an element of controversy concerning the radiative corrections presented here and in the literature [105, 107, 445]. Kong and Ravndal [449] suggested that the Euler-Heisenberg Lagrangian[§] [454, 455]

$$\mathcal{L}_{\text{EH}} = -\mathcal{F} - \frac{1}{8\pi^2} \int_0^\infty \frac{ds}{s^3} e^{-m^2 s} \left[(es)^2 \mathcal{G} \frac{\text{Re} \cosh esX}{\text{Im} \cosh esX} - 1 - \frac{2}{3}(es)^2 \mathcal{F} \right],$$

$$(13.34)$$

which describes the scattering of light by light, may be regarded as an effective Lagrangian, for calculation of higher-loop corrections to the single

---

[‡] A typical value for a light quark is $\mu a \sim 0.05$.

[§] It may be of interest to note that the Casimir-Euler-Heisenberg effective action was considered in Ref. [453], in which $\mathcal{L}_{\text{EH}}$ was derived in a space in which one dimension is compactified with antiperiodic boundary conditions.

photon loop process. Here the invariant field strength combinations are

$$\mathcal{F} = \frac{1}{4}F^2 = \frac{1}{2}(\mathbf{H}^2 - \mathbf{E}^2), \quad \mathcal{G} = \frac{1}{4}F^*F = \mathbf{E} \cdot \mathbf{H}, \tag{13.35}$$

$^*F_{\mu\nu} = \frac{1}{2}\epsilon_{\mu\nu\alpha\beta}F^{\alpha\beta}$ being the dual field strength tensor, and the argument of the hyperbolic cosine in (13.34) is given in terms of

$$X = [2(\mathcal{F} + i\mathcal{G})]^{1/2} = [(\mathbf{H} + i\mathbf{E})^2]^{1/2}. \tag{13.36}$$

Since (13.34) is derived under the assumption that the external photon lines carry zero momentum, and involves only a single electron loop, it is *a priori* difficult to see why this should be at all relevant to computing high-loop corrections; indeed, the fact that use of (13.34) leads to qualitatively different behavior than that found by a full calculation demonstrates the inapplicability of this intriguing idea.¶

Ravndal and Thomassen [457] now attribute this discrepancy to a previously omitted surface term in the effective theory, which has an unknown numerical coefficient which can be adjusted to reproduce the result (13.1)—however, this apparently only reproduces the $\mu a \gg 1$ limit. Melnikov [458] proposes instead to add a new bulk term with a judiciously chosen coefficient depending on the plate separation to reproduce (13.2). Both "solutions" seem rather contrived, to save a description which cannot be regarded as fundamental.

¶For additional discussion of why the effective Lagrangian approach is ineffective, see Ref. [445]. For another example of the failure of $\mathcal{L}_{EH}$ to capture the relevant higher-loop corrections, see Ref. [456].

Chapter 14
Conclusions and Outlook

In this book, we have tried to survey the subject of the Casimir effect, which can be thought of as describing "macroscopic" manifestations of zero-point fluctuations in quantum fields. (The word *macroscopic* is in quotes because applications of these ideas have been made to hadrons, of a size on order of fm, 10^{-13} cm, and to cosmology, where the fundamental scale is the Planck length, $\sim 10^{-33}$ cm.) The phenomenon probes the most fundamental quantum idea, that of the uncertainty principle and the associated zero-point energy of a quantum oscillator. It is quite remarkable that in the 75 years since the birth of quantum mechanics, so little understanding of the role of zero-point energy exists. Textbooks even used to assert that these fluctuations are unobservable [459]. It was the genius of Casimir in the years immediately after the Second World War to recognize that these quantum fluctuations, having their action-at-a-distance counterpart in the van der Waals forces between neutral molecules, could be observed in the laboratory as forces between macroscopic objects [1]. With the fundamentally identical theory of Lifshitz [7] verified experimentally in the 1970s [66], no one could seriously doubt the reality of the effect, although, for psychological reasons, what convinced most physicists seemed to be the recent measurements between conducting lenses and plates [68, 71, 75].

So what have we learned in the more than 50 years since Casimir's brilliant observation? Remarkably little! This has not been for want of theoretical activity, as recounted a few years ago by Bordag [460]. The yearly number of papers per year citing the Casimir effect now must approach 100. (Only a select subset of those papers have been cited in this book.) There have been some remarkable milestones: Boyer's discovery [13] that the Casimir force on a conducting spherical shell in *repulsive*; the study of the dimensional dependence of the Casimir effect by Bender and Milton [32]; the discovery of the unique finite part of the Casimir energy for a dilute dielectric sphere by Barton and others [40, 39], and the equivalence of that energy with the renormalized sum of van der Waals energies [38], which brings the saga full circle to its starting point. But we lack fundamental understanding. Why should a sphere give a repulsive rather than an attractive self-stress? Why is the stress on an even-dimensional hypersphere infinite? It seems to do little good to argue, as does Barton [219], that the leading infinite term is, of course, attractive, so we have no hope of understanding the sign of the subleading finite terms, because this begs the issue in cases when the results are manifestly finite, for spherical and cylindrical shells and the like. And new puzzles keep emerging. Why should a dilute dielectric cylinder have a zero Casimir self-stress, when a dilute sphere experiences a repulsion?

Applications are burgeoning. There is much recent work on applying Casimir energies to cosmological models, and the time is probably ripe to revisit applications at the hadronic scale. The Casimir effect's possible role in explaining sonoluminescence remains intriguing. Yet what is required to confront phenomena is further development of the theory.

The subject of radiative corrections to the Casimir effect is in its infancy [107]. A proper calculation with interacting quarks and gluons could have significant implications in our understanding of nuclear physics. As for cosmology (and perhaps for sonoluminescence) what is required are true dynamical calculations, where the boundaries (or geometry) are not changing perturbatively or adiabatically.

The recent impressive experiments with atomic force microscopes [71, 72, 175], and the very recent micromachined torsional measurement [75] (like the corresponding experiments testing Newton's law to the 100 μm level [362]) suggest we could be on the verge of significant new experimental input to the field. For example, if a 2-dimensional experiment could be devised to measure the stress on a circle, we might hope to understand the breakdown of the theory in even dimensions. Moreover, it is not fanciful to imagine that we are on the threshold of practical applications of the Casimir effect. (Some applications are suggested in Ref. [75].) It is well to remember that physics is an experimental science. After all, we recall that Schwinger could not have carried the formalism of QED to its logical conclusion without the impetus of the postwar experiments, announced at the Shelter Island Conference, which overcame prewar paralysis by showing that the quantum corrections "were neither infinite nor zero, but finite and small, and demanded understanding" [461].

So this book does not represent a survey of a completed field, but a status report of a fundamental aspect of quantum field theory, where the most important questions have yet to be addressed. This book will have served its role if it helps stimulate the next stage in that development.

Appendix A

Relation of Contour Integral Method to Green's Function Approach

Most of this book is devoted to the calculation of the Casimir effect in different situations using the Green's function approach, which is deemed the most physical. However, it cannot be denied that zeta-function techniques and related methods are very efficacious, and an example of such a method was given in Sec. 7.2. Other examples in the text occur in the evaluation of the Casimir energy for a dilute dielectric sphere in Sec. 5.7, and of the Casimir energy in Kaluza-Klein spaces in Sec. 10.1.4. Here we offer a derivation of the starting point of the first of these examples from the Green's function approach. We will not discuss the basis of zeta-function evaluation, however, since treatises describing such methods are readily available [271, 272].*

Specifically, we sketch the relation of the formula (7.50) to the Green's function formalism. We will confine our remarks to the case of a massless scalar field, as the generalization to, say, electromagnetism is rather immediate. We take the scalar Green's function to satisfy the differential equation (2.13)

$$\left(\frac{\partial^2}{\partial t^2} - \nabla^2\right) G(x, x') = \delta(x - x'), \qquad (A1)$$

subject to appropriate boundary conditions. The stress tensor (2.21) (use

---

*Beneventano and Santangelo [462] claim discrepancies between ζ-function and cutoff regularization, yet both methods reproduce the known results for massive fields in a d-dimensional box. See also the paper of Elizalde [463] responding in particular to complementary claims of Svaiter and Svaiter [464].

of the conformal stress tensor has the same effect) is

$$T^{\mu\nu} = \partial^\mu \phi \partial^\nu \phi - \frac{1}{2} g^{\mu\nu} \partial_\lambda \phi \partial^\lambda \phi, \tag{A2}$$

so, since the Green's function is given as a vacuum expectation value of a time-ordered product of fields (2.23),

$$G(x, x') = i\langle T\phi(x)\phi(x')\rangle, \tag{A3}$$

the energy density is

$$u = \langle T^{00} \rangle = -\frac{i}{2}[\partial^0 \partial^{0\prime} + \boldsymbol{\nabla} \cdot \boldsymbol{\nabla}']G(x, x')\bigg|_{x=x'}. \tag{A4}$$

To compute the total energy, we integrate u over all space; then we can integrate by parts on one of the gradient terms, and use the differential equation (A1), omitting the delta function because the limit of point coincidence is understood:

$$\boldsymbol{\nabla} \cdot \boldsymbol{\nabla}' \to -\nabla^2 \to -\frac{\partial^2}{\partial t^2}. \tag{A5}$$

(The net effect is that the Lagrangian term in $T^{\mu\nu}$ does not contribute.) In terms of the Fourier transform of the Green's function

$$G(x, x') = \int \frac{d\omega}{2\pi} e^{-i\omega(t-t')} \mathcal{G}_\omega(\mathbf{r}, \mathbf{r}'), \tag{A6}$$

the Casimir energy is

$$E = -i \int \frac{d\omega}{2\pi} e^{-i\omega(t-t')} \int (d\mathbf{r})\, \omega^2 \mathcal{G}_\omega(\mathbf{r}, \mathbf{r}')\bigg|_{x=x'}. \tag{A7}$$

Now introduce eigenfunctions of the differential operator ∇^2 subject to the same boundary conditions as $\mathcal{G}_\omega$:

$$\nabla^2 \psi_p(\mathbf{r}) = -k_p^2 \psi_p(\mathbf{r}), \tag{A8a}$$

$$\sum_p \psi_p(\mathbf{r})\psi_p^*(\mathbf{r}') = \delta(\mathbf{r} - \mathbf{r}'), \tag{A8b}$$

$$\int (d\mathbf{r})\psi_p^*(\mathbf{r})\psi_{p'}(\mathbf{r}) = \delta_{pp'}. \tag{A8c}$$

Then the Green's function has an eigenfunction expansion

$$\mathcal{G}_\omega(\mathbf{r}, \mathbf{r}') = -\sum_p \frac{\psi_p(\mathbf{r})\psi_p^*(\mathbf{r}')}{\omega^2 - k_p^2 + i\epsilon}. \tag{A9}$$

(The Green's function is the causal, or Feynman, propagator.) Carrying out the volume integration,

$$\int (d\mathbf{r}) \mathcal{G}_\omega(\mathbf{r}, \mathbf{r}) = -\sum_p \frac{1}{\omega^2 - k_p^2 + i\epsilon}, \tag{A10}$$

we find that the energy can be written in the form

$$E = i \int_{-\infty}^{\infty} \frac{d\omega}{2\pi} e^{-i\omega\tau} \frac{\omega}{2} \sum_p \left(\frac{1}{\omega - k_p + i\epsilon} + \frac{1}{\omega + k_p - i\epsilon} \right). \tag{A11}$$

Here we have retained a time splitting, $\tau = t - t' \to 0$, which is a technique to regulate the divergent expression. What does this integral mean? Since the energy must be real, when τ is set equal to zero, the i is to be interpreted as an instruction to pick out the negative imaginary part of the integral. Because the path of integration in (A11) passes above the poles on the positive real axis, and below the poles on the negative real axis, that means that we can replace the integration path by a contour C which encircles all the poles on the real axis, the positive poles by a contour closed in the counterclockwise sense, and the negative poles by a contour closed in the clockwise sense (see Sec. 10.1 and Fig. 10.1):

$$E = -\frac{i}{8\pi} \oint_C \omega \, d\omega \ln \prod_p (\omega - k_p)(\omega + k_p). \tag{A12}$$

(We may verify the sign by noting that (7.43) is formally reproduced.) This is the content of (7.50). The equivalent mathematical result

$$\frac{1}{2\pi i} \oint_C \omega \frac{d}{d\omega} \ln g(\omega) \, d\omega = \sum \omega_0 - \sum \omega_\infty, \tag{A13}$$

where ω_0 denotes the zeroes, and ω_∞ the poles, of $g(\omega)$ lying within C, is called the argument principle—see for example Ref. [182], p. 428. It is the mode-sum analog of the Sommerfeld-Watson transformation of partial wave sums [465, 466].

The argument principle was apparently first used in this context by van Kampen, Nijboer, and Schram in 1968 [149], who rederived the nonretarded part of the Lifshitz interaction between dielectric slabs. Further examples of evaluating such mode sums, particularly for a string with massive ends (see Appendix B) are given in Ref. [467].

Appendix B

Casimir Effect for a Closed String

Here we offer another elementary example of the calculation of the Casimir energy, for the simple situation of a uniform closed string. Brevik and collaborators [468, 469, 470, 471, 472, 473, 474] generalized this by considering a piecewise uniform string with a uniform sound speed but different string tensions. See also Ref. [475]. A nonuniform sound speed has also been considered [476].

The uniform closed string may be regarded as a circle, of radius a, characterized by a Green's function $G(\theta - \theta', t - t')$, satisfying the differential equation

$$\left(\frac{\partial^2}{\partial t^2} - \frac{1}{a^2} \frac{\partial^2}{\partial \theta^2} \right) G(\theta - \theta', t - t') = \frac{1}{a} \delta(\theta - \theta') \delta(t - t'). \tag{B1}$$

The solution may be written in eigenfunction form as

$$G(\theta - \theta', t - t') = \int \frac{d\omega}{2\pi} e^{-i\omega(t-t')} \frac{1}{2\pi} \sum_{m=-\infty}^{\infty} e^{im(\theta-\theta')} \frac{a}{m^2 - \omega^2 a^2}. \tag{B2}$$

As usual, the energy density may be obtained from this by differentiation:

$$\langle T^{00} \rangle = \frac{1}{i} \partial^0 \partial'^0 G(0, t - t') \Big|_{t=t'}$$

$$= \frac{a}{2\pi i} \int_{-\infty}^{\infty} \frac{d\omega}{2\pi} \sum_{m=-\infty}^{\infty} \frac{\omega^2}{m^2 - \omega^2 a^2}. \tag{B3}$$

We now carry out the m sum by use of (2.45), perform the usual Euclidean rotation, $\omega \to i\zeta$, and omit terms which are nonnegative powers of a (can-

273

celled by contact terms). We are left with

$$\langle T^{00}\rangle = -\frac{1}{\pi}\int_0^\infty d\zeta\,\zeta\frac{1}{e^{2\pi\zeta a}-1} = -\frac{1}{24\pi a^2}. \tag{B4}$$

This is the energy per unit length of the string. The energy is obtained by multiplying by the circumference $L = 2\pi a$ of the string,

$$E = -\frac{1}{12a} = -\frac{\pi}{6L}. \tag{B5}$$

Compare this with an open string, with Dirichlet boundary conditions at the ends. That is just the Lüscher energy (1.35) [83, 477] [see also (6.2) and (9.21)],

$$E_{\text{open string}} = -\frac{\pi}{24L}. \tag{B6}$$

Periodic boundary conditions give an enhancement of a factor of four, as a consequence of (2.49).

Brevik and Nielsen [468] have considered the case when the material of the string consists of two parts, with lengths L_1 and L_2, $L_1 + L_2 = L$. For example, in the special case when one of the string tensions is zero (but with uniform sound velocity) the energy is shifted by the simple formula

$$\Delta E = -\frac{\pi}{24L}\left(\frac{L_1}{L_2}+\frac{L_2}{L_1}-2\right). \tag{B7}$$

If $L_1 = L_2$, $\Delta E = 0$, which also holds true for any value of the string tension. In general, numerical results are given in Refs. [468, 469]. In Refs. [470, 472, 471] the division of the string into $2N$, and into 3, pieces, respectively, is considered. The various cases are summarized in Ref. [473], while an interesting scaling property for the $2N$ string, that $E_N(x)/E_N(0)$ is nearly independent of N, where $x = T_1/T_2$ is the ratio of string tensions in the two portions of string material, was discovered in Ref. [474]. The decay spectrum of a two-piece string was considered in Ref. [478]. A recent paper discusses the thermodynamics of the piecewise uniform string [479]. An elegant regularization method was given by Li, Shi, and Zhang [475]. The difficulties inherent in the more realistic case of a nonuniform string, where the sound speed is different in different sectors, is addressed in Ref. [476].

B.1 Open Strings

D'Hoker and Sikivie [480] considered the Casimir energy between beads on an open string. The Lüscher energy (B6) is recovered in the limit of large masses. The interaction (always nonnegative) between beads on d-dimensional membranes due to quantum fluctuations of the membrane were considered by D'Hoker, Skivie, and Kanev [481].

Actor, Bender, and Reingruber [482] considered the Casimir effect on a one-dimensional lattice, recovering the continuum result (B6), the effect of masses [contained in (2.54)] and the effect of a lattice potential. Elizalde and Odintsov [483] calculated quantum corrections to the Lüscher potential, which renormalized the string tension; they interpreted the phenomenon as a kind of nonlocal Casimir effect.

Kleinert, Lambiase, and Nesterenko [484, 485] considered the interquark potential in the model of the Nambu-Goto string with point masses μ at the ends. The Lüscher term is recovered in either the $\mu = 0$ or $\mu = \infty$ limits. The deconfinement temperature is not affected by finite quark masses, however [486].

Bibliography

[1] H. B. G. Casimir. *Proc. Kon. Ned. Akad. Wetensch.*, 51:793, 1948.

[2] J. Schwinger. *Lett. Math. Phys.*, 1:43, 1975.

[3] F. Sauer. PhD thesis, Göttingen, 1962.

[4] J. Mehra. *Physica*, 37:145, 1967.

[5] J. Mehra. PhD thesis, Université de Neuchâtel, Switzerland, 1963.

[6] J. Mehra. *Acta Physica Austriaca*, 27:341, 1968.

[7] E. M. Lifshitz. *Zh. Eksp. Teor. Fiz.*, 29:94, 1956. [English transl.: *Soviet Phys. JETP* 2:73, 1956].

[8] I. D. Dzyaloshinskii, E. M. Lifshitz, and L. P. Pitaevskii. *Usp. Fiz. Nauk*, 73:381, 1961. [English transl.: *Soviet Phys. Usp.* 4:153, 1961].

[9] L. D. Landau and E. M. Lifshitz. *Electrodynamics of Continuous Media.* Pergamon, Oxford, 1960.

[10] C. M. Hargreaves. *Proc. Kon. Ned. Akad. Wetensch. B*, 68:231, 1965.

[11] J. Schwinger, L. L. DeRaad, Jr., and K. A. Milton. *Ann. Phys. (N.Y.)*, 115:1, 1978.

[12] H. B. G. Casimir. *Physica*, 19:846, 1956.

[13] T. H. Boyer. *Phys. Rev.*, 174:1764, 1968.

[14] R. Balian and B. Duplantier. *Ann. Phys. (N.Y.)*, 112:165, 1978.

[15] K. A. Milton, L. L. DeRaad, Jr., and J. Schwinger. *Ann. Phys. (N.Y.)*, 115:388, 1978.

[16] K. A. Milton. *Ann. Phys. (N.Y.)*, 127:49, 1980.

[17] K. Johnson. In B. Margolis and D. G. Stairs, editors, *Particles and Fields 1979*, page 353, New York, 1980. AIP.

[18] K. A. Milton. *Phys. Rev. D*, 22:1441, 1980.

[19] K. A. Milton. *Phys. Rev. D*, 22:1444, 1980.

[20] K. A. Milton. *Phys. Rev. D*, 25:3441 (E), 1982.

[21] K. A. Milton. *Phys. Lett. B*, 104:49, 1981.

[22] K. A. Milton. *Phys. Rev. D*, 27:439, 1983.

[23] K. A. Milton. *Ann. Phys. (N.Y.)*, 150:432, 1983.

[24] L. L. DeRaad, Jr. and K. A. Milton. *Ann. Phys. (N.Y.)*, 136:229, 1981.

[25] R. Kantowski and K. A. Milton. *Phys. Rev. D*, 35:549, 1987.

[26] R. Kantowski and K. A. Milton. *Phys. Rev. D*, 36:3712, 1987.

[27] D. Birmingham, R. Kantowski, and K. A. Milton. *Phys. Rev. D*, 38:1809, 1988.

[28] T. Appelquist and A. Chodos. *Phys. Rev. Lett.*, 50:141, 1983.

[29] P. Candelas and S. Weinberg. *Nucl. Phys. B*, 237:397, 1984.

[30] K. A. Milton and Y. J. Ng. *Phys. Rev. D*, 42:2875, 1990.

[31] K. A. Milton and Y. J. Ng. *Phys. Rev. D*, 46:842, 1992.

[32] C. M. Bender and K. A. Milton. *Phys. Rev. D*, 50:6547, 1994.

[33] K. A. Milton. *Phys. Rev. D*, 55:4940, 1997.

[34] B. P. Barber, R. A. Hiller, R. Löfstedt, S. J. Putterman, and K. Weniger. *Phys. Rep.*, 281:65, 1997.

[35] J. Schwinger. *Proc. Natl. Acad. Sci. USA*, 90:958, 2105, 4505, 7285, 1993.

[36] J. Schwinger. *Proc. Natl. Acad. Sci. USA*, 91:6473, 1994.

[37] K. A. Milton and Y. J. Ng. *Phys. Rev. E*, 55:4207, 1997.

[38] K. A. Milton and Y. J. Ng. *Phys. Rev. E*, 57:5504, 1998.

[39] I. Brevik, V. N. Marachevsky, and K. A. Milton. *Phys. Rev. Lett.*, 82:3948, 1999.

[40] G. Barton. *J. Phys. A*, 32:525, 1999.

[41] M. Bordag, K. Kirsten, and D. Vassilevich. *Phys. Rev. D*, 59:085011, 1999.

[42] J. S. Høye and I. Brevik. *J. Stat. Phys.*, 100:223, 2000.

[43] K. A. Milton, A. V. Nesterenko, and V. V. Nesterenko. *Phys. Rev. D*, 59:105009, 1999.

[44] A. Romeo. *private communication*, 1998.

[45] D. ter Haar. *Elements of Statistical Mechanics*. Holt, Reinhart and Winston, New York, 1954.

[46] J. D. van der Waals. *Over de Continueïteit van den Gas- en Vloeistoftoestand*. PhD thesis, Leiden, 1873.

[47] J. C. Maxwell. *Nature*, 10:477, 1874.

[48] R. Eisenschitz and F. London. *Z. Physik*, 60:491, 1930.

[49] F. London. *Z. Physik*, 63:245, 1930.

[50] H. B. G. Casimir and D. Polder. *Phys. Rev.*, 73:360, 1948.

[51] H. B. G. Casimir. In *Colloque sur la theorie de la liaison chimique*, Paris, 12–17 April, 1948. Published in *J. Chim. Phys.*, 46:407, 1949.

[52] H. B. G. Casimir. In M. Bordag, editor, *The Casimir Effect 50 Years Later: The Proceedings of the Fourth Workshop on Quantum Field Theory Under the Influence of External Conditions, Leipzig, 1998*, page 3, Singapore, 1999. World Scientific.

[53] H. Rechenberg. In M. Bordag, editor, *The Casimir Effect 50 Years Later: The Proceedings of the Fourth Workshop on Quantum Field Theory Under the Influence of External Conditions, Leipzig, 1998*, page 10, Singapore, 1999. World Scientific.

[54] I. I. Abrikosova and B. V. Deriagin (Derjaguin). *Dokl. Akad. Nauk SSSR*,

90:1055, 1953.

[55] B. V. Deriagin (Derjaguin) and I. I. Abrikosova. *Zh. Eksp. Teor. Fiz.*, 30:993, 1956. [English transl.: *Soviet Phys. JETP* 3:819, 1957].

[56] B. V. Deriagin (Derjaguin) and I. I. Abrikosova. *Zh. Eksp. Teor. Fiz.*, 31:3, 1956. [English transl.: *Soviet Phys. JETP* 4:2, 1957].

[57] B. V. Derjaguin, I.I. Abrikosova, and E. M. Lifshitz. *Quart. Rev.*, 10:295, 1956.

[58] A. Kitchener and A. P. Prosser. *Proc. Roy. Soc. (London) A*, 242:403, 1957.

[59] M. Y. Sparnaay. *Physica*, 24:751, 1958.

[60] W. Black, J. G. V. de Jongh, J. Th. G. Overbeck, and M. J. Sparnaay. *Trans. Faraday Soc.*, 56:1597, 1960.

[61] A. van Silfhout. *Proc. Kon. Ned. Akad. Wetensch. B*, 69:501, 1966.

[62] D. Tabor and R. H. S. Winterton. *Nature*, 219:1120, 1968.

[63] D. Tabor and R. H. S. Winterton. *Proc. Roy. Soc. (London) A*, 312:435, 1969.

[64] R. H. S. Winterton. *Contemp. Phys.*, 11:559, 1970.

[65] J. N. Israelachivili and D. Tabor. *Proc. Roy. Soc. (London) A*, 331:19, 1972.

[66] E. S. Sabisky and C. H. Anderson. *Phys. Rev. A*, 7:790, 1973.

[67] I. D. Dzyaloshinskii, E. M. Lifshitz, and L. P. Pitaevskii. *Zh. Eksp. Teor. Fiz.*, 37:229, 1959. [English transl.: *Soviet Phys. JETP* 10:161, 1960].

[68] S. K. Lamoreaux. *Phys. Rev. Lett.*, 78:5, 1997.

[69] S. K. Lamoreaux. *Phys. Rev. Lett.*, 81:5475(E), 1998.

[70] S. K. Lamoreaux. *Phys. Rev. A*, 59:R3149, 1999.

[71] U. Mohideen and A. Roy. *Phys. Rev. Lett.*, 81:4549, 1998.

[72] A. Roy, C.-Y. Lin, and U. Mohideen. *Phys. Rev. D*, 60:R111101, 1999.

[73] B. W. Harris, F. Chen, and U. Mohideen. *Phys. Rev. A*, 62:052109, 2000.

[74] T. Erdeth. *Phys. Rev. A*, 62:062104, 2000.

[75] H. B. Chan, V. A. Aksyuk, R. N. Kleiman, D. J. Bishop, and F. Capasso. *Science*, 291:1941, 2001. (10.1126/science.1057984).

[76] B. Davies. *J. Math. Phys.*, 13:1324, 1972.

[77] W. Lukosz. *Physica*, 56:109, 1971.

[78] W. Lukosz. *Z. Phys.*, 258:99, 1973.

[79] W. Lukosz. *Z. Phys.*, 262:327, 1973.

[80] J. R. Ruggiero, A. H. Zimerman, and A. Villani. *Rev. Bras. Fis.*, 7:663, 1977.

[81] J. R. Ruggiero, A. H. Zimerman, and A. Villani. *J. Phys. A*, 13:761, 1980.

[82] J. Ambjørn and S. Wolfram. *Ann. Phys. (N.Y.)*, 147:1, 1983.

[83] M. Lüscher, K. Symanzik, and P. Weisz. *Nucl. Phys. B*, 173:365, 1980.

[84] A. Chodos, R. L. Jaffe, K. Johnson, C. B. Thorn, and V. Weisskopf. *Phys. Rev. D*, 9:3471, 1974.

[85] A. Chodos, R. L. Jaffe, K. Johnson, and C. B. Thorn. *Phys. Rev. D*, 10:2599, 1974.

[86] T. DeGrand, R. L. Jaffe, K. Johnson, and J. Kiskis. *Phys. Rev. D*, 12:2060,

1975.

[87] K. Johnson. *Acta Phys. Pol.*, B6:865, 1975.

[88] J. F. Donoghue, E. Golowich, and B. R. Holstein. *Phys. Rev. D*, 12:2875, 1975.

[89] R. E. Schrock and S. B. Treiman. *Phys. Rev. D*, 19:2148, 1979.

[90] G. A. Vilkovisky. In S. C. Christensen, editor, *Quantum Theory of Gravity*, Bristol, England, 1984. Hilger.

[91] B. S. DeWitt. In I. A. Batalin, C. J. Isham, and G. A. Vilkovisky, editors, *Quantum Field Theory and Quantum Statistics*, Bristol, England, 1987. Hilger.

[92] R. B. Laughlin. *Phys. Rev. Lett.*, 50:1395, 1983.

[93] F. D. M. Haldane. *Phys. Rev. Lett.*, 51:605, 1985.

[94] W. P. Su. *Phys. Rev. B*, 30:1069, 1984.

[95] R. Tao and Y. S. Wu. *Phys. Rev. B*, 31:6859, 1985.

[96] R. B. Laughlin. *Science*, 242:525, 1988.

[97] Y. H. Chen, B. I. Halperin, F. Wilczek, and E. Witten. *Int. J. Mod. Phys.*, B3:1001, 1989.

[98] X. G. Wen and A. Zee. *Phys. Rev. B*, 41:240, 1990.

[99] J. Schwinger, L. L. DeRaad, Jr., K. A. Milton, and W.-y. Tsai. *Classical Electrodynamics*. Perseus Books, Reading, Massachusetts, 1998.

[100] L. S. Brown and G. J. Maclay. *Phys. Rev.*, 184:1272, 1969.

[101] N. D. Birrell and P. C. W. Davies. *Quantum Fields in Curved Space*. Cambridge University Press, Cambridge, 1982.

[102] D. Deutsch and P. Candelas. *Phys. Rev. D*, 20:3063, 1979.

[103] J. Schwinger. *Lett. Math. Phys.*, 24:59, 227, 1992.

[104] J. Schwinger. *Proc. Natl. Acad. Sci. USA*, 89:4091, 11118, 1992.

[105] M. Bordag, D. Robaschik, and E. Wieczorek. *Ann. Phys. (N.Y.)*, 165:192, 1985.

[106] K. Langfeld, F. Schmüser, and H. Reinhardt. *Phys. Rev. D*, 51:765, 1995.

[107] M. Bordag and J. Lindig. *Phys. Rev. D*, 58:045003, 1998.

[108] E. T. Whittaker and G. N. Watson. *A Course in Modern Analysis*. Cambridge University Press, London, 1997.

[109] I. S. Gradshteyn and I. M. Ryzhik. *Table of Integrals, Series, and Products*. Academic Press, New York, 1965.

[110] A. P. Prudnikov, Yu. A. Brychkov, and O. I. Marichev. *Integrals and Series*. Gordon and Breach, New York, 1986. three volumes.

[111] M. Abramowitz and I. A. Stegun, editors. *Handbook of Mathematical Functions*. National Bureau of Standards, Washington, D.C., 1964.

[112] G. Plunien, B. Müller, and W. Greiner. *Phys. Rep.*, 134:87, 1986.

[113] V. M. Mostepanenko and N. N. Trunov. *Usp. Fiz. Nauk*, 156:385, 1988. [English transl.: *Soviet Phys. Usp.*, 31:965, 1988].

[114] V. M. Mostepanenko and N. N. Trunov. *The Casimir Effect and its Applications*. Oxford Science Publications, Oxford, 1997.

[115] F. S. Levin and D. A. Micha. *Long-Range Casimir Forces: Theory and*

Recent Experiments on Atomic Systems. Plenum, New York, 1993.

[116] M. Krech. *Casimir Effect in Critical Systems.* World Scientific, Singapore, 1994.

[117] P. Milonni. *The Quantum Vacuum: An Introduction to Quantum Electrodynamics.* Academic Press, Boston, 1994.

[118] M. Bordag, U. Mohideen, and V. M. Mostepanenko. New Developments in the Casimir Effect. *Physics Reports*, in press. quant-ph/0106045.

[119] J. Schwinger. *Particles, Sources, and Fields*, volume I. Addison-Wesley, Reading, Mass., 1970.

[120] C. G. Callan, Jr., S. Coleman, and R. Jackiw. *Ann. Phys. (N.Y.)*, 59:42, 1970.

[121] J. Schwinger. *Phys. Rev.*, 82:644, 1951.

[122] M. V. Cougo-Pinto, C. Farina, A. J. Seguí-Santonja. *Lett. Math. Phys.*, 30:169, 1994.

[123] M. V. Cougo-Pinto, C. Farina, A. J. Seguí-Santonja. *Lett. Math. Phys.*, 31:309, 1994.

[124] M. V. Cougo-Pinto, C. Farina, and A. Tort. *Lett. Math. Phys.*, 37:159, 1996.

[125] M. V. Cougo-Pinto, C. Farina, and A. Tort. *Lett. Math. Phys.*, 38:97, 1996.

[126] M. V. Cougo-Pinto, C. Farina, and A. Tort. *Lett. Math. Phys.*, 38:337, 1996.

[127] M. Bordag, D. Hennig, and D. Robaschik. *J. Phys. A*, 25:4483, 1992.

[128] P. Hays. *Ann. Phys. (N.Y.)*, 121:32, 1979.

[129] M. Fierz. *Helv. Phys. Acta*, 33:855, 1960.

[130] E. M. Lifshitz. *private letter to J. Schwinger*, 27 April 1978.

[131] R. Balian and B. Duplantier. *Ann. Phys. (N.Y.)*, 104:300, 1977.

[132] M. Revzen, R. Opher, M. Opher, and A. Mann. *J. Phys. A*, 30:7783, 1997.

[133] H. Mitter and D. Robaschik. *Eur. Phys. J. B*, 13:335, 2000.

[134] T. H. Boyer. *Phys. Rev. A*, 9:2078, 1974.

[135] F. C. Santos, A. Tenorio, and A. C. Tort. *Phys. Rev. D*, 60:105022, 1999.

[136] M. V. Cougo-Pinto, C. Farina, F. C. Santos, and A. C. Tort. *J. Phys. A*, 32:4463, 1999.

[137] A. Kenneth and S. Nussinov. *Phys. Rev. D*, 63:121701(R), 2001. preprint: hep-th/9802149.

[138] G. Barton. *J. Phys. A*, 24:991, 1991.

[139] G. Barton. *J. Phys. A*, 24:5533, 1991.

[140] D. Robaschik and E. Wieczorek. *Ann. Phys. (N.Y.)*, 236:43, 1994.

[141] D. Robaschik and E. Wieczorek. *Phys. Rev. D*, 52:2341, 1995.

[142] M.-T. Jaekel and S. Reynaud. *Phys. Lett. A*, 167:227, 1992.

[143] M.-T. Jaekel and S. Reynaud. In J. P. Dowling, editor, *Proceedings of NATO Advanced Study Institute: Electron Theory and Quantum Electrodynamics—100 Years Later, Edirne, Turkey, 5–6 September 1995*, pages 55–65, New York, 1995. Plenum Press.

[144] K. Scharnhorst. *Phys. Lett. B*, 236:354, 1990.

[145] G. Barton. *Phys. Lett. B*, 237:559, 1990.

[146] G. Barton and K. Scharnhorst. *J. Phys. A*, 26:2037, 1993.

[147] J. Mehra and K. A. Milton. *Climbing the Mountain: The Scientific Biography of Julian Schwinger*. Oxford University Press, Oxford, 2000.

[148] T. H. Boyer. *Ann. Phys. (N.Y.)*, 56:474, 1970.

[149] N. G. van Kampen, B. R. A. Nijboer, and K. Schram. *Phys. Lett. A*, 26:307, 1968.

[150] M. Schaden, L. Spruch, and F. Zhou. *Phys. Rev. A*, 57:1108, 1998.

[151] M. Boström and Bo E. Sernelius. *Phys. Rev. Lett.*, 84:4757, 2000.

[152] M. Boström and Bo E. Sernelius. *Phys. Rev. A*, 61:052703, 2000.

[153] S. K. Lamoreaux. *preprint*, quant-ph/0007029.

[154] V. B. Svetovoy and M. V. Lokhanin. *Mod. Phys. Lett. A*, 15:1437, 2000.

[155] V. B. Svetovoy and M. V. Lokhanin. *Phys. Lett. A*, 280:177, 2001.

[156] M. Bordag, B. Geyer, G. L. Klimchitskaya, and V. M. Mostepanenko. *Phys. Rev. Lett.*, 85:503, 2000.

[157] M. L. Levin and S. M. Rytov. *The Theory of Thermal Equilibrium Fluctuations in Electrodynamics*. Nauka, Moscow, 1967.

[158] F. Ravndal and D. Tollefsen. *Phys. Rev. D*, 40:4191, 1989.

[159] M. Schaden and L. Spruch. *preprint*, quant-ph/0012082.

[160] V. M. Mostepanenko and N. N. Trunov. *Sov. J. Nucl. Phys.*, 42:812, 1985.

[161] A. Lambrecht, M.-T. Jaekel, and S. Reynaud. *Phys. Lett. A*, 225:188, 1997.

[162] C. I. Sukenik, M. G. Boshier, D. Cho, V. Sundoghar, and G. A. Hinds. *Phys. Rev. Lett.*, 70:560, 1993.

[163] G. Barton. *Proc. Roy. Soc. London*, 410:175, 1987.

[164] L. H. Ford and N. F. Svaiter. *Phys. Rev. A*, 62:062105, 2000.

[165] M. Schaden and L. Spruch. *Phys. Rev. A*, 58:935, 1998.

[166] M. Schaden and L. Spruch. *Phys. Rev. Lett.*, 84:459, 2000.

[167] B. V. Deryagin (Derjaguin). *Kolloid Z.*, 69:155, 1934.

[168] B. V. Deryagin (Derjaguin) et al. *J. Colloid. Interface Sci.*, 53:314, 1975.

[169] M. J. Sparnaay. In A. Sarlemijn and M. J. Sparnaay, editors, *Physics in the Making: Essays on Developments in 20th Century Physics in Hounor of H.B.G. Casimir on the Occasion of his 80th Birthday*, page 235, Amsterdam, 1989. North-Holland.

[170] J. Blocki, J. Randrup, W. J. Świątecki, and C. F. Tsang. *Ann. Phys. (N.Y.)*, 105:427, 1977.

[171] U. Mohideen and A. Roy. *Phys. Rev. Lett.*, 83:3341, 1999.

[172] A. Lambrecht and S. Reynaud. *Eur. Phys. J. D*, 8:309, 2000.

[173] A. Lambrecht and S. Reynaud. *Phys. Rev. Lett.*, 84:5672, 2000.

[174] S. K. Lamoreaux. *Phys. Rev. Lett.*, 84:5673, 2000.

[175] A. Roy and U. Mohideen. *Phys. Rev. Lett.*, 82:4380, 1999.

[176] G. L. Klimchitskaya, A. Roy, U. Mohideen, and V. M. Mostepanenko. *Phys. Rev. A*, 60:3487, 1999.

[177] M. Bordag, B. Geyer, G. L. Klimchitskaya, and V. M. Mostepanenko. *Phys. Rev. D*, 58:075003, 1998.

[178] G. L. Klimchitskaya, U. Mohideen, and V. M. Mostepanenko. *Phys. Rev. A*, 61:062107, 2000.

[179] A. A. Maradudin and P. Mazur. *Phys. Rev. B*, 22:1677, 1980.

[180] V. B. Bezerra, G. L. Klimchitskaya, and C. Romero. *Mod. Phys. Lett. A*, 12:2613, 1997.

[181] H. C. Corben and J. Schwinger. *Phys. Rev.*, 58:953, 1940.

[182] P. M. Morse and H. Feshbach. *Methods of Theoretical Physics*. McGraw-Hill, New York, 1953. Part II, Sec. 13.3.

[183] J. D. Jackson. *Classical Electrodynamics*. Wiley, New York, third edition, 1998. Chapter 9.

[184] J. A. Stratton. *Electromagnetic Theory*. McGraw-Hill, New York, 1941.

[185] C. R. Hagen. *Phys. Rev. D*, 61:065005, 2000.

[186] I. Brevik, H. Skurdal, and R. Sollie. *J. Phys. A*, 27:6853, 1994.

[187] S. Leseduarte and A. Romeo. *Europhys. Lett.*, 34:79, 1996.

[188] S. Leseduarte and A. Romeo. *Ann. Phys. (N.Y.)*, 250:448, 1996.

[189] V. V. Nesterenko and I. G. Pirozhenko. *Phys. Rev. D*, 57:1284, 1998.

[190] M. E. Bowers and C. R. Hagen. *Phys. Rev. D*, 59:0250007, 1999.

[191] G. Lambiase, V. V. Nesterenko, and M. Bordag. *J. Math. Phys.*, 40:6254, 1999.

[192] G. Esposito, A. Yu. Kamenshchik, and K. Kirsten. *Int. J. Mod. Phys.*, 14:281, 1999.

[193] G. Esposito, A. Yu. Kamenshchik, and K. Kirsten. *Phys. Rev. D*, 62:085027, 2000.

[194] C. Eberlein. *J. Phys. A*, 25:3015, 1992.

[195] M. Bordag, E. Elizalde, K. Kirsten, and S. Leseduarte. *Phys. Rev. D*, 56:4896, 1997.

[196] M. Revzen, R. Opher, M. Opher, and A. Mann. *Europhys. Lett.*, 38:245, 1997.

[197] J. Feinberg, A. Mann, and M. Revzen. *Ann. Phys. (N.Y.)*, 288:103, 2001.

[198] K. Johnson. *private communication*.

[199] E. Elizalde, M. Bordag, and K. Kirsten. *J. Phys. A*, 31:1743, 1998.

[200] G. Cognola, E. Elizalde, and K. Kirsten. *preprint*, hep-th/9906228.

[201] P. Candelas. *Ann. Phys. (N.Y.)*, 167:257, 1986.

[202] W. Pauli. *Theory of Relativity*. Pergamon, Oxford, 1958. §33.

[203] I. Brevik. *Phys. Rep.*, 52:133, 1979.

[204] I. Brevik. *Ann. Phys. (N.Y.)*, 138:36, 1982.

[205] I. Brevik and H. Kolbenstvedt. *Phys. Rev. D*, 25:1731, 1982.

[206] I. Brevik and H. Kolbenstvedt. *Phys. Rev. D*, 26:1490, 1982. (E).

[207] I. Brevik and H. Kolbenstvedt. *Ann. Phys. (N.Y.)*, 143:179, 1982.

[208] I. Brevik and H. Kolbenstvedt. *Ann. Phys. (N.Y.)*, 149:237, 1983.

[209] I. Klich. *Phys. Rev. D*, 61:025004, 2000.

[210] I. H. Brevik, V. V. Nesterenko, and I. G. Pirozhenko. *J. Phys. A.*, 31:8661, 1998.

[211] I. Brevik. *J. Phys. A*, 20:5189, 1987.

[212] I. Brevik and G. Einevoll. *Phys. Rev. D*, 37:2977, 1988.

[213] P. Candelas. *Ann. Phys. (N.Y.)*, 143:241, 1982.

[214] I. Klich, J. Feinberg, A. Mann, and M. Revzen. *Phys. Rev. D*, 62:045017, 2000.

[215] I. Brevik and T. A. Yousef. *J. Phys. A*, 33:5819, 2000.

[216] I. Brevik and V. N. Marachevsky. *Phys. Rev. D*, 60:085006, 1999.

[217] G. Lambiase, G. Scarpetta, and V. V. Nesterenko. *preprint*, hep-th/9912176 v2, 2000.

[218] V. V. Nesterenko, G. Lambiase, and G. Scarpetta. *Phys. Rev.D*, 64:025013, 2001.

[219] G. Barton. *J. Phys. A*, 34:4083, 2001.

[220] P. Hasenfratz and J. Kuti. *Phys. Rep.*, 40:75, 1978.

[221] T. H. Hansson, K. Johnson, and C. Peterson. *Phys. Rev. D*, 26:2069, 1982.

[222] P. M. Fishbane, S. G. Gasiorowicz, and P. Kaus. *Phys. Rev. D*, 37:2623, 1988.

[223] C. M. Bender and P. Hays. *Phys. Rev. D*, 14:2622, 1976.

[224] J. Baacke and Y. Igarashi. *Phys. Rev. D*, 27:460, 1983.

[225] A. Romeo. *Phys. Rev. D*, 52:7308, 1995.

[226] R. Balian and C. Bloch. *Ann. Phys. (N.Y.)*, 64:271, 1971.

[227] R. Balian and C. Bloch. *Ann. Phys. (N.Y.)*, 84:559, 1974.

[228] S. K. Blau, M. Visser, and A. Wipf. *Nucl. Phys. B*, 310:163, 1988.

[229] J. F. Donoghue and K. Johnson. *Phys. Rev. D*, 21:1975, 1980.

[230] C. W. Wong. *Phys. Rev. D*, 24:1416, 1981.

[231] K. F. Liu and C. W. Wong. *Phys. Lett. B*, 113:1, 1982.

[232] A. Chodos and C. B. Thorn. *Phys. Rev. D*, 12:2733, 1975.

[233] H. Hosaka and H. Toki. *Phys. Rep.*, 277:65, 1996. See references therein.

[234] A. I. Vainshtein, V. I. Zakharov, and M. A. Shifman. *Pis'ma Zh. Eksp. Teor. Fiz*, 27:60, 1978. English transl.: *JETP Lett.* 27:55, 1978.

[235] A. I. Vainshtein, V. I. Zakharov, and M. A. Shifman. *Nucl. Phys. B*, 147:385, 448, 519, 1979.

[236] E. Fahri, N. Graham, R. L. Jaffe, and H. Weigel. *Nucl. Phys. B*, 585:443, 2000.

[237] N. Graham, R. L. Jaffe, M. Quandt, and H. Weigel. *preprint*, hep-th/0103010.

[238] R. Hofmann, M. Schumann, and R. D. Violler. *Eur. Phys. J. C*, 11:153, 1999.

[239] R. Hofmann, T. Gutsche, M. Schumann, and R. D. Violler. *Eur. Phys. J. C*, 16:677, 2000.

[240] M. Schumann, R.J. Lindebaum, R.D. Viollier. *Eur. Phys. J. C*, 16:331, 2000.

[241] I. Cherednikov, S. Fedorov, M. Khalili, and K. Sveshnikov. *Nucl. Phys. A*, 676:339, 2000.

[242] I. O. Cherednikov. *Phys. Lett. B*, 498:40, 2001.

[243] P. Gosdzinsky and A. Romeo. *Phys. Lett. B*, 441:265, 1998.

[244] V. V. Nesterenko and I. G. Pirozhenko. *J. Math. Phys.*, 41:4521, 2000.

[245] M. Scandurra. *J. Phys. A*, 33:5707, 2000.

[246] S. Leseduarte and A. Romeo. *Commun. Math. Phys.*, 193:317, 1998.

[247] Yu. A. Sitenko and A. Yu. Babansky. *Mod. Phys. Lett. A*, 13:379, 1998.

[248] F. Caruso, N. P. Neto, B. F. Svaiter, and N.F. Svaiter. *Phys. Rev. D*, 43:1300, 1991.

[249] F. Caruso, R. De Paola, N.F. Svaiter. *Int. J. Mod. Phys. A*, 14:2077, 1999.

[250] A. A. Actor. *Ann. Phys. (N.Y.)*, 230:303, 1994.

[251] A. A. Actor and I. Bender. *Phys. Rev. D*, 52:3581, 1995.

[252] X. Li, H. Cheng, and X. Zhai. *Phys. Rev. D*, 56:2155, 1997.

[253] K. Kirsten. *J. Phys. A*, 24:3281, 1991.

[254] I. Brevik and M. Lygren. *Ann. Phys. (N.Y.)*, 251:157, 1996.

[255] I. Brevik, M. Lygren, and V. N. Marachevsky. *Ann. Phys. (N.Y.)*, 267:134, 1998.

[256] I. Brevik and G. H. Nyland. *Ann. Phys. (N.Y.)*, 230:321, 1994.

[257] T. M. Helliwell and D. A. Konkowski. *Phys. Rev. D*, 34:1918, 1986.

[258] J. S. Dowker. *Phys. Rev. D*, 36:3095, 1987.

[259] I. Brevik, G. E. A. Matsas, and E. S. Moreira, Jr. *Phys. Rev. D*, 58:027502, 1998.

[260] I. Brevik and K. Petterson. *preprint*, quant-ph/0101114.

[261] G. E. A. Matsas. *Phys. Rev. D*, 41:3846, 1990.

[262] E. S. Moreira, Jr. *Nucl. Phys. B*, 451:365, 1995.

[263] A. N. Aliev. *Phys. Rev. D*, 55:3903, 1997.

[264] V. V. Nesterenko and I. G. Pirozhenko. *J. Math. Phys.*, 38:6265, 1997.

[265] T. D. Lee. *Particle Physics and Introduction to Field Theory.* Harwood, New York, 1981.

[266] I. Brevik. *J. Phys. A*, 15:L369, 1982.

[267] I. Brevik. *Can. J. Phys.*, 61:493, 1983.

[268] I. Brevik and H. Kolbenstvedt. *Can J. Phys.*, 62:805, 1984.

[269] I. Brevik and H. Kolbenstvedt. *Can J. Phys.*, 63:1409, 1985.

[270] Louis de Broglie. *Problèms de Propagations Guidées des Ondes Electromagnétiques.* Gauthier-Villars, Paris, 1941.

[271] E. Elizalde, S. D. Odintsov, A. Romeo, A. A. Bytsenko, and S. Zerbini. *Zeta Regularization Techniques with Applications.* World Scientific, Singapore, 1994.

[272] E. Elizalde. *Ten Physical Applications of Spectral Zeta Functions.* Springer, Berlin, 1995.

[273] I. Klich and A. Romeo. *Phys. Lett. B*, 476:369, 2000.

[274] V. V. Nesterenko and I. G. Pirozhenko. *Phys. Rev. D*, 60:125007, 1999.

[275] V. V. Nesterenko, G. Lambiase, and G. Scarpetta. *J. Math. Phys.*, 42:1974, 2001.

[276] J. S. Dowker. *preprint*, hep-th/0006138.

[277] M. Bordag and I. G. Pirozhenko. *Phys. Rev. D*, 64:025019, 2001.

[278] W. Siegel. *Nucl. Phys. B*, 156:135, 1979.

[279] R. Jackiw and S. Templeton. *Phys. Rev. D*, 23:2291, 1981.

[280] J. Schonfeld. *Nucl. Phys.*, B185:157, 1981.

[281] S. Deser, R. Jackiw, and S. Templeton. *Phys. Rev. Lett*, 48:975, 1982.

[282] S. Deser, R. Jackiw, and S. Templeton. *Ann. Phys. (N.Y.)*, 140:372, 1982.

[283] A. Niemi and G. Semenoff. *Phys. Rev. Lett.*, 51:2077, 1983.

[284] A. N. Redlich. *Phys. Rev. Lett.*, 52:18, 1984.

[285] R. B. Laughlin. *Rev. Mod. Phys.*, 71:863, 1999.

[286] K. Ishikawa. *Phys. Rev. Lett.*, 53:1615, 1984.

[287] K. Ishikawa. *Phys. Rev. D*, 31:143, 1985.

[288] K. Ishikawa and K. Matsuyama. *Z. Phys. C*, 33:41, 1986.

[289] K. Ishikawa and K. Matsuyama. *Nucl. Phys. B*, 280:523, 1987.

[290] N. Imai, K. Ishikawa, and T. Matsuyama. *Phys. Rev. B*, 42:10610, 1990.

[291] Q. Niu, D. J. Thouless, and Y.-S. Wu. *Phys. Rev. B*, 31:3372, 1985.

[292] J. Goryo and K. Ishikawa. *Phys. Lett. A*, 260:294, 1999.

[293] S. Deser and A. N. Redlich. *Phys. Rev. Lett.*, 61:1541, 1988.

[294] N. E. Mavromatos and M. Ruiz-Altaba. *Phys. Lett.*, A142:419, 1989.

[295] A. S. Goldhaber. *Phys. Rev. Lett.*, 26:1390, 1971.

[296] N. M. Kroll. *Phys. Rev. Lett.*, 26:1395, 1971.

[297] P. C. W. Davies and S. D. Unwin. *Phys. Lett.*, 98B:274, 1981.

[298] M. Bordag and D. V. Vassilevich. *Phys. Lett. A*, 268:75, 2000.

[299] S. Sen. *Phys. Rev. D*, 24:869, 1981.

[300] S. Sen. *J. Math. Phys.*, 22:2968, 1981.

[301] J. C. Slater and N. H. Frank. *Electromagnetism*. McGraw-Hill, New York, 1947. p. 153.

[302] A. Romeo. *Phys. Rev. D*, 53:3392, 1995.

[303] T. Kaluza. *Sitz. Preuss. Akad. Wiss. Phys. Math.*, K1:966, 1921.

[304] O. Klein. *Nature*, 118:516, 1926.

[305] O. Klein. *Z. Phys.*, 37:895, 1926.

[306] H. Mandel. *Z. Phys.*, 39:136, 1926.

[307] A. Einstein and P. Bergmann. *Ann. Math.*, 39:683, 1938.

[308] B. S. DeWitt. *Phys. Rep. C*, 19:297, 1975.

[309] B. DeWitt. *Relativity, Groups, and Topology*. Gordon and Breach, New York, 1964.

[310] J. H. Schwarz, editor. *Superstrings*. World Scientific, Singapore, 1985. two volumes.

[311] M. B. Green, J. H. Schwarz, and E. Witten. *Superstring Theory*. Cambridge University Press, Cambridge, 1987. two volumes.

[312] M. Kaku. *Introduction to Superstrings and M Theory*. Springer, New York, 1999. 2nd ed.

[313] N. Arkani-Hamed, S. Dimopoulos, and G. Dvali. *Phys. Lett. B*, 429:263, 1998.

[314] N. Arkani-Hamed, S. Dimopoulos, G. Dvali, and N. Kaloper. *Phys. Rev. Lett.*, 84:586, 2000.

[315] S. Dimopoulos and G. Guidice. *Phys. Lett. B*, 379:105, 1996.

[316] R. Sundrum. *J. High Energy Phys.*, 9907:001, 1999.
[317] L. Randall and R. Sundrum. *Phys. Rev. Lett.*, 83:4690, 1999.
[318] L. Randall and R. Sundrum. *Phys. Rev. Lett.*, 83:3370, 1999.
[319] P. M. Garnavich et al. *Astrophys. J.*, 493:L53, 1998.
[320] A. G. Reiss et al. *Astron. J.*, 116:1009, 1998.
[321] S. Perlmutter et al. *Astrophys. J.*, 517:565, 1999.
[322] Particle Data Group. *Eur. Phys. J. C*, 15:1, 2000.
[323] A. Balbi et al. *Astrophys. J. Lett.*, 545:L1, 2000.
[324] S. Hanany et al. *Astrophys. J. Lett.*, 545:L5, 2001.
[325] The MaxiBoom Collaboration (J. Richard Bond et al.). In *Conference on Cosmology and Particle Physics (CAPP 2000), Verbier, Switzerland, 17-28 Jul 2000.* astro-ph/0011379.
[326] P. de Bernadis et al. *Nature*, 404:955, 2000.
[327] A. E. Lange et al. *Phys. Rev. D*, 63:042001, 2001.
[328] C. B. Netterfield et al. *preprint*, astro-ph/0104460.
[329] C. Pryke et al. *preprint*, astro-ph/0104490.
[330] T. Appelquist and A. Chodos. *Phys. Rev. D*, 28:772, 1983.
[331] S. Weinberg. *Phys. Lett.*, 125B:265, 1983.
[332] T. Appelquist, A. Chodos, and E. Myers. *Phys. Lett. B*, 127:51, 1983.
[333] M. A. Rubin and B. Roth. *Nucl. Phys. B*, 226:44, 1983.
[334] M. A. Rubin and B. Roth. *Phys. Lett. B*, 127:55, 1983.
[335] C. R. Ordóñez and M. A. Rubin. *Nucl. Phys. B*, 260:456, 1985.
[336] M. A. Rubin and C. R. Ordóñez. *J. Math. Phys.*, 25:2888, 1984.
[337] M. A. Rubin and C. R. Ordóñez. *J. Math. Phys.*, 26:65, 1985.
[338] M. H. Sarmadi. *Nucl. Phys. B*, 263:187, 1986.
[339] A. Chodos and E. Myers. *Ann. Phys. (N.Y.)*, 156:412, 1984.
[340] A. Chodos and E. Myers. *Phys. Rev. D*, 31:3064, 1985.
[341] E. Myers. *Phys. Rev. D*, 33:1563, 1986.
[342] J. S. Dowker. *Class. Quant. Grav.*, 1:359, 1984.
[343] J. S. Dowker. *Phys. Rev. D*, 29:2773, 1984.
[344] G. Kunstatter and H. P. Leivo. *Phys. Lett. B*, 166:321, 1986.
[345] S. R. Huggins, G. Kunstatter, H. P. Leivo, and D. J. Toms. *Phys. Rev. Lett.*, 58:296, 1987.
[346] E. Myers. *Phys. Rev. Lett.*, 59:165, 1987.
[347] S. Weinberg. *Rev. Mod. Phys.*, 61:1, 1989.
[348] S. Weinberg. In *Dark Matter 2000, Marina del Rey, CA.* astro-ph/0005265.
[349] G. A. Vilkovisky. *Nucl. Phys. B*, 234:125, 1984.
[350] A. O. Barvinsky and G. A. Vilkovisky. *Phys. Rep.*, 119:1, 1985.
[351] H. T. Cho and R. Kantowski. *Phys. Rev. D*, 62:124003, 2000.
[352] C.-I Kuo and L. H. Ford. *Phys. Rev. D*, 47:4510, 1993.
[353] N. G. Phillips and B. L. Hu. *Phys. Rev. D*, 62:084017, 2000.
[354] I. Brevik, K. A. Milton, S. D. Odintsov, and K. E. Osetrin. *Phys. Rev. D*, 62:064005, 2000.
[355] I. Brevik, K. A. Milton, S. Nojiri, and S. D. Odintsov. *Nucl. Phys. B*,

599:305, 2001.

[356] A. Bytsenko and S. D. Odintsov. *Class. Quantum Grav.*, 9:391, 1992.

[357] A. Bytsenko and S. Zerbini. *Class. Quantum Grav.*, 9:1365, 1992.

[358] N. Shtykov and D. V. Vassilevich. *Mod. Phys. Lett. A*, 10:755, 1995.

[359] K. Kirsten and E. Elizalde. *Phys. Lett. B*, 365:72, 1996.

[360] R. C. Myers. *Phys. Rev. D*, 60:046002, 1999.

[361] R. Garatini. *Class. Quantum Grav.*, 17:3335, 2000.

[362] C. D. Hoyle, U. Schmidt, B. R. Heckel, E. G. Adelberger, J. H. Gundlach,
       D. J. Kapner, and H. E. Swanson. *Phys. Rev. Lett.*, 86:1418, 2001.

[363] J. C. Long, H. W. Chan, and J. C. Price. *Nucl. Phys. B*, 539:23, 1999.

[364] J. C. Long, A. B. Churnside, and J. C. Price. In *Proceedings of 9th Mar-
       cel Grossmann Meeting on Recent Developments in Theoretical and Experi-
       mental General Relativity, Gravitation and Relativistic Field Theories (MG
       9), Rome, Italy, 2–9 July, 2000*. hep-ph/0009062.

[365] M. Bordag, B. Geyer, G. L. Klimchitskaya, and V. M. Mostepanenko. *Phys.
       Rev. D*, 62:011701(R), 2000.

[366] A. A. Actor. *J. Phys. A*, 28:5737, 1995.

[367] B Guberina, R. Meckbach, R. D. Peccei, and R. Rückl. *Nucl. Phys. B*,
       184:476, 1981.

[368] R. Fukuda. *Phys. Rev. D*, 21:485, 1980.

[369] R. Fukuda and Y. Kazama. *Phys. Rev. Lett.*, 45:1142, 1980.

[370] G. K. Savvidy. *Phys. Lett. B*, 71:113, 1977.

[371] H. Pagels and E. Tomboulis. *Nucl. Phys. B*, 143:485, 1978.

[372] S. Weinberg. *Trans. N. Y. Acad. Sci., Series II*, 38:185, 1977.

[373] K. Olaussen and F. Ravndal. *Phys. Lett. B*, 100:497, 1981.

[374] K. Olaussen and F. Ravndal. *Nucl. Phys. B*, 192:237, 1981.

[375] A. Romeo and A. A. Saharian. *preprint*, hep-th/0007242.

[376] A. A. Saharian. *Phys. Rev. D*, 63:125007, 2001.

[377] A. Romeo and A. A. Saharian. *Phys. Rev. D*, 63:105019, 2001.

[378] H. Frenzel and H. Shultes. *Z, Phys. Ch. Abt. B*, 27:421, 1934.

[379] N. Marinesco and J. J. Trillat. *C. R. Acad. Sci. Paris*, 196:858, 1933.

[380] D. F. Gaitan, L. A. Crum, C. C. Church, and R. A. Roy. *J. Acoust. Soc.
       Am.*, 91:3166, 1992.

[381] B. P. Barber and S. J. Putterman. *Nature*, 352:318, 1991.

[382] B. P. Barber, R. Hiller, K. Arisaka, H. Fetterman, and S. J. Putterman. *J.
       Acoust. Soc. Am.*, 91:3061, 1992.

[383] R. G. Holt, D. G. Gaitan, A. A. Atchley, and J. Holzfuss. *Phys. Rev. Lett.*,
       72:1376, 1994.

[384] R. Hiller, S. J. Putterman, and B. P. Barber. *Phys. Rev. Lett.*, 69:1182,
       1992.

[385] B. P. Barber and S. J. Putterman. *Phys. Rev. Lett.*, 69:3839, 1992.

[386] R. A. Hiller and S. J. Putterman. *Phys. Rev. Lett.*, 75:3549, 1995.

[387] R. A. Hiller and S. J. Putterman. *Phys. Rev. Lett.*, 77:2345, 1996.

[388] B. Gompf, R. Günther, G. Nich, R. Pecha, and E. Eisenmenger. *Phys. Rev.*

Lett., 79:1405, 1997.

[389] R Hiller, S. Putterman, and S. Weninger. *Phys. Rev. Lett.*, 80:1090, 1998.

[390] M. J. Moran and D. Sweider. *Phys. Rev. Lett.*, 80:4987, 1998.

[391] J. Holzfuss, M. Rüggeberg, and A. Billo. *Phys. Rev. Lett.*, 81:5434, 1998.

[392] D. Lohse and S. Hilgenfeldt. *J. Chem Phys.*, 107:6986, 1997.

[393] D. Lohse, M. Brenner, T. Dupont, S. Hilgenfeldt, and B. Johnston. *Phys. Rev. Lett.*, 78:1359, 1997.

[394] T. J. Matula and L. A. Crum. *Phys. Rev. Lett.*, 80:865, 1998.

[395] J. A. Kettering and R. E. Apfel. *Phys. Rev. Lett.*, 81:4991, 1998.

[396] E. B. Flint and K. S. Suslik. *J. Am. Chem. Soc.*, 111:6987, 1989.

[397] C. C. Wu and P. H. Roberts. *Phys. Rev. Lett.*, 70:3424, 1993.

[398] C. C. Wu and P. H. Roberts. *Proc. Roy. Soc. London A*, 445:323, 1994.

[399] H. P. Greenspan and A. Nadim. *Phys. Fluids A*, 5:1065, 1993.

[400] S. Hilgenfeldt, S. Grossmann, and D. Lohse. *Phys. Rev. Lett.*, 70:3424, 1993.

[401] M. P. Brenner, S. Hilgenfeldt, D. Lohse, and R. R. Rosales. *Phys. Rev. Lett.*, 77:3467, 1996.

[402] H. Kwak and H. Yang. *J. Phys. Soc. Japan*, 64:1980, 1995.

[403] H. Kwak and J. Na. *Phys. Rev. Lett.*, 77:4454, 1996.

[404] J. B. Young, T. Schmiehl, and W. Kang. *Phys. Rev. Lett.*, 77:4816, 1996.

[405] L. Frommhold and W. Meyer. In M. Zoppi and L. Ulivi, editors, *Spectral Line Shapes*, volume 9, New York, 1997. AIP. 13th ICSLS, Firenze, Italy, June 1996, AIP Conf. Proc. 386.

[406] L. Frommhold and A. A. Atchley. *Phys. Rev. Lett.*, 73:2883, 1994.

[407] L. Motyka and M. Sadzikowski. *preprint.* physics/9912013.

[408] P. Mohanty and S. V. Khare. *Phys. Rev. Lett.*, 80:189, 1998.

[409] P. Mohanty. *preprints.* cond-mat/9912271, cond-mat/0005233.

[410] M. Buzzacchi, E. Del Guidice, and G. Preparata. *preprint.* quant-ph/9804006.

[411] K. A. Milton. In M. Bordag, editor, *Proceedings of the Third Workshop on Quantum Field Theory Under the Influence of External Conditions, Leipzig, 1995*, page 13, Stuttgart, 1996. Teubner.

[412] E. Sassaroli, Y. N. Srivastava, and A. Widom. *Phys. Rev. A*, 50:1027, 1994.

[413] C. Eberlein. *Phys. Rev. A*, 53:2772, 1996.

[414] C. Eberlein. *Phys. Rev. Lett.*, 76:3842, 1996.

[415] A. Chodos. In B. Kursonolglu, S. Mintz, and A. Perlmutter, editors, *Orbis Scientiae 1996, Miami Beach*, New York, 1996. Plenum.

[416] A. Chodos and S. Groff. *Phys. Rev. E*, 59:3001, 1999.

[417] S. Liberati, M. Visser, F. Belgiorno, and D. W. Sciama. *Phys. Rev. Lett.*, 83:678, 1999.

[418] S. Liberati, M. Visser, F. Belgiorno, and D. W. Sciama. *Phys. Rev. D*, 61:085023, 2000.

[419] S. Liberati, M. Visser, F. Belgiorno, and D. W. Sciama. *Phys. Rev. D*, 61:085024, 2000.

[420] S. Liberati, F. Belgiorno, M. Visser, and D. W. Sciama. *J. Phys. A*, 33:2251, 2000.

[421] F. Belgiorno, S. Liberati, M. Visser, and D. W. Sciama. *Phys. Lett. A*, 271:308, 2000.

[422] C. E. Carlson, C. Molina-París, J. Pérez-Mercader, and M. Visser. *Phys. Lett. B*, 395:76, 1997.

[423] C. E. Carlson, C. Molina-París, J. Pérez-Mercader, and M. Visser. *Phys. Rev. D*, 56:1262, 1997.

[424] C. Molina-París and M. Visser. *Phys. Rev. D*, 56:6629, 1997.

[425] B. Jensen and I. Brevik. *Phys. Rev. E*, 61:6639, 2000.

[426] G. T. Moore. *J. Math. Phys.*, 11:2679, 1970.

[427] P. C. W. Davies. *J. Phys. A*, 8:365, 1975.

[428] W. G. Unruh. *Phys. Rev. D*, 14:870, 1976.

[429] S. A. Fulling and P. C. W. Davies. *Proc. R. Soc. London, Ser. A*, 348:393, 1976.

[430] C. S. Unnikrishnan and S. Mukhopadhyay. *Phys. Rev. Lett.*, 77:4690, 1996.

[431] A. Lambrecht, M.-T. Jaekel, and S. Reynaud. *Phys. Rev. Lett.*, 78:2267, 1997.

[432] I. Brevik and I. Clausen. *Phys. Rev. D*, 39:603, 1989.

[433] L. H. Ford and A. Vilenkin. *Phys. Rev. A*, 25:2569, 1982.

[434] S. A. Fulling. *Aspects of Quantum Field Theory in Curved Space-Time*. Cambridge University Press, Cambridge, 1989.

[435] S. Liberati, F. Belgiorno, and M. Visser. *preprint*, hep-th/0010140.

[436] R. Schützhold, G. Plunien, and G. Soff. *Phys. Rev. A*, 57:2311, 1998.

[437] R. Schützhold, G. Plunien, and G. Soff. *Phys. Rev. A*, 58:1783, 1998.

[438] G. Plunien, R. Schützhold, and G. Soff. *Phys. Rev. Lett.*, 84:1882, 2000.

[439] P. Davis. *Nature*, 382:761, 1996.

[440] G. Barton and C. Eberlein. *Ann. Phys. (N.Y.)*, 227:222, 1993.

[441] M.-T. Jaekel and S. Reynaud. *J. Physique*, 2:149, 1992.

[442] R. Golestanian and M. Kardar. *Phys. Rev. Lett.*, 78:3421, 1997.

[443] A. Kenneth and S. Nussinov. *preprint*, hep-th/9912291.

[444] D. Robaschik, K Scharnhorst, and E. Wieczorek. *Ann. Phys. (N.Y.)*, 174:401, 1987.

[445] M. Bordag and K. Scharnhorst. *Phys. Rev. Lett.*, 81:3815, 1998.

[446] S.-S. Xue. *Commun. Theor. Phys. (Wuhan)*, 11:243, 1989.

[447] Tai-Yu Zheng and S.-S. Xue. *Chin. Sci. Bull.*, 38:631, 1993.

[448] C. Peterson, T. H. Hansson, and K. Johnson. *Phys. Rev. D*, 26:415, 1982.

[449] X. Kong and F. Ravndal. *Phys. Rev. Lett.*, 79:545, 1997.

[450] X. Kong and F. Ravndal. *Nucl. Phys. B*, 526:627, 1998.

[451] J. Schwinger. *Particles, Sources, and Fields*, volume II. Addison-Wesley, Reading, Mass., 1973.

[452] L. C. Albuquerque. *Phys. Rev. D*, 55:7754, 1997.

[453] M. V. Cougo-Pinto, C. Farina, A. Tort, and J. Rafelski. *Phys. Lett. B*, 434:388, 1998.

[454] W. Heisenberg and H. Euler. *Z. Phys.*, 98:714, 1936.

[455] V. Weisskopf. *Kgl. Danske Videnskab. Selskab. Mat-fys. Medd.*, 14(6), 1936.

[456] L. Gamberg, G. R. Kalbfleisch, and K. A. Milton. *Found. Phys.*, 30:543, 2000.

[457] F. Ravndal and J. B. Thomassen. *Phys. Rev. D*, 63:113007, 2001.

[458] K. Melnikov. *Phys. Rev. D*, 64:045002, 2001.

[459] W. Heitler. *The Quantum Theory of Radiation.* Oxford University Press, Oxford, 1954. 3rd edition.

[460] M. Bordag. In M. Bordag, editor, *The Casimir Effect 50 Years Later: The Proceedings of the Fourth Workshop on Quantum Field Theory Under the Influence of External Conditions, Leipzig, 1998*, page v, Singapore, 1999. World Scientific.

[461] J. Schwinger. *Renormalization Theory of Quantum Electrodynamics: An Individual View*, page 329. Cambridge University Press, 1983.

[462] C. G. Beneventano and E. M. Santangelo. *Int. J. Mod. Phys. A*, 11:2871, 1996.

[463] E. Elizalde. *J. Phys. A*, 27:L229, 1994.

[464] B. F. Svaiter and N.F. Svaiter. *Phys. Rev. D*, 47:4581, 1993.

[465] G. N. Watson. *Proc. Roy. Soc. London*, 95:83, 1918.

[466] A. Sommerfeld. *Partial Differential Equations in Physics.* Academic Press, New York, 1949.

[467] T. J. Allen, M. G. Olsson, and J. R. Schmidt. *Phys. Rev. D*, 62:066006, 2000.

[468] I. Brevik and H. B. Nielsen. *Phys. Rev. D*, 41:1185, 1990.

[469] I. Brevik and E. Elizalde. *Phys. Rev. D*, 49:5319, 1994.

[470] I. Brevik and H. B. Nielsen. *Phys. Rev. D*, 51:1869, 1995.

[471] I. Brevik, H. B Nielsen, and S. D. Odintsov. *Phys. Rev. D*, 53:3224, 1996.

[472] I. Brevik and R. Sollie. *J. Math. Phys.*, 38:2774, 1997.

[473] M. H. Berntsen, I. Brevik, and S. D. Odintsov. *Ann. Phys. (N.Y.)*, 257:84, 1997.

[474] I. Brevik, E. Elizalde, R. Sollie, and J. B. Aarseth. *J. Math. Phys.*, 40:1127, 1999.

[475] X. Li, X. Shi, and J. Zhang. *Phys. Rev. D*, 44:560, 1991.

[476] L. Hadasz, G. Lambiase, and V. V. Nesterenko. *Phys. Rev. D*, 62:025011, 2000.

[477] M. Lüscher. *Nucl. Phys. B*, 180:317, 1981.

[478] I. Brevik, A. A. Bytsenko and A. E. Gonçalves. *Phys. Lett. B*, 453:217, 1999.

[479] I. Brevik and A. Bytsenko. In *Proceedings of the Londrina Winter School, Mathematical Methods in Physics, Londrina-Parana, Brazil, 1999*, page 44, 2000. hep-th/0002064.

[480] E. D'Hoker and P. Sikivie. *Phys. Rev. Lett.*, 71:1136, 1993.

[481] E. D'Hoker, P. Sikivie, and Y. Kanev. *Phys. Lett. B*, 347:56, 1995.

[482] A. Actor, I. Bender, and J. Reingruber. *Fortsch. Phys.*, 48:303, 2000.

[483] E. Elizalde and S. Odintsov. *Class. Quantum Grav.*, 12:2881, 1995.

[484] G. Lambiase and V. V. Nesterenko. *Phys. Rev. D*, 54:6387, 1996.

[485] H. Kleinert, G. Lambiase, and V. V. Nesterenko. *Phys. Lett. B*, 384:213, 1996.

[486] G. Lambiase and V. V. Nesterenko. *Phys. Lett. B*, 398:335, 1997.

Index

zeta-f° regularisat°

my mother's biggest blindspot is
her inability to see that she
has any blindspots